The Formation of the Ikhwānian Corpus

Islamic Philosophy, Theology and Science

TEXTS AND STUDIES

Founding Editor

Hans Daiber

Edited by

Anna Akasoy
Emilie Savage-Smith
Ayman Shihadeh

VOLUME 132

The titles published in this series are listed at *brill.com/ipts*

The Formation of the Ikhwānian Corpus

A Philological Study of the Epistles and Two Summaries Attributed to the Brethren of Purity

By

Janne Mattila

BRILL

LEIDEN | BOSTON

Funding ERC project (2017–2023) AdG 740618 directed by Professor Godefroid de Callataÿ at the University of Louvain (UCLouvain) and the Helsinki Collegium for Advanced Studies.

Cover illustration: Painting (1236–1237 CE) by Yaḥyā Ibn Maḥmūd al-Wāsiṭī portraying a group of scholars in a library. The illustration appears in a manuscript of al-Ḥarīrī's (d. 1122) *Maqāmāt*. The image is reproduced from MS Arabe 5847, fol. 5v of Bibliothèque nationale de France.

The Library of Congress Cataloging-in-Publication Data is available online at https://catalog.loc.gov

Typeface for the Latin, Greek, and Cyrillic scripts: "Brill". See and download: brill.com/brill-typeface.

ISSN 0169-8729
ISBN 978-90-04-69013-4 (hardback)
ISBN 978-90-04-69014-1 (e-book)
DOI 10.1163/9789004690141

This book is printed on acid-free paper and produced in a sustainable manner.

Contents

Acknowledgements

This book is the main result of the four years I spent as a researcher at the Université catholique de Louvain in Belgium. The ERC project I worked in—"The Origin and Early Development of Philosophy in Tenth-century al-Andalus: the Impact of Ill-Defined Materials and Channels of Transmission" (2017–2023)—focused on the genesis of philosophy in Islamic Spain, including the sources that influenced that genesis, such as the *Epistles* of the Brethren of Purity. I was assigned the task of researching the formal questions concerning the *Epistles* and two further works attributed to the Brethren of Purity. I would first and foremost like to express my gratitude to Godefroid de Callataÿ, the main innovator and director of the *PhilAnd* project. I thank him for the warmth by which he welcomed me to Belgium. I had little prior experience of manuscript work; my previous work on the Brethren of Purity had concerned their philosophical ideas rather than the more technical aspects of the *Epistles*. I might also have initially felt intimidated by some aspects of the work at hand, such as collecting the more than 500 cross-references in the *Epistles*. In the end, I feel that the project made me a better scholar by forcing me to improve my philological skills. I am also grateful to Godefroid de Callataÿ for his innovative design and implementation of the project. Despite my initial fears, I quickly came to view my research task as a fascinating puzzle that has intrigued me ever since. Whether this book is successful in resolving that puzzle or not, I most certainly would not have written the book without the assignment.

I also wish to thank the many wonderful people who worked in, or were associated with, the *PhilAnd* project. I found the University of Louvain to be an exceptionally inspiring environment for research. Many aspects of this book were discussed not only in project seminars and workshops but also during lunch and coffee. I would especially like to thank the project's vice-director Sébastien Moureau who was instrumental for the project's operation and provided his valuable insights in both official contexts and informal discussions. I am particularly grateful to Cécile Bonmariage not only for the valuable comments produced by her sharp mind but also for the many reading groups and other occasions that enhanced my understanding of Islamic philosophy, as well as reduced the tedium of quarantine during the pandemic in Belgium. The other members who formed a part of the *PhilAnd* circle at one time or another were equally important for the genesis of this book. I am grateful for all the fruitful discussions with José Bellver, Charles Burnett, Marion Dapsens, Agnieszka Erdt, Michele Petrone, Liana Saif, Elaine Van Dalen, and Gregory Vandamme.

Besides the project events in Belgium, I gave presentations related to this book in conferences and seminars in Princeton and Helsinki. I wish to thank the many participants in those events that provided me with comments. I am also grateful for the suggestions that the two anonymous reviewers offered for the manuscript. Finally, I want to thank the wonderful institution of the Helsinki Collegium for Advanced Studies at the University of Helsinki for the opportunity of revising the final version of this book.

I wish to dedicate this book to the memory of my doctoral supervisor, Jaakko Hämeen-Anttila. He was the one who first introduced me to the intricacies of Arabic philology. I suspect he would have appreciated the philological nature of this study.

Tables

Introduction

The anonymous group of philosophers known as the "Brethren of Purity and Friends of Loyalty" (Ikhwān al-Ṣafāʾ wa-Khullān al-Wafāʾ) has presented a formidable puzzle for medieval authors and modern scholars alike. At some point during the ninth to tenth centuries, the group authored an Arabic encyclopedia of philosophical sciences known as the *Epistles of the Brethren of Purity and Friends of Loyalty* (*Rasāʾil Ikhwān al-Ṣafāʾ wa-Khullān al-Wafāʾ*), which consists of fifty-two treatises in the modern editions.[1] Besides the main work, two lengthy summaries of the *Epistles*, under the titles *Comprehensive Epistle* (*al-Risāla al-jāmiʿa*) and *Super-Comprehensive Epistle* (*Risālat Jāmiʿat al-jāmiʿa*), are attributed to the same authors.[2] These three works constitute what I call the Ikhwānian corpus.

A major part of modern scholarship on the Brethren of Purity has focused on the problem of their identity. Yet, after two centuries of academic debate, there is no scholarly agreement on the central questions regarding the authors or the works they supposedly wrote. Who were the authors behind the pseudonym, or

1 The references in this study are primarily to the Arabic pages of the critical editions of the *Epistles* published by the Oxford University Press and the Institute of Ismaili Studies in the series *Epistles of the Brethren of Purity* edited by Nader El-Bizri. In the case of the epistles for which no critical edition is yet available, the references are to the Beirut edition: Ikhwān al-Ṣafāʾ, *Rasāʾil Ikhwān al-Ṣafāʾ wa-Khullān al-Wafāʾ*, 4 vols., ed. Buṭrus al-Bustānī (Beirut: Dār Ṣādir, 1957). Many manuscripts also bear the title *Book of the Brethren of Purity* (*Kitāb Ikhwān al-Ṣafāʾ*). For the suggestion that this is the work's original title, see Susanne Diwald, *Arabische Philosophie und Wissenschaft in der Enzyklopädie: Kitāb Iḫwān aṣ-Ṣafāʾ (III): Die Lehre von Seele und Intellekt* (Wiesbaden: Otto Harrassowitz, 1975), 16–17. It is, however, a common convention to refer to the work as the *Epistles*, which is also the title by which the work refers to itself.

2 There are two semi-critical editions for the *Comprehensive Epistle*: Ikhwān al-Ṣafāʾ, *al-Risāla al-jāmiʿa: Tāj Rasāʾil Ikhwān al-Ṣafāʾ wa-Khullān al-Wafāʾ*, 2 vols., ed. Jamīl Ṣalībā (Damascus: Maṭbaʿat al-taraqqī, 1949–1951); *al-Risāla al-jāmiʿa: Tāj Rasāʾil Ikhwān al-Ṣafāʾ wa-Khullān al-Wafāʾ*, ed. Muṣṭafā Ghālib (Beirut: Dār al-Andalus, 1974). There is now also a properly critical edition in the form of the second part of a doctoral dissertation: Mourad Kacimi, "Edición crítica árabe y estudio de la *Risālat al-ŷāmiʿa ḏāt al-fawāʾid al-nāfiʿa, tāŷ Rasāʾil Ijwān al-Ṣafāʾ* (finales del s. IV H./X e.C., principios del s. V H./XI e.C.): Tomo II" (doctoral thesis, University of Alicante, 2015). A critical edition of the *Comprehensive Epistle*, based on Kacimi's doctoral thesis and accompanied by an annotated English translation by Carmela Baffioni, is currently in preparation for the Oxford University Press and the Institute of Ismaili Studies series of editions. In this study, references are given to both the Ghālib and Kacimi editions. For the *Super-Comprehensive Epistle*, the only available, semi-critical, edition will be employed: Ikhwān al-Ṣafāʾ, *Risālat Jāmiʿat al-jāmiʿa li-Ikhwān al-Ṣafāʾ wa-Khullān al-Wafāʾ*, ed. ʿĀrif Tāmir (Beirut: Dār Maktabat al-ḥayāt, 1970).

was there perhaps only one author? Was the Brethren of Purity a real society, or a fiction invented by the authors? What was the authors' doctrinal affiliation, and for what end were the works written? Why did the author or authors decide to remain anonymous? Were all three works authored by the same individual or group, or were the summaries produced by different authors? When were the works written, and were the *Epistles* the result of many redactions by successive generations of authors?

The continuing lack of scholarly agreement on the questions of dating and authorship of the *Epistles* and the related works perhaps shows that these problems might be unsolvable. At the same time, Ikhwānian scholarship has recently made great progress. This is the case, in particular, due to the series of critical editions of the *Epistles* produced by Oxford University Press in association with the Institute of Ismaili Studies, currently approaching its completion. Until recent times, research was overwhelmingly based on uncritical modern editions, mainly the Beirut edition, which give the false impression of an entirely uniform text. The new editions make visible the major variation in the manuscripts of the texts. In the most extreme cases, the manuscripts provide completely different textual variants for some epistles. This supports the view earlier taken by Yves Marquet, and more recently by Carmela Baffioni and Godefroid de Callataÿ, that the text of the *Epistles* we now have is the result of a long period of redaction.

The aim of this book is to take full use of this scholarly progress in order to address the puzzle of the Brethren of Purity from a novel perspective. The novelty consists in the book's focus on the question of the evolution of the Ikhwānian texts, which has been the object of much less scholarly interest than the problem of identity. Thus, the methodological approach differs from previous studies in two main respects. First, the book relies on internal, rather than external, evidence. Second, while most previous scholarship on these questions has been based on a strong hypothesis of the authors' identity or doctrinal alignment, such as the Ismāʿīlī hypothesis, the starting point here is neutral with regard to the question of authorship.

As regards the first aspect, the reliance on internal evidence in practice consists of an analysis of those textual components that provide reasonably objective evidence for the textual evolution of the Ikhwānian corpus. For the *Epistles*, this means that the book is not so much concerned with the work's doctrinal contents as with internal contradictions that attest to its development in time. This evidence notably includes manuscript variation, and besides the critical editions, the book employs a total of twenty manuscripts of the *Epistles*. For the *Comprehensive Epistle*, the evidence consists of a comparative analysis of the two texts, and this also extends to doctrinal questions. To substantiate its claims, the book provides nineteen tables that document the evidence. The book's one major aim is simply to present this evidence. Although the book will ultimately construct a narrative of the Ikhwānian corpus, hopefully that evidence will provide more objective tools for claims made by future scholars, even if they were to arrive at completely different conclusions from the current book.

This study is, then, concerned with formal questions of the Ikhwānian corpus, rather than the substance of its ideas and doctrines.[3] It should be emphasized that this is not to say that research on the doctrinal content of the *Epistles* would not be at least equally important. On the contrary, a long line of excellent research has demonstrated the work's intrinsic value within Islamic intellectual history. It is also clear that the *Epistles* exercised considerable influence on later thinkers of diverse intellectual orientations, most notably Ismāʿīlī theologians, but also the Andalusian author of the magical treatise *Goal of the Sage* (*Ghāyat al-ḥakīm*), which became well-known in medieval Europe through its Latin translation as *Picatrix*, as well as such later illustrious mystical and philosophical authors as Ibn ʿArabī (d. 1240) and Mullā Ṣadrā (d. ca. 1636). It is also true that the work attracted criticism from some famous figures of Islamic intellectual history, such as al-Ghazālī (d. 1111) and Ibn Taymiyya (d. 1328).[4] This shows that they considered the work relevant enough to be an object of attack, and at least in al-Ghāzālī's case, the *Epistles* provided an intellectual influence.[5] Precisely because of the prestige and notoriety that the *Epistles* came to enjoy later, a better understanding of the nature and formation of the Ikhwānian corpus is indispensable.

As for the second aspect, the aim is not to present yet another hypothesis of the identity of the Brethren of Purity. A plethora of such theories exists

---

3 For overviews of the philosophical and doctrinal contents of the *Epistles*, see Godefroid de Callataÿ, *Ikhwan al-Safa': a Brotherhood of Idealists on the Fringe of Orthodox Islam* (New York: Oneworld Publications, 2006); Nader El-Bizri (ed.), *The Ikhwān al-Ṣafāʾ and their Rasāʾil: An Introduction* (Oxford: Oxford University Press and the Institute of Ismaili Studies, 2008); Yves Marquet, *La philosophie des Iḫwān al-Ṣafāʾ: Nouvelle édition augmentée* (Paris and Milan: S.É.H.A and Archè, 1999); Ian Richard Netton, *Muslim Neoplatonists: An Introduction to the Thought of the Brethren of Purity (Ikhwān al-Ṣafāʾ)*, 2nd ed. (Edinburgh: Edinburgh University Press, 1991).

4 For Ibn Taymiyya's criticism of the Brethren of Purity, see Yahya Jean Michot, "Misled and Misleading … Yet Central in Their Influence: Ibn Taymiyya's Views on the Ikhwān al-Ṣafāʾ," in *The Ikhwān al-Ṣafāʾ and Their Rasāʾil: An Introduction*, ed. Nader El-Bizri (Oxford: Oxford University Press and the Institute of Ismaili Studies, 2008), 139–179.

5 The influence of the Brethren of Purity on al-Ghazālī is suggested in, for example, Taneli Kukkonen, "Al-Ghazālī on the Origins of Ethics," *Numen* 63 (2016): 280.

already, and it seems impossible to either verify or falsify any of them. Instead, the book's starting point is one of neutrality with regard to the questions of both identity and ideological affiliation. I will, then, make no precise claims about who the Brethren of Purity were. However, the problem of whether the *Comprehensive Epistle* was written by the same authors as the *Epistles* is unavoidably linked with the question of ideological affiliation. Thus, in the second part of the book, the Ismāʿīlī hypothesis of authorship will in fact assume a prominent role in the discussion.

Despite the reluctance to take a firm stand on the question of authorship, the book will, then, formulate a "soft" hypothesis of the Brethren of Purity's identity. The purpose of that hypothesis, however, is to provide a plausible narrative that explains the redaction chronology established by internal evidence. The claim is that in its initial form, the *Epistles* constituted a philosophical encyclopedia that fits doctrinally with the philosophical circle of al-Kindī (d. after 870 CE). The philosophical work was gradually expanded by inclusion of further epistles addressing religious and esoteric concerns. The two summaries constitute the final stage of the evolution of the Ikhwānian corpus, the adoption of the Brethren of Purity by Ismāʿīlī authors. This book claims, then, that the story of the Ikhwānian corpus is one of gradual transformation of an initially philosophical encyclopedia to a theological treatise. This narrative involves two claims about the Brethren of Purity's ideological commitment: (1) the authors of the *Epistles* were doctrinally close to al-Kindī's school; and (2) unlike the author of the *Comprehensive Epistle*, the authors of the *Epistles* probably were not Ismāʿīlīs.

The book's two major aims are, then, to present a theory of (1) the redaction history of the *Epistles* based on internal evidence; and (2) of the relationship of the *Comprehensive Epistle*, as well as of the *Super-Comprehensive Epistle*, with the *Epistles* based on textual comparison. To this end, the book is divided into two main parts, preceded by this introduction and followed by an epilogue.

In the remaining sections of the introduction, I will offer an overview of the *Epistles* and their two summaries, as well review the academic discussions that are relevant for the present study. The main purpose of the introduction is to situate this book within the context of our current scholarly understanding of the Brethren of Purity, and thus provide the necessary prerequisites for the arguments presented in the research chapters. This means that I will also address the theories of the Brethren of Purity's identity and doctrinal affiliation, as well as the dating of the *Epistles*. Even though the purpose is not to resolve these questions, they are relevant for various parts of the book: (1) theories of absolute dating relate to the claims of the relative chronology of the epistles; (2) theories of authorship relate to the claim that the *Epistles* have their origin within al-Kindī's circle; and (3) the Ismāʿīlī hypothesis relates to the claim that the two summaries represent an Ismāʿīlī adaptation of the *Epistles*.

The first part of the book is devoted to the *Epistles* of the Brethren of Purity. It serves two main objectives: (1) to present and analyze the evidence for the claim that the *Epistles* evolved in time and (2) to formulate a theory of the methods and chronology of the redaction process. These serve the final aim of (3) presenting a narrative by which the redaction history can be explained. Taken together, the three chapters of the first part provide seven kinds of evidence for the gradual evolution of the *Epistles*.

In Chapter 1, the second and third sections advance the argument that the current configuration of epistles in part contradicts, first, the rationale for the classification of philosophy explicitly adopted by the authors and, second, the passages in various epistles on the proper classification of knowledge in general, and of the epistles in particular. It is mainly these two sections that lay the ground for the initial hypothesis that an early variant of the *Epistles* was a philosophical work whose structure roughly corresponded to the curricular order of philosophical sciences followed in al-Kindī's circle. The other two sections of the chapter address, respectively, the configuration of epistles in known manuscripts and the *Fihrist*, that is, the table of contents that precedes the epistles. The former section supports the general conclusion by showing the textual instability of the manuscript tradition of the *Epistles*. The latter section demonstrates the same for the *Fihrist*, which is supposed to attest to the authors' pre-established plan in their composition of the *Epistles*.

Chapter 2 presents two kinds of technical evidence as further support for the thesis of the gradual evolution of the *Epistles*: (1) variation of epistle titles in manuscripts and cross-references provides evidence that many epistles previously formed a part of a larger epistle, and that the method of splitting epistles into two or more parts was therefore employed in the redaction process; (2) contradictions in transition passages between epistles show that many epistles previously occupied a different position within the work. Chapter 3 analyzes the cross-references between epistles for the end of constructing a theory of relative chronology of the fifty-two epistles. The final section of the chapter combines the evidence presented in the three chapters into a narrative of the redaction history of the *Epistles*.

The second part of the book aims to establish the relationship between the *Comprehensive Epistle* and the *Epistles* on both technical and doctrinal grounds. Its three chapters amount to an overall argument that an assumption of the shared authorship of the two works is not plausible.

Chapter 4 makes this argument on two specific philological grounds: (1) the second section of the chapter shows that most references from the *Epistles* to the *Comprehensive Epistle* are interpolations, and thus of no value for the theory of common authorship; (2) the third section demonstrates a textual dependence between the *Comprehensive Epistle* and a specific variant of the *Fihrist*, which almost certainly postdates the *Epistles* (as shown in Chapter 1).

Chapters 5 and 6 address the structural and doctrinal relationship, respectively, between the two works. The former shows the evolution from the *Epistles* to the *Comprehensive Epistle* at the higher level of subjects and central themes addressed by the two works. The latter focuses on questions that relate specifically to Ismāʿīlī doctrine to show that, while ideas of Ismāʿīlī inspiration were present in the *Epistles*, the *Comprehensive Epistle* manifests much closer alignment with Ismāʿīlī thought at the level of both terminology and doctrine, and, at times, even contradicts the *Epistles* in these doctrinal questions. The results of the book's second part are taken together in the final section of Chapter 6: (1) the theory of common authorship is incorrect; and (2) the *Comprehensive Epistle* is an Ismāʿīlī adaptation of the *Epistles*.

The epilogue finally addresses the *Super-Comprehensive Epistle*. It shows that the work has a particular textual relation with the *Comprehensive Epistle*. At the same time, the final chapter completes the book's general narrative of the Ikhwānian corpus from an encyclopedia of philosophical sciences, through an esoteric commentary of that encyclopedia, to an Ismāʿīlī theological treatise with little relation to the original philosophical work.

1 The Nature of the *Epistles*

The *Epistles* of the Brethren of Purity are a comprehensive work of philosophical sciences in fifty-one or fifty-two epistles (*risāla*), each of which is devoted to a single science or subject. While the format of presenting all parts of philosophy within a single work was later to become more common in both the Arabic and Latin traditions, the *Epistles* appear to form the first complete philosophical encyclopedia in this sense.[6] The epistles are divided into four

6 By 'encyclopedia,' I mean a book that aims to gather within its covers all relevant knowledge about a given field. Thus, it is not necessarily a reference work that consists of a large number of entries. In this sense, the Brethren of Purity's *Epistles* qualify as a philosophical encyclopedia, as perhaps do also al-Fārābī's (d. 950) *Enumeration of Sciences* (*Iḥṣāʾ al-ʿulūm*) and Avicenna's (Ibn Sīnā; d. 1037) larger works, such as the *Healing* (*al-Shifāʾ*) and the *Book of Knowledge* (*Dānishnāma-yi ʿAlāʾī*). All of these take as their classificatory principle the curricular order of the parts of philosophy and aim for comprehensiveness, albeit for different ends: al-Fārābī's *Enumeration* presents a concise classificatory account of the theoretical sciences and their aims; Avicenna's *Book of Knowledge* and the *Epistles* offer introductions to the doctrinal content of philosophical sciences aimed at non-specialists; and Avicenna's

parts: (1) propaedeutic-mathematical (*riyāḍiyya taʿlīmiyya*), consisting of thirteen or fourteen epistles on mainly mathematics and logic; (2) physical (*jismāniyya ṭabīʿiyya*), consisting of seventeen epistles; (3) psychical-intelligible (*nafsāniyya ʿaqliyya*), with ten epistles; and (4) nomic-theological (*nāmūsiyya ilāhiyya*), with eleven epistles. In the case of many epistles, the reason for their position within one of the four sections is far from obvious. Neither is it immediately clear why the authors chose this particular four-division, since it diverges both from the standard division of philosophical sciences of their time and the classification the authors themselves outline in the seventh epistle.[7]

The *Epistles* contain certain characteristic features as regards both style and content that distinguish the work from other philosophical treatises of this period. First, with regard to stylistic features, the authors, with a few notable exceptions, refer to themselves in the first-person plural, giving the impression that the *Epistles* were produced by the collective effort of a group of authors calling themselves the "Brethren of Purity." Second, the authors consistently employ a number of formulaic expressions.[8] The most common is the direct address to the reader, often at the beginning of an epistle or a chapter, with some variant of the formula: "Know, O (pious and compassionate) brother, may God support you and us with His spirit, that …" (*iʿlam ayyuhā al-akh al-bārr al-raḥīm ayyadaka Allāh wa-iyyānā bi-rūḥ minhu*). Often at the end of an epistle, the authors employ a variant of another formula: "May your soul awake from the sleep of negligence and the slumber of ignorance" (*laʿalla nafsaka tantabih min nawm al-ghafla wa-raqdat al-jahāla*). These set expressions act as signa-

Healing is in essence a comprehensive philosophical commentary of the Aristotelian corpus written for philosophers. For the applicability of the term 'encyclopedia' to the pre-modern Islamic world, and for Arabic non-philosophical and philosophical encyclopedias in general, see *Organizing Knowledge: Encyclopædic Activities in the Pre-Eighteenth Century Islamic World*, ed. Gerhard Endress (Leiden: Brill, 2006); Hinrich Biesterfeldt, "Medieval Arabic Encyclopedias of Science and Philosophy," in *The Medieval Hebrew Encyclopedias of Science and Philosophy: Proceedings of the Bar-Ilan University Conference*, ed. Steven Harvey (Dordrecht: Springer, 2000), 77–98. Among the articles in *Organizing Knowledge*, see, in particular, Gerhard Endress, "The Cycle of Knowledge: Intellectual Traditions and Encyclopædias of the Rational Sciences in Arabic Islamic Hellenism"; Dimitri Idem, "The Greek and Persian Background of Early Arabic Encyclopedism"; and Josef van Ess, "Encyclopædic Activities in the Islamic World: A Few Questions, and No Answers."

7 The order of the epistles and the authors' classification of sciences will be discussed in detail in Chapter 1.

8 For these Ikhwānian 'shibboleths,' see Godefroid de Callataÿ, "From Ibn Masarra to Ibn ʿArabī: References, Shibboleths and Other Subtle Allusions to the *Rasāʾil Ikhwān al-Ṣafāʾ* in the Literature of al-Andalus," in *Labor limae: Atti in onore di Carmela Baffioni*, ed. Antonella Straface, Carlo De Angelo, and Andrea Manzo (Naples: Centro di studi maġrebini, 2014–2015), 228–230.

tures by which the authors identify the text as pertaining to the Brethren of Purity. Beyond this, they point to certain essential doctrinal features of the *Epistles*. The first formulaic address brings readers and authors together into an ingroup of "pure brothers," who are guided by, or at least plead for the guidance of, God. The second exhortation indicates that the *Epistles* aim for something beyond instructing the reader on the technicalities of philosophy and science. Their goal is to rouse readers from ignorance to awareness concerning the true nature of reality and of themselves.

The distinctive nature of the *Epistles* as a philosophical work is also clear when the epistles are analyzed as independent philosophical treatises. Each epistle is devoted to a specific science, such as geometry or astronomy; or subject, such as pleasure or love; or a work of the Aristotelian curriculum, such as the *Posterior Analytics* or *Generation and Corruption*. Compared to similar treatises and commentaries written by philosophers and natural scientists during the ninth and tenth centuries, the Ikhwānian treatises are distinctive in both style and content. Each epistle contains various layers. The authors often specify the pedagogical goal of each epistle at its beginning, and more systematically in the introductory *Fihrist* (table of contents), which precedes the first epistle. For most epistles, the expressed goal consists of two layers: (1) the technical content of the epistle's subject and (2) an inner meaning related to the central cosmological, ethical, or eschatological themes of the *Epistles*.

As regards the first layer, each epistle aims to explain the technical content of the subject at hand; that is, to provide philosophical or scientific knowledge to the reader. The authors do not aim for exhaustive technical accounts, but the *Epistles* often state that the purpose of each epistle is to function as an introduction to, or a summary of the essential points of, a given subject.[9]

9 Thus, in Epistle 2 (*Geometry*), the authors state (*On Arithmetic & Geometry: An Arabic Critical Edition and English Translation of Epistles 1–2*, ed. and trans. Nader El-Bizri (Oxford: Oxford University Press and the Institute of Ismaili Studies, 2013), 76): "We have made to each kind of these sciences an epistle which is like an introduction and preamble (*shibh al-madkhal wa-l-muqaddimāt*)." In Epistle 7 (*Theoretical Sciences*), they say (*On Composition and the Arts: An Arabic Critical Edition and English Translation of Epistles 6–8*, ed. and trans. Nader El-Bizri and Godefroid de Callataÿ (Oxford: Oxford University Press and the Institute of Ismaili Studies, 2018), 79): "We made for each of these disciplines an epistle where we stated its necessary essential points (*nukat mā yuḥtāj ilayhā*) but dispensed with lengthy exposition;" and in Epistle 15 (*Matter and Form*) (*On the Natural Sciences: An Arabic Critical Edition and English Translation of Epistles 15–21*, ed. and trans. Carmela Baffioni (Oxford: Oxford University Press and the Institute of Ismaili Studies, 2013), 55): "Because of that, we made these epistles, and summarized in them the discussion as a kind of introduction and preamble, so that it may be close to a student's understanding (*yaqrub ʿalā al-mutaʿallim fahmuhā*) and facilitate a beginner's study of the subject (*yashul ʿalā al-mubtadiʾīna al-naẓar fīhā*)." The introductory nature of the

The introductory nature is also apparent in the authors' methodology, in the sense that the epistles present philosophical and scientific doctrines in a dogmatic manner and rarely provide systematic arguments to support them. Both features stand out when the epistles are compared with contemporary philosophical and scientific treatises written by specialists on the same subjects. Although the term 'popular' philosophy or science might be inappropriate,[10] it is clearly true that the epistles are not primarily written for mathematicians, astronomers, or philosophers but for educated laymen.

Second, each epistle contains what might be called an esoteric layer. Here, the term 'esoteric' is meant in the specific sense that the subject at hand is given an inner meaning. Thus, for example, the technical subject of Epistle 1 (*Arithmetic*) is number theory. The tractate relates this further to the Pythagorean idea that natural numbers correspond to degrees of reality.

Beyond these two layers of the explicit and implicit meanings of the epistle subjects, each epistle contains what may be called an exhortative layer. This takes the form of moral appeals to readers to liberate their souls from their material state by acquisition of knowledge and virtue. These moral exhortations permeate the *Epistles* as a whole and are mostly unrelated to an epistle's specific subject. The exhortative character is reinforced by the repeated use of allegorical stories, the longest and most famous of which concerns a trial between animals and human beings, occupying most of Epistle 22 on zoology.[11] All of this gives the *Epistles* their peculiar character in which philosophical-scientific discussions are seamlessly blended with eschatological themes and moralizing sermons and fables.

As regards the doctrinal content of the *Epistles* as a whole, scholars have often employed the terms eclectic or syncretistic to characterize the work.[12] These terms refer to the tendency of the authors to draw on a wide variety of classical and contemporary philosophical, scientific, religious, and literary sources, which they mix to construct their particular philosophical-religious

epistles is further supported by the fact that the titles of certain epistles in many manuscripts include the word 'introduction.'

10 The term is attributed to the *Epistles* in, for example, Ulrich Rudolph, *Islamische Philosophie: von den Anfängen bis zur Gegenwart* (Munich: C.H. Beck, 2004), 39; the term's application to the *Epistles* is rejected in Carmela Baffioni, "Ikhwân al-Safâ'," in *The Stanford Encyclopedia of Philosophy*, ed. Edward N. Zalta (2008); Diwald, *Arabische Philosophie*, 4.

11 For the pedagogical function of stories in the *Epistles*, see Shatha Almutawa, "Education through Narrative in *Rasāʾil Ikhwān al-Ṣafāʾ*," in *Knowledge and Education in Classical Islam: Religious Learning between Continuity and Change*, ed. Sebastian Günther (Leiden and Boston: Brill, 2020), 389–403.

12 See, for example, de Callataÿ, *Ikhwan al-Safa'*, 73–87; Netton, *Muslim Neoplatonists*, 7–8, 39.

system. The philosophical and scientific stratum draws on classical authorities that laid the foundation for the scientific worldview that emerged during the first Islamic centuries: Plato, Aristotle, and the Neoplatonists for philosophy; Ptolemy for astronomy; Euclid and Nicomachus of Gerasa for mathematics; and Galen for medicine.[13] Especially in their numerological, astrological, and occult inclinations, the *Epistles* are further influenced by Neopythagorean, Hermetic, and Jābirean sources.[14] As regards astrology and mythical stories, in particular, the *Epistles* draw further on Persian and Indian heritage.[15]

---

13 For a comprehensive survey of the explicit citations of classical authors in the *Epistles*, see Carmela Baffioni, *Frammenti e testimonianze di autori antichi nelle Epistole degli Iḫwān aṣ-Ṣafāʾ* (Rome: Istituto Nazionale per la Storia Antica, 1994).

14 For the Brethren of Purity as Pythagoreans, see Yves Marquet, *Les "Frères de la pureté", pythagoriciens de l'Islam: La marque du pythagorisme dans la rédaction des Épîtres des Iḫwān aṣ-Ṣafāʾ* (Paris: Textes et travaux de Chrysopaiea, 2006). For a rebuttal of Marquet's thesis of the Brethren of Purity's Pythagoreanism, and, in particular, its alleged channel of transmission through the Sabians of Ḥarrān, see Daniel De Smet, "Yves Marquet, les Iḫwān al-Ṣafāʾ et le pythagorisme," *Journal Asiatique* 295 (2007). For the Hermetic influence in the *Epistles*, see de Callataÿ, *Ikhwan al-Safa'*, 76–77; Netton, *Muslim Neoplatonists*, 50–52. The 'Hermetic' corpus refers to the body of texts attributed to Hermes Trismegistus, dealing, in particular, with cosmology, astrology, alchemy, and magic, and spanning from classical Greek through medieval Arabic to early modern Latin traditions. For the complex history of the Arabic Hermes, and of the ideas and works attributed to him, see Martin Plessner, "Hermes Trismegistus and Arab Science," *Studia Islamica* 2 (1954); Kevin van Bladel, *The Arabic Hermes: From Pagan Sage to Prophet of Science* (Oxford and New York: Oxford University Press, 2009). Only a few of the so-called Hermetic works can be shown to have a Greek predecessor, and it is less than clear that they share any coherent body of doctrines. To a large extent, then, the 'Hermetic' strand in the *Epistles* represents the influence of astrological, alchemical, and magical works of Arabic origin. The terms 'Pythagorean' and 'Hermetic' are therefore vague and misleading as an indication of specific sources of textual influence. However, the Brethren of Purity identify themselves as Pythagoreans and cite Hermes as an authority; this is manifested in the importance of numerology, on the one hand, and of astrology, alchemy, and magic, on the other. These ideas were transmitted to the *Epistles* by means of late ancient Neopythagorean texts, such as *The Golden Verses* (*Xrusa epē*) and Nicomachus of Gerasa's *Introduction to Arithmetic* (*Arithmetikē eisagōgē*), or texts supposedly transmitting Hermetic wisdom, such as the *Book of the Apple* (*Kitāb al-Tuffāḥa*). In their alchemical views, the *Epistles* are probably also influenced by the corpus attributed to Jābir Ibn Ḥayyān, dated by Kraus at between the mid-ninth and mid-tenth centuries. For the Jābirean influence, see Yves Marquet, *La philosophie des alchimistes et l'alchimie des philosophes: Jâbir ibn Ḥayyân et les "Frères de la Pureté"* (Paris: Maisonneuve and Larose, 1988). For the Ikhwānian citations of Nicomachus, see also Carmela Baffioni, "Citazioni di autori antichi nelle Rasail degli Ikhwan al-Safa: il caso di Nicomacho di Gerasa," in *The Ancient Tradition in Christian and Islamic Hellenism: Studies on the Transmission of Greek Philosophy and Sciences Dedicated to H.J. Drossaart Lulofs on His Ninetieth Birthday*, ed. Gerhard Endress and Remke Kruk (Leiden: Research School CNWS, 1997), 3–27.

15 See de Callataÿ, *Ikhwan al-Safa'*, 77–79; Netton, *Muslim Neoplatonists*, 89–94.

Syncretism is equally apparent in the religious stratum. The authors obviously identify themselves as Muslims, citing the Qur'ān extensively and prophetic traditions occasionally. But they also repeatedly quote the Bible and the Torah,[16] and even discuss the Sabian astral cult approvingly.[17] The authors manifest a similarly eclectic tendency toward the many interpretations of Islam, which is precisely why it has been so difficult to pinpoint their doctrinal affiliation. The Brethren of Purity's eclecticism, moreover, appears to be self-conscious in the sense that they emphasize that all religions and doctrines can provide at least the starting point for the pursuit of truth.[18]

A second general feature characterizing the *Epistles* is the work's esoteric character. The term 'esoteric' may seem somewhat vague since its referent is often ambiguous.[19] But the term may be defined in a specific sense. The first, literal, sense of inner meaning was mentioned above. The *Epistles*, for example, ascribe the inner meaning of a Pythagorean correspondence between numbers and reality to the technical subject matter of mathematics. This represents a special case of the authors' tendency to emphasize a binary division between exoteric (*ẓāhir*) and esoteric (*bāṭin*) aspects of revelation, reality, and knowledge. The idea is, then, that a specific Qur'ānic passage or some aspect of the sensible reality contains an immediately apparent external meaning which points toward a deeper inner meaning.

In a second sense, esotericism refers to the gnostic idea of an elect group in possession of secret saving knowledge that must be guarded from the uninitiated. The authors of the *Epistles* portray themselves as precisely such a divinely guided group. They claim that they hold regular philosophical-religious sessions, and caution against divulging their secrets to the unworthy. Beyond this, the authors connect themselves to a larger scheme of religious history alternating between the rule of the virtuous and the vicious and predict the appearance of a messianic figure of final days reinstating virtuous rule on the Earth.

---

16 For the Judeo-Christian layer in the *Epistles*, see Netton, *Muslim Neoplatonists*, 53–77.

17 See Janne Mattila, "Sabians, the School of al-Kindī, and the Brethren of Purity," in *Religious Identities in Antiquity and the Early Middle Ages*, ed. Ilkka Linstedt, Nina Nikki, and Riikka Tuori (Leiden: Brill, 2021), 92–114.

18 See Janne Mattila, "The Ikhwān al-Ṣafā' on Religious Diversity," *Journal of Islamic Studies* 28, no. 2 (2017).

19 For the meaning of 'esoteric' in the medieval Islamic context, see Daniel De Smet, "Esotericism and exotericism," in *Encyclopaedia of Islam Three*, ed. Kate Fleet, Gudrun Krämer, and Denis Matringe (Leiden: Brill, 2015); Paul Walker, "Bāṭiniyya," in *Encyclopaedia of Islam Three*, ed. Kate Fleet, Gudrun Krämer, and Denis Matringe (Leiden: Brill, 2009).

Esotericism in either of these two senses is obviously not unique to the *Epistles*. Rather, it relates the work to Shīʿīte groups, Ṣūfī mystics, or alchemists and magicians who have been branded as esoteric either by their contemporaries or by modern scholars. At least as regards the Peripatetic philosophers of Baghdād, however, the esotericism of the *Epistles* distinguishes the Brethren of Purity from their philosophical contemporaries.

These distinct aspects that characterize the *Epistles* bring up a fundamental question: for what purpose did the authors write the *Epistles*? The question is intimately connected with the problem of the authors' identity and ideological commitment. But before turning to the various theories of authorship, it may be worthwhile to summarize what the preceding analysis of the *Epistles* says of that purpose. The very first sentence of the very first epistle emphasizes the authors' theoretical ambitions to acquire a comprehensive knowledge of everything that exists in our world.[20] As an encyclopedic work, the *Epistles* provide a synthesis of the scientific knowledge of their time. Thus, the work's first aim is clearly the philosophical-scientific one of educating the reader in mathematics, astronomy, logic, physics, and other sciences. At the same time, the authors consistently attribute inner meanings related to ethical and eschatological themes even to this scientific content. Clearly, then, the *Epistles* also serve the religious purpose of disseminating both the knowledge and moral instructions that are necessary for the soul's salvation. Hence, based on the contents of the *Epistles*, the work serves the dual purpose of disseminating philosophical knowledge and encouraging the reader to choose the path toward the soul's salvation in the afterlife.

2 Authorship

The authors of the *Epistles* were careful to conceal their identity. To solve the riddle of the work's authorship, medieval and modern scholars alike have come up with an impressive number of theories. As for the pseudonym Ikhwān al-Ṣafāʾ, it has commonly been translated as the "Brethren of Purity," but sometimes also as the "Sincere Brethren." In the famous animal fable of two jackal brothers, *Kalīla and Dimna* (*Kalīla wa-Dimna*), the phrase is used of a group

20 *On Arithmetic & Geometry*, 8–9: "Know, O pious and compassionate brother, may God support you and us with His spirit, that it is the doctrine (*madhhab*) of our noble brothers, may God support them, to investigate (*naẓar*) all knowledge of existents (*ʿulūm al-mawjūdāt*) that are in the world."

of animals who escape their captors through sincere co-operation; it has been suggested that the authors took their name from the story.[21] This is supported by the multiple references that the *Epistles* make to the animal fable.[22] While *Kalīla and Dimna* might have provided the immediate inspiration, the name should probably be primarily understood as a literal reference to the work's ethical and eschatological objectives. The *Epistles* emphasize the importance of the co-operation of like-minded "brothers" in pursuit of knowledge and virtue, and ultimately the salvation of the soul.[23] This ethical-eschatological end coincides with the "purity" of the soul, understood in the Platonic sense as the soul's separation from the material world; to this may be added the religious connotations of purity with piety and right belief.[24] The epithet "Brethren of Purity," then, refers to the community, real or imagined, of virtuous and pious men aiding each other to save their souls. The term's precise object of reference appears somewhat ambiguous, however, in that the authors employ it to refer to themselves, their addressees, a specific religious-philosophical society with regular

21 This was first proposed in Ignaz Goldziher, "Über die Benennung der 'Ichwān al-ṣafā,'" *Der Islam* 1 (1910).

22 See, in particular, the reference in Epistle 2 (*Geometry*), which occurs in a context that emphasizes the necessity of co-operation of the virtuous in seeking salvation (*On Arithmetic & Geometry*, 118): "O brother, take heed of the tale of the ring-dove that is mentioned in the book *Kalīla and Dimna*, and how it escaped from the net, in order that you grasp the truth of what we have said." For *Kalīla and Dimna* in the context of Arabic animal fables, see Robert Irwin, "The Arabic Beast Fable," *Journal of the Warburg and Courtauld Institutes* 55 (1992); Ewald Wagner, "Sprechende Tiere in der arabischen Prosa," *Asiatische Studien: Zeitschrift der Schweizerischen Asiengesellschaft* 48, no. 3 (1994). The fable might also have provided an inspiration for the story about a trial between human beings and animals in Epistle 22; both feature speaking animals and a trial, and the Brethren of Purity's fable also refers to the *Kalīla and Dimna* (*The Case of the Animals versus Man Before the King of the Jinn*, ed. and trans. Lenn E. Goodman and Richard McGregor (Oxford: Oxford University Press and the Institute of Ismaili Studies, 2010), 95). For the suggestion that the literary circle in which Ibn al-Muqaffaʿ (d. ca. 758), the author-translator of *Kalīla and Dimna*, participated was an early model for the society of the Brethren of Purity, see also Gustav Richter, *Studien zur Geschichte der älteren arabischen Fürstenspiegel* (Leipzig: Heinrichs, 1932), 32. Both societies were located in Baṣra. However, it must also be noted that the term *ikhwān al-ṣafāʾ* can be found in Arabic poetry pre-dating the *Epistles* and could have been in common enough use for the authors to have come across it in other contexts besides *Kalīla and Dimna*.

23 See, in particular, Ian Richard Netton, "Brotherhood versus Imāmate: Ikhwān al-Ṣafāʾ and the Ismāʿīlīs," *Jerusalem Studies in Arabic and Islam* 2 (1980). Netton argues against the Ismāʿīlī theory of authorship by claiming that the concept of imāmate in the *Epistles* is replaced by that of brotherhood.

24 See, in particular, the discussion of the various philosophical and religious contexts of the term purity in Diwald, *Arabische Philosophie*, 17–21.

meetings, and righteous people in general.[25] While some scholars have therefore denied that the term refers to the authors at all,[26] the *Epistles*' employment of the first person plural and external evidence support the contrary view.

As for the identity of the author or authors hiding behind the pseudonym, scholarly discussion has revolved around two main points: (1) the implication of contemporary reports on the identity of the authors and (2) the authors' ideological or religious affiliation. Regarding the first point, the central testimony is offered by the Baghdādian literate-philosopher Abū Ḥayyān al-Tawḥīdī (d. 1023) in the *Book of Pleasure and Conviviality* (*Kitāb al-Imtāʿ wa-l-muʾānasa*).[27] Al-Tawḥīdī locates the Brethren of Purity in Baṣra and names four or five of their members.[28] Al-Tawḥīdī then describes the aims of their society:

> This is a group (*ʿiṣāba*) that has been united by companionship and sincere friendship (*qad taʾālafat bi-l-ʿishra wa-taṣāfat bi-l-ṣadāqa*), and agrees on sanctity (*quds*), purity (*ṭahāra*), and friendly advice (*naṣīḥa*). They have founded amongst themselves a school of thought (*waḍaʿū baynahum madhhaban*) that they claim brings them closer to success in God's favor (*al-fawz bi-riḍwān Allāh*) and the way to paradise (*al-maṣīr ilā jannatihi*). For they say: religion has become stained by ignorance and mixed with error (*al-sharīʿa qad dunnisat bi-l-jahālāt wa-ukhtuliṭat bi-l-ḍalālāt*); the only way to cleanse and purify it is by means of philosophy, for it comprises the wisdom of beliefs and benefit of actions (*al-ḥikma al-iʿtiqādiyya wa-l-maṣlaḥa al-ijtihādiyya*). They also claim that when Greek philosophy and Arabic religion are brought together, perfection is attained.[29]

---

25 For a more in-depth analysis of the semantic field of the term "Brethren of Purity," see pp. 192–211.

26 Thus, for example, Stern ("New Information about the Authors of the 'Epistles of the Sincere Brethren,'" *Islamic Studies* 3 (1964): 420–421) claims that the term refers neither to the authors nor to any concrete society but is instead a subtle reference to the Ismāʿīlīs. Similarly, scholars who identify a single author for the *Epistles* would have to deny that the epithet refers to a group that authored the epistles.

27 Al-Tawḥīdī, *Kitāb al-Imtāʿ wa-l-muʾānasa*, 3 vols., ed. Aḥmad Amīn and Aḥmad al-Zayn (Beirut: Dār Maktabat al-ḥayāt, 1939–1944), vol. 2, 4–11. Al-Tawḥīdī transmits thirty-seven discussions held at the court of Ibn Saʿdān in Baghdād during the years 983–985, when the latter acted as a Būyid vizier. Al-Tawḥīdī brings up the Brethren of Purity incidentally in the seventeenth discussion.

28 Al-Tawḥīdī's famous list (Ibid, 4–5), which he says is not comprehensive, consists of Abū Sulaymān al-Bustī al-Maqdisī, Abū al-Ḥasan al-Zanjānī, Abū Aḥmad al-Mihrajānī, and al-ʿAwfī, as well as Zayd Ibn Rifāʿa as an associate member.

29 Ibid, 5. This and all subsequent translations are mine unless otherwise indicated.

The passage makes several claims about the Brethren of Purity's nature: (1) they constitute a society united by friendship, which (2) founded a doctrine that they claim leads to the soul's salvation; (3) they allege that Islam has become mixed with error which can only be corrected by Greek philosophy; and (4) their aim is therefore to combine Greek philosophy with Islamic religion. Al-Tawḥīdī's account would, then, seem to confirm that the Brethren of Purity were an actual society that was active in Baṣra. The account also presents that society's aim as both religious and philosophical: their objective is to construct a synthesis between Greek philosophy and Islam. As we saw in the previous section, this is a fairly accurate description of the *Epistles*' contents.

To attain this end, al-Tawḥīdī states that the group authored fifty epistles and a table of contents (*fihrist*), consisting of all parts of both theoretical and practical philosophy, which they called the *Epistles of the Brethren of Purity and Friends of Loyalty*.[30] The authors then concealed their names, and divulged the *Epistles* in order to liberate people from the false beliefs and immoral actions that harm their souls. Neither al-Tawḥīdī nor his mentor, the philosopher Abū Sulaymān al-Sijistānī (d. ca. 985), known as the "logician" (*al-manṭiqī*), accord the *Epistles* much value as either a philosophical or religious work. According to al-Tawḥīdī, what is correct (*ṣawāb*) in it is muddled with the predominance of error (*ghalaṭ*).[31] According to al-Sijistānī, the premise of fusing philosophy with religion is ill-founded to begin with: philosophy is founded on reason and religion on revelation, and they pertain to distinct spheres of knowledge. If it was the case that revelation can be corrected with philosophy, then revelation would be subject to philosophy, and there would be no need for revelation in the first place.[32]

---

30 Ibid.

31 Ibid, 5–6.

32 Ibid, pp. 6–23. The passage constitutes a highly interesting discussion of the relation between religion and philosophy, for which see also Joel L. Kraemer, *Philosophy in the Renaissance of Islam: Abū Sulaymān al-Sijistānī and his Circle* (Leiden: Brill, 1986), 230–243. Al-Tawḥīdī's interlocutor, the vizier, is surprised by al-Sijistānī's position (*Kitāb al-Imtāʿ*, 18), given that he is known as the "logician" and a student of Greek philosophy. Al-Tawḥīdī replies that al-Sijistānī's claim is that both philosophy and religion are true, but they simply have nothing to do with each other. All the examples of philosophical truths mentioned in this context pertain to mathematics, astronomy, logic, and physics, suggesting that al-Sijistānī's position is perhaps explained by the focus on the religiously neutral parts of philosophy. In contrast, al-Fārābī (d. 950), whom al-Tawḥīdī does not mention, had previously devoted much of his political philosophy to the problem of the relation between religion and philosophy, and resolved the question by explicitly subjugating religion to philosophy.

Two further reports complement the picture painted by al-Tawḥīdī's account. Al-Sijistānī, al-Tawḥīdī's mentor, attributes fifty-one epistles to al-Maqdisī.[33] Al-Maqdisī is also included in al-Tawḥīdī's list. But the difference is that al-Sijistānī seems to claim that the *Epistles* were authored by a single person, unless he means, for example, that al-Maqdisī was responsible for the final redaction of the *Epistles*.[34] The second report is provided by the famous Muʿtazilite theologian, ʿAbd al-Jabbār (d. 1025). He polemizes against a group of Ismāʿīlīs in Baṣra, mentioning by name all but al-Maqdisī from al-Tawḥīdī's list, and adding to the list the name of Ibn Abī Baghl.[35] Interestingly, he adds that the group includes other followers outside Baṣra; this coincides with the claims of some epistles about the wide geographical spread of the Brethren of Purity. However, ʿAbd al-Jabbār does not identify these people with the Brethren of Purity nor does he mention the *Epistles*. Yet, when al-Tawḥīdī's account is combined with that of ʿAbd al-Jabbār, it could be concluded that the authors were a group in Baṣra affiliated with the Ismāʿīlīs.

The three contemporary reports give us seemingly valuable information about who the authors were and why they wrote the *Epistles*. However, there has been little scholarly agreement on how these reports should be interpreted. Most scholars have accepted al-Tawḥīdī's account as a more or less accurate description of the identity of the Brethren of Purity.[36] This has still left plenty

33 Al-Sijistānī, *The Muntakhab Ṣiwān al-ḥikmah of Abū Sulaimān as-Sijistānī: Arabic Text, Introduction and Indices*, ed. D.M. Dunlop (Hague, Paris, and New York: Mouton, 1979), §§ 309–310, 161–163. Since we only possess an abridgement (*muntakhab*) of al-Sijistānī's original work, *Repository of Wisdom* (*Ṣiwān al-ḥikma*), it is not clear whether all of the text can be attributed to al-Sijistānī.

34 For this view, see Diwald, *Arabische Philosophie*, 13–14. See also al-Bayhaqī, *Tārikh ḥukamāʾ al-islām*, ed. Muḥammad Kurd ʿAlī (Damascus: Maṭbaʿat al-taraqqī, 1946), § 18, 35–37. In this work, known as the "complement" of the *Repository of Wisdom* (*tatimmat Ṣiwān al-ḥikma*), al-Bayhaqī (d. 1174), perhaps to harmonize the two accounts, states that al-Maqdisī was responsible for the "language" (*alfāẓ*) of the *Epistles*. Al-Shahrazūrī (fourteenth cent.), on the other hand, states in *Rawḍat al-afrāḥ* that al-Maqdisī was the spokesman for the group that authored the *Epistles*. It is, however, al-Zanjānī, rather than al-Maqdisī, whom al-Tawḥīdī identifies as the "the school's leader" (*ṣāḥib al-madhhab*) in another passage (*Kitāb al-Imtāʿ*, vol. 2, 157), in which he attributes to al-Zanjānī a story that also appears in the *Epistles*. ʿAbd al-Jabbār (Stern, "New Information," 411) similarly calls al-Zanjānī "one of their leaders" (*raʾīs min ruʾasāʾihim*). Since neither passage mentions the Brethren of Purity, it is not clear which *madhhab* is meant; nor is it necessary that the leader of the society of pure brothers and the final redactor of the *Epistles* is the same person.

35 ʿAbd al-Jabbār, *Tathbīt dalāʾil al-nubuwwa*, vol. 1, ed. ʿAbd al-Karīm ʿUthmān (Beirut: Dār al-ʿarabiyya, 1962), 355–356. The relevant passages are cited in Arabic and translated into English in Stern, "New Information," 407–411.

36 See, in particular, Diwald, *Arabische Philosophie*, 8–16; Joel L. Kraemer, *Humanism in the*

of room for wildly divergent theories about their doctrinal affiliation. Some scholars have nevertheless dismissed the account's historical value entirely, either based on al-Tawḥīdī's alleged untrustworthiness,[37] or because internal grounds suggest a date or an author for the *Epistles* that is incompatible with al-Tawḥīdī's claims.

The medieval Arabic sources are not particularly helpful, either. First, it is striking that there are so few contemporary references to the Brethren of Purity; this suggests that, in the eastern part of the Islamic world, the *Epistles* may not have become widely known until after a considerable time had passed since their completion.[38] Second, the disagreement of subsequent Arabic authors suggests that they had no access to more information about the authors' identity than we have. Many writers derive their information from al-Tawḥīdī or al-Sijistānī; others offer an incredible range of mostly implausible theories that suggest various Shīʿite imāms, Muʿtazilite theologians, Jābir Ibn Ḥayyān, the Ṣūfī martyr al-Ḥallāj (d. 922), and the Ashʿarite theologian al-Ghazālī (d. 1111) as potential authors.[39]

The attribution to the Andalusian astronomer al-Majrīṭī (d. 1007)[40] is more noteworthy, because some manuscripts of the *Epistles*, and most manuscripts

---

Renaissance of Islam: The Cultural Revival during the Buyid Age (Leiden: Brill, 1986), 165–178; Yves Marquet, "Ikhwān al-Ṣafāʾ," in *Encyclopaedia of Islam, 2nd Edition* (Leiden: Brill, 1968); Samuel Miklos Stern, "The Authorship of the Epistles of the Ikhwān aṣ-Ṣafāʾ," *Islamic Culture* 20 (1946); Stern, "New Information."

37 See, in particular, Abbas Hamdani, "Abū Ḥayyān al-Tawḥīdī and the Brethren of Purity," *International Journal of Middle East Studies* 9, no. 3 (1978). Hamdani argues that al-Tawḥīdī's purpose is polemical in that he wants to bring the courtier Zayd Ibn Rifāʿa into disrepute by associating him with a heterodox group. For al-Tawḥīdī's untrustworthiness as a reporter, see also Kraemer, *Philosophy*, 31–40, 44; Ismail K. Poonawala, "Why We Need an Arabic Critical Edition with an Annotated English Translation of the *Rasāʾil Ikhwān al-Ṣafāʾ*," in *The Ikhwān al-Ṣafāʾ and Their Rasāʾil: An Introduction*, ed. Nader El-Bizri (Oxford: Oxford University Press and the Institute of Ismaili Studies, 2008), 33–57, at 52–56.

38 Thus, the otherwise well-informed Baghdādian bibliographer Ibn al-Nadīm (d. ca. 995) does not mention the Brethren of Purity.

39 The entry on the Brethren of Purity in Ibn al-Qifṭī's (d. 1248) bibliographical history of science (Ibn al-Qifṭī, *Tārīkh al-ḥukamāʾ*, ed. Julius Lippert (Leipzig: Dieterichische Verlagsbuchhandlung, 1903), 82–88) first states that some people attribute the work to a Shīʿite imām, while they disagree on which one, and others to an unnamed early Muʿtazilite theologian, and then reproduces al-Tawḥīdī's report. Al-Ṣafadī (d. 1363) cites Yāqūt's (d. 1229) summary of al-Tawḥīdī in the entry on Zayd Ibn Rifāʿa (al-Ṣafadī, *Kitāb al-Wāfī bi-l-wafiyyāt*, 2nd ed., vol. 15, ed. Bernd Radtke (Beirut: Institut der Deutschen Morgenländischen Gesellschaft, 1991), § 60, 48–49), and then mentions al-Maqdisī, the Shīʿite imāms ʿAlī (d. 662) and Jaʿfar al-Ṣādiq (d. 765), as well as al-Ḥallāj, al-Ghazālī, and Jābir Ibn Ḥayyān as potential authors.

40 For the attribution of the *Epistles* to al-Majrīṭī in al-Andalus, see Godefroid de Callataÿ and

of the *Comprehensive Epistle*, are also attributed to al-Majrīṭī. Moreover, early external sources connect the *Epistles* with al-Andalus in general and al-Majrīṭī, in particular. The Andalusian historian of science Ṣāʿid al-Andalusī (d. 1070) reports that al-Majrīṭī's pupil al-Kirmānī (d. 1066) was to his knowledge the first to introduce the *Epistles* into the Iberian peninsula.[41] Also, the author of the alchemical treatise *Degree of the Sage* (*Rutbat al-ḥakīm*) refers explicitly to either fifty or fifty-one epistles, and even gives the impression that he is their author; at the same time, he affirms that people remain ignorant of who authored the epistles.[42] The magical treatise *Goal of the Sage* (*Ghāyat al-ḥakīm*), later famous in Europe as the *Picatrix*, employs the *Epistles* as one of its principal sources without mentioning them explicitly.[43] While it has been known for some time that the traditional attribution of these works to al-Majrīṭī is doubtful, Fierro has convincingly argued that the real author of the two treatises is the *ḥadīth* scholar Maslama al-Qurṭubī al-Majrīṭī (d. 964).[44] This makes the two Andalusian works quite probably the earliest external testimonies on the *Epistles*, antedating those by al-Sijistānī and al-Tawḥīdī.[45]

Sébastien Moureau, "Again on Maslama Ibn Qāsim al-Qurṭubī, the Ikhwān al-Ṣafāʾ and Ibn Khaldūn: New Evidence from Two Manuscripts of *Rutbat al-ḥakīm*," *Al-Qanṭara* 37, no. 2 (2016).

41 Ṣāʿid al-Andalusī, *Kitāb Ṭabaqāt al-umam*, ed. Louis Cheikho (Beirut: al-Maṭbaʿa al-kathūlīkiyya li-l-ābāʾ al-yasūʿiyyīn, 1912), 70–71.

42 The treatise has been edited in Pseudo-Majrīṭī, *Rutbat al-ḥakīm: The Book of the Rank of the Sage*, ed. Wilferd Madelung (Zürich: Living Human Heritage Publications, 2016). Godefroid de Callataÿ and Sébastien Moureau are currently preparing a new edition. For the references that the treatise makes to the *Epistles*, see Godefroid de Callataÿ, "Magia en al-Andalus: *Rasāʾil Ijwān al-Ṣafāʾ*, *Rutbat al-ḥakīm* y *Gāyat al-ḥakīm* (*Picatrix*)," *Al-Qanṭara* 34, no. 2 (2013): 317–326.

43 The tractate has been edited in Pseudo-Majrīṭī, *Das Ziel des Weisen*, ed. Helmut Ritter (Leipzig and Berlin: B.G. Teubner, 1933). For the tractate's employment of the *Epistles*, see de Callataÿ, "Magia," 326–333; Pseudo-Majrīṭī, *"Picatrix:" Das Ziel des Weisen von Pseudo-Mağrīṭī*, trans. Helmut Ritter and Martin Plessner (London: The Warburg Institute, University of London, 1962), lix–lxxv. The footnotes of the German translation by Ritter and Plessner provide a detailed analysis of the treatise's use of the *Epistles*.

44 Maribel Fierro, "Bāṭinism in al-Andalus: Maslama b. Qāsim al-Qurṭubī (d. 353/964), Author of the *Rutbat al-Ḥakīm* and the *Ghāyat al-Ḥakīm* (*Picatrix*)," *Studia Islamica* 84, no. 2 (1996).

45 Even if the attribution to Maslama al-Qurṭubī were not accepted, the author explicitly states that he wrote the *Aim of the Sage* in 343–348 AH (954–960 CE) after he had finished the *Degree of the Sage* (Pseudo-Majrīṭī, *"Picatrix,"* 1). That these works provide the earliest testimony for the *Epistles* was thus already suggested in Diwald, *Arabische Philosophie*, 15–16. The problem is that some manuscripts give the dates as 434–448, which would completely change the testimonial value that the two works have for the *Epistles*.

It is worth noting, then, that many medieval authors do not take seriously the claim that the *Epistles* were composed by a group of authors and instead attribute the work to a single author. Modern scholars, in contrast, have rarely disputed the idea of group authorship. A notable exception is Diwald, who argues that nothing in the work contradicts the idea that it was written by a single author. The work's stylistic unity and systematic arrangement, on the other hand, support this view.[46] The employment of the first person plural does not necessarily entail group authorship either, since it could merely be a literary device, and some passages also employ the first person singular. More recently, certain scholars have gone further to suggest specific candidates for the authorship. Thus, de Vaulx d'Arcy argues, based on doctrinal similarities between the *Epistles* and the school of al-Kindī, that the author is al-Kindī's student al-Sarakhsī (d. 899).[47] Carusi and Madelung have put forward the hypothesis that the above-mentioned Andalusian author al-Qurṭubī joined the Brethren of Purity while travelling in the east, and was thereafter responsible for the final redaction of the *Epistles*.[48] Madelung relates al-Tawḥīdī's account to an earlier redaction of the *Epistles* which no longer exists; de Vaulx d'Arcy dismisses al-Tawḥīdī's testimony altogether.

In the absence of external evidence to confirm these theories, they remain entirely speculative. What seems clear, however, is that neither internal nor external grounds make the assumption of group authorship self-evident. The history of classical Arabic or Persian literature, let alone of philosophy, apparently knows of no other examples of collaborative works. If the *Epistles* in fact were a collaborative work, it is singular in this respect also. Alternatively, the stylistic unity, systematic arrangement, and al-Sijistānī's early attribution of

---

46 Diwald, *Arabische Philosophie*, 12–15.

47 Guillaume de Vaulx d'Arcy, *Les Épîtres des Frères en Pureté* (*Rasāʾil Ikhwān al-Ṣafāʾ*)*: Mathématique et philosophie: Présentation et traduction de six épîtres* (Paris: Les Belles lettres, 2019), 22–48; Idem, "Aḥmad b. al-Ṭayyib al-Saraḫsī, réviseur de l'*Introduction Arithmétique* de Nicomaque de Gérase et rédacteur des *Rasāʾil Iḫwān al-Ṣafāʾ*," *Arabic Sciences and Philosophy* 29, no. 2 (2019).

48 Paola Carusi, "Alchimia islamica e religione: la legittimazione difficile di una scienza della natura," *Oriente Moderno* 19, no. 3 (2000): 500; Wilferd Madelung, "Maslama al-Qurṭubī's Contribution to the Shaping of the Encyclopedia of the Ikhwān al-Ṣafāʾ," in *Labor limae: Studi in onore di Carmela Baffioni*, ed. Antonella Straface, Carlo De Angelo, and Andrea Manzo, Studi magrebini: nuova serie (Naples: Università degli studi di Napoli, 2015), 403–417. That al-Majrīṭī re-redacted the *Epistles* and then divulged them in his own name was already suggested in Gustav Flügel, "Ueber Inhalt und Verfasser der arabischen Encyclopädie رسائل اخوان الصفا وخلان الوفا d.i. die Abhandlungen der aufrichtigen Brüder und treuen Freunde: Nebst Andeutungen über die Einrichtungen des Bundes der Verbrüderten," *Zeitschrift der Deutschen Morgenländischen Gesellschaft* 13, no. 1/2 (1859): 25.

the authorship to only one person could support the hypothesis that a single person was responsible for the *Epistles*' final redaction, while the epistles were originally written by a group of authors.

The range of scholarly opinions has been equally wide regarding the question of the authors' doctrinal affiliation. As we have seen, medieval authors were far from unanimous in this question. Al-Tawḥīdī does not explicitly claim that the authors adhered to any specific religious persuasion; ʿAbd al-Jabbār identifies the authors in al-Tawḥīdī's list as Ismāʿīlīs without mentioning the *Epistles*. Later authors attribute the *Epistles* not only to the Ismāʿīlī creed but also to Shīʿism more generally, as well as to authors of Ṣūfī, Muʿtazilite, or even Ashʿarite persuasions. Some modern scholars have similarly argued that the *Epistles* are closely related to Muʿtazilite theology, supported by their emphasis on divine justice and the unity of God's essence, the two main principles by which the Muʿtazilite school was known, as well as by some further doctrinal similarities.[49] Many of the oldest manuscripts explicitly relate the *Epistles* to Sufism,[50] and Diwald associates the work with the early stages of Ṣūfī thought when the technical mystical terminology had not yet been developed.[51] The authors, however, repeatedly polemize against theologians in general; this makes it unlikely that they were formally affiliated with the Muʿtazilite school.[52] Moreover, the ethical-eschatological end the authors attribute to human beings seems to derive more from Neoplatonic thought than from Ṣūfī mysticism.[53] It seems rather to be the case that the authors, true to their eclec-

49 Muʿtazilite orientation of the *Epistles* has been argued for especially in Adel Awa, *L'esprit critique des "Frères de la Pureté": Encyclopédistes arabes du IVe/Xe siècle* (Beirut: Imprimerie catholique, 1948). In the *Epistles*, the introductory *Fihrist* employs the attribute "people of justice" (*ahl al-ʿadl*) three times (*Rasāʾil*, vol. 1, 21, 43, 47) as an attribute for the Brethren of Purity. However, none of these instances is present in the older, main, variant of the *Fihrist*, nor is the appellation employed elsewhere in the *Epistles*.

50 Thus, in MS Tübingen MaVI86 (dated 1206 CE), the title of the work is given as *Kitāb Ikhwān al-Ṣafawiyya al-Ṣūfiyya* (*Book of the Pure Ṣūfī Brothers*). In many manuscripts, including the oldest one of MS Atıf Efendi 1681, each epistle title ends with the stock phrase *fī tahdhīb al-nafs wa-iṣlāḥ al-akhlāq min kalām al-ṣūfiyya* ("on purification of the soul and rectification of character traits pertaining to the Ṣūfī discourse"). The fact that in some other manuscripts the epistle titles are complemented with exactly the same phrase excluding the last part of *min kalām al-ṣūfiyya* suggests, however, that, in the manuscript tradition, the characterization of the goal of the *Epistles* as ethical has only later been identified as being of a specifically Ṣūfī nature.

51 Diwald, *Arabische Philosophie*, 19–23.

52 See Stern, "New Information," 419.

53 Moreover, Epistle 7 (*Theoretical Sciences*) (*On Composition*, 74) does not accord Sufism a particularly prominent position; the "science of remembrance, exhortations, asceticism,

tic inclinations, are influenced by both Muʿtazilite theology and early Ṣūfī thought,[54] but they did not adhere to either group in any confessional sense.

Most of the scholarly debate, however, has revolved around the Ismāʿīlī hypothesis of authorship. A great number of scholars have held that the authors of the *Epistles* were affiliated with the Ismāʿīlī branch of Shīʿism in one way or another. Corbin, Marquet, and Hamdani have maintained that the *Epistles* represent the official propaganda of the Fāṭimids, either before or after the establishment of the Fāṭimid state in North Africa in 909.[55] More recently, Baffioni has also argued for the authors' Ismāʿīlī commitment and, in less certain terms, for their possible affiliation with the Fāṭimid mission.[56] Other scholars have claimed either that the authors of the *Epistles* were unaffiliated Ismāʿīlīs, or that, on the contrary, they were affiliated with anti-Fāṭimid Ismāʿīlī groups, including the Qarmaṭians, who found their own state in the eastern Arabian peninsula in 899.[57] Yet further scholars have argued that while the *Epistles*

---

and Sufism" (*ʿilm al-tidhkār wa-l-mawāʿiẓ wa-l-zuhd wa-l-taṣawwuf*) is given as the fifth religious science, preceding all philosophical sciences in the curricular order.

54 For doctrinal affinities with Sufism, see, for example, Carmela Baffioni, "From Sense Perception to the Vision of God: A Path Toward Knowledge According to the Iḫwān al-Ṣafāʾ," *Arabic Sciences and Philosophy* 8 (1998): 230–231; Idem, "An Essay on Terminological Research in Philosophy: The 'Friends of God' in the *Rasāʾil Iḫwān al-Ṣafāʾ*," in *Scritti in onore di Giovanni Oman*, ed. Carmela Baffioni, Giuseppina Igonetti, and Vincenzo Strika, Studi Maġrebini (Naples: Istituto universitario orientale, 2002), 23–43. The authors show particular fondness for Ṣūfī ascetic practices, which in Epistle 27 (*Spread of Particular Souls into Physical Bodies*) (*Rasāʾil*, vol. 3, 8) they view to have similar beneficial effects as Christian monasticism.

55 See, in particular, Henry Corbin, *Histoire de la philosophie islamique, I: Des origines jusqu'à la mort d'Averroës (1198)* (Paris: Gallimard, 1964), 190–194; Abbas Hamdani, "An Early Fāṭimid Source on the Time and Authorship of the 'Rasāʾil Iḫwān al-Ṣafāʾ'," *Arabica* 26, no. 1 (1979); Idem, "Brethren of Purity, a Secret Society for the Establishment of the Fatimid Caliphate: New Evidence for the Early Dating of their Encyclopaedia," in *L'Egypte fatimide: son art et son histoire*, ed. Marianne Barrucand (Paris: Presses de l'Université Paris-Sorbonne, 1999), 73–82; Yves Marquet, "Iḫwān al-Ṣafāʾ, ismaïliens et qarmaṭes," *Arabica* 24 (1977); Idem, "Les epîtres des Ikhwân as-Safâ', œuvre ismaïlienne," *Studia Islamica* 61 (1985). Among earlier generations of scholars, the *Epistles* were already connected with Ismāʿīlī thought in Paul Casanova, "Notice sur un manuscrit de la secte des Assasins," *Journal Asiatique* 11 (1898). In Wladimir A. Ivanow, *A Guide to Ismaili Literature* (London: Royal Asiatic Society, 1933), 31, the work was argued to be of specifically Fāṭimid character.

56 Carmela Baffioni, "Epistle 48 of the Ikhwān al-Ṣafāʾ and their Ismāʿīlī Commitment," in *Case Studies in Transmission*, ed. Jaakko Hämeen-Anttila et al. (Münster: Ugarit Verlag, 2015), 11–21.

57 For the argument that the *Epistles* represent a non-Fāṭimid, idealized, reconstruction of the Ismāʿīlī doctrine not connected with any specific faction, see Stern, "The Authorship;"

manifest a definite Ismāʿīlī influence, this should be regarded as one more constituent in the authors' eclectic mixture of distinct sources.[58]

The Ismāʿīlī hypothesis has been dominant, because there is considerable external and internal evidence to support it. We have seen that ʿAbd al-Jabbār's contemporary testimony relates the authors in al-Tawḥīdī's list to the Ismāʿīlī movement. As it happens, Qarmaṭian Ismāʿīlīs had strong presence in the southern Iraq where the *Epistles* in all likelihood were redacted.[59] Beyond this, the Ṭayyibite faction of the later Fāṭimid Ismāʿīlīs adopted the *Epistles* as one of their central works, attributing its authorship to the hidden imām Aḥmad, although, as far as we know, not before the mid-twelfth century.[60] This leaves a gap of at least two centuries between the redaction of the *Epistles* and their known adoption by the Ismāʿīlīs. The fact that classical Fāṭimid authors men-

---

Idem, "New Information." Bausani (*L'enciclopedia dei Fratelli della Purità: riassunto, con introduzione e breve commento dei 52 trattati o epistole degli Ikhwān aṣ-Ṣafāʾ* (Naples: Istituto Universitario Orientale, 1978), 13–16) similarly views that the authors manifest an "Ismāʿīlī tendency" but do not pertain to any particular group. Ebstein (*Mysticism and Philosophy in al-Andalus: Ibn Masarra, Ibn al-ʿArabī and the Ismāʿīlī Tradition* (Leiden and Boston: Brill, 2014), 180) states that the Brethren of Purity expressed their Ismāʿīlī beliefs in a "unique way." For a careful and more affirmative claim that the *Epistles* pertain to the Qarmaṭians, see Farhad Daftary, *The Ismāʿīlīs: Their History and Doctrines* (Cambridge: Cambridge University Press, 1992), 28, 234–236; Wilferd Madelung, "Ḳarmaṭī," in *Encyclopaedia of Islam, 2nd edition* (Leiden Brill, 1998), 663. The major dividing line between the two factions is that for the Fāṭimids the chain of imāms continues in their dynastic line, while for the Qarmaṭians it ends in the seventh imām.

58 de Callataÿ, *Ikhwan al-Safa'*, xi; Kraemer, *Humanism*, 175–178; Netton, *Muslim Neoplatonists*, 95–104. Nasr (*An Introduction to Islamic Cosmological Doctrines: Conceptions of Nature and Methods Used For Its Study by the Ikhwān al-Ṣafāʾ, al-Bīrūnī, and Ibn Sīnā* (Cambridge: Harvard University Press, 1964), 35–36) also connects the *Epistles* with what he calls the "Ismāʿīlī gnosis," mixed with generically Shīʿite and Ṣūfī tendencies. Tibawi ("Ikhwān al-Ṣafā and their *Rasāʾil*: A Critical Review of a Century and a Half of Research," *Islamic Quarterly* 2 (1955): 33, 43) admits Ismāʿīlī sympathies but inclines toward the view that a closer affiliation between the authors and Ismāʿīlī groups is a later development.

59 Daftary, *The Ismāʿīlīs*, 236.

60 See Daftary, *The Ismāʿīlīs*, 236; Daniel De Smet, "L'auteur des *Rasāʾil Ikhwān al-Ṣafāʾ* selon les sources ismaéliennes ṭayyibites," *Shii Studies Review* 1, no. 1–2 (2017); Husain F. Hamdani, "*Rasāʾil Ikhwān aṣ-Ṣafāʾ* in the Literature of the Ismāʿīlī Ṭayibī Daʿwat," *Zeitschrift für Geschichte und Kultur des Islamischen Orients* 20 (1932). The Yamanī Ṭayyibite *dāʿī* Ibrāhīm Ibn al-Ḥusayn al-Ḥāmidī (d. 1162) is the first Ismāʿīlī author known to have explicitly referred to and cited from the *Epistles* in his *Kitāb Kanz al-walad*. As shown by De Smet, al-Ḥāmidī introduces the citations as attributed either to the "sage" (*ḥakīm*) or the "eminent person, author of the epistles" (*al-shakhṣ al-fāḍil ṣāḥib al-rasāʾil*), and once to the "virtuous imām" (*al-imām al-fāḍil*). The specific attribution to imām Aḥmad seems to be of later provenance.

tion neither the *Epistles* nor the Brethren of Purity seems to contradict the theory that the work was an official Fāṭimid treatise.[61]

Baffioni has shown, however, that, among Fāṭimid *dāʿīs*, Ḥamīd al-Dīn al-Kirmānī (d. 1021) was quite possibly influenced by the *Epistles* in the mineralogical section of his *Rest of the Intellect* (*Rāḥat al-ʿaql*), while Nāṣir-i Khosrow (d. ca. 1088) cites from the *Epistles* in the *Book of Two Wisdoms* (*Jāmiʿat al-ḥikmatayni*), even if neither author explicitly acknowledges his debt on the *Epistles*.[62] Clearly, then, it is not the case that the *Epistles* had no influence at all on classical Fāṭimid authors.[63] It is also true that neither the Fāṭimids' employment nor later adoption of the *Epistles* proves that the Brethren of Purity were themselves Ismāʿīlīs.[64] It is also worth noting that some later Sunnite authors, such as Ibn Taymiyya (d. 1328), count the *Epistles* as an Ismāʿīlī work,[65] while others polemizing against the Ismāʿīlīs, such as al-Baghdādī (d. 1037) and al-Shahrastānī (d. 1153), ignore the *Epistles* completely;[66] al-Ghazālī, in turn, attacks both the Ismāʿīlīs and the Brethren of Purity without connecting the two.[67]

---

61 See, for example, Stern, "New Information," 416–417.

62 Carmela Baffioni, "L'influence des Ikhwān al-Ṣafāʾ sur la minéralogie de Ḥamīd al-Dīn al-Kirmānī," in *Une lumière venue d'ailleurs: héritages et ouvertures dans les encyclopédies d'Orient et d'Occident au Moyen Âge*, ed. Godefroid de Callataÿ and Baudouin Van den Abeele (Turnhout: Brepols, 2008), 319–331; Idem, "The Religious Approach to Natural Sciences: The Case of Mineralogy in the Ikhwān al-Ṣafāʾ and in Ḥamīd al-Dīn al-Kirmānī," in *Islamic Thought in the Middle Ages: Studies in Text, Transmission and Translation, in Honour of Hans Daiber*, ed. Anna Akasoy and Wim Raven (Leiden and Boston: Brill, 2008), 181–194; Idem, "Nāṣir-i Khosrow, Translator of the Ikhwān al-Ṣafāʾ?" in *Sources and Approaches across Disciplines in Near Eastern Studies: Proceedings of the 24th Congress, Union Européenne des Arabisants et Islamisants, Leipzig 2008*, ed. Verena Klemm and Nuha al-Shaʿar (Leuven: Peeters, 2013), 319–331.

63 This is claimed in, for example, Daftary, *The Ismāʿīlīs*, 236.

64 See also Diwald, *Arabische Philosophie*, 27. Diwald already recognizes that Khosrow's *Rawsha Nāʾīnāma* and *Zād al-musāfirīn* draw on the *Epistles*, but emphasizes that this does not prove the Ismāʿīlī affiliation for the authors of the *Epistles*.

65 Michot, "Misled and Misleading," 143–147.

66 Diwald, *Arabische Philosophie*, 27.

67 Al-Ghazālī mentions the "author of the *Book of the Brethren of Purity*" (*ṣāḥib Kitāb Ikhwān al-Ṣafāʾ*) three times in his spiritual autobiography, *Deliverer from Error* (al-Ghazālī, *al-Munqidh min al-ḍalāl*, ed. Jamīl Ṣalība and Kāmil ʿAyyār (Beirut: Dār al-Andalus, 1967), 88, 89, 98). The first two references are included in the section addressing philosophers, where al-Ghazālī's point is that the fact that the Brethren of Purity employ the Qurʾān, Prophetic sayings, and Ṣūfī wisdom means neither that these should be rejected because the author happens to be wrong in everything else, nor that all that is untrue in the work should be accepted because it is mixed together with truthful statements. The third reference does occur in the section devoted to the Ismāʿīlīs. But al-Ghazālī brings up the Brethren

As for internal evidence, the authors' generic Shīʿīte persuasion is clear, for example, in their identifying Prophet Muḥammad's spiritual deputyship (*wilāya*) with ʿAlī[68] and the Prophet's family,[69] decrying the massacre of imām Ḥusayn and his family in Karbalāʾ,[70] and advocating the idea of imāmate in general.[71] The authors explicitly refute the doctrine of hidden imām,[72] which would seem to place them outside Twelver-Shīʿism.

Regarding specifically Ismāʿīlī views: first, the general doctrinal content of the *Epistles* is reminiscent of the highly philosophical Ismāʿīlī theology of the tenth century. In particular, the *dāʿī* Abū Yaʿqūb al-Sijistānī (d. after 971) shares

---

of Purity there only as a side remark on the philosophical worthlessness of Pythagoras on whom some anonymous Ismāʿīlī thinker draws. This coincides with the low worth of the *Epistles* that follow Pythagoras. Al-Ghazālī, then, clearly includes the Brethren of Purity among philosophers, rather than among the Ismāʿīlīs. See also Tibawi, "Ikhwān al-Ṣafāʾ," 4, for Ibn Sabʿīn's (d. 1270) surprising accusation that al-Ghazālī borrowed most of his philosophical ideas from the *Epistles*, which are as weak as the original on which he draws.

68 *The Call to God: An Arabic Critical Edition and English Translation of Epistle 48*, ed. and trans. Abbas Hamdani and Abdallah Soufan (Oxford: Oxford University Press and the Institute of Ismaili Studies, 2019), 144–145: "Among the things that unite us to you, O brother, is love of the pure people of our Prophet's family (*maḥabbat ahl bayt nabiyyinā al-ṭāhirīn*), and the guardianship (*wilāya*) of the commander of the believers, ʿAlī Ibn Abī Ṭālib, the best of guardians (*khayr al-waṣiyyīn*)." However, as shown by the critical apparatus of the edition, this whole chapter, entitled in many manuscripts as the "address of the Shīʿītes" (*mukhāṭabat al-mutashayyiʿīna*), is missing in many manuscripts. If the chapter is an interpolation, then it provides evidence of later efforts to make the *Epistles* more Shīʿīte, rather than of the original authors' affiliation. For other references to the "commander of the faithful" (*amīr al-muʾminīna*), see Ikhwān al-Ṣafāʾ, *Rasāʾil*, vol. 4, 362; *On Companionship and Belief: An Arabic Critical Edition and English Translation of Epistles 43–45*, ed. and trans. Toby Mayer, Ian Richard Netton, and Samer F. Traboulsi (Oxford: Oxford University Press and the Institute of Ismaili Studies, 2017), 38.

69 See *Rasāʾil*, vol. 4, 375. The passage distinguishes between "great guardianship" (*wilāya ʿaẓīma*) restricted to the Prophet, and the guardianship after the Prophet attributed (*makhṣūṣa*) to the prophetic family (*ahl bayt al-risāla*). For the Shīʿīte concepts of *walī* and *wilāya*, see Daftary, *The Ismāʿīlīs*, 83–84.

70 *Rasāʾil*, vol. 4, 75; *The Case of the Animals*, 158–159; *On Music: An Arabic Critical Edition and English Translation of Epistle 5*, ed. and trans. Owen Wright (Oxford: Oxford University Press and the Institute of Ismaili Studies, 2011), 9; *On Companionship and Belief*, 74; *On God and the World: An Arabic Critical Edition and English Translation of Epistles 49–51*, ed. and trans. Wilferd Madelung et al. (Oxford: Oxford University Press and the Institute of Ismaili Studies, 2019), 283.

71 See, in particular, *Rasāʾil*, vol. 2, 478; vol. 3, 483; *On Composition*, 74, 93; *The Call to God*, 165–167; *On Ethics and Character Traits: An Arabic Critical Edition and English Translation of Epistle 9*, ed. and trans. Omar Alí-de-Unzaga (Oxford: Oxford University Press and the Institute of Ismaili Studies, 2024), 73, 83–84. These passages attribute governance (*siyāsa*), secret knowledge, and interpretation (*taʾwīl*) of the revelation to imāms.

72 *The Call to God*, 165–166.

with the *Epistles* the Plotinian cosmological scheme of God, Intellect, and Soul to explain the creation of the world by gradual emanation of existence.[73] While most Muslim philosophers similarly draw on Neoplatonism in their accounts of creation, the Brethren of Purity's system and Ismāʿīlī thought are distinguished by two additional elements: (1) the first created being of the Intellect is the result of God's active creation (*ibdāʿ*) or command (*amr*), rather than an involuntary and necessary process of emanation;[74] and (2) the reascent of existence, the eschatological counterpart of creation by descent, is related to the historical event of Judgment Day, versus being an eternal and timeless process. Other Ismāʿīlī-like features include the prominence of the distinction between exoteric and esoteric knowledge,[75] the doctrine of prophetic cycles determined by celestial conjunctions, and the identification of the seventh era with the appearance of a messianic figure (*qāʾim*).[76] The *Epistles*, moreover, employ Ismāʿīlī terminology related to creation, cosmology, eschatology, and prophets, although not as systematically as Ismāʿīlī works.[77]

There are also doctrinal features in the *Epistles* that do not fit well with the authors' supposed Ismāʿīlī identity. Some passages manifest a reverent attitude toward the first three caliphs, which raises the question whether the authors were even Shīʿīte, given the Shīʿīte belief in the illegitimacy of their rule.[78] Netton has further argued that the *Epistles* substitute the idea of imāmate for brotherhood; it is, then, individual reason rather an imām that acts as the spir-

---

73 For an overview of al-Sijistānī's thought, see Paul Walker, *Early Philosophical Shiism: The Ismaili Neoplatonism of Abū Yaʿqūb al-Sijistānī* (Cambridge: Cambridge University Press, 2003). For doctrinal parallels between the *Epistles* and al-Kirmānī's *Rest of the Intellect* (*Rāhat al-ʿaql*), see Daniel De Smet, *La quiétude de l'intellect: Néoplatonisme et gnose ismaélienne dans l'œuvre de Ḥamîd al-Dîn al-Kirmânî* (Leuven: Peeters, 1995), 115–116, 199–200, 258–260, 300–301, 318–333, 337–338, 340–343.

74 See Carmela Baffioni, "Ibdāʿ, Divine Imperative and Prophecy in the *Rasāʾil Ikhwān al-Ṣafāʾ*," in *Fortresses of the Intellect: Ismaili and Other Islamic Studies in Honour of Farhad Daftary*, ed. Omar Alí-de-Unzaga (London and New York: I.B. Tauris and the Institute of Ismaili Studies, 2011); Marquet, *La philosophie des Iḫwān al-Ṣafāʾ*, 56–58, 63–72. The *Epistles* seem to affirm both God's creation ex-nihilo (*ibdāʿ*), command (*amr*), and creative word (*kalima*/*kun*) as resulting in the first created being (*mubdaʿ*) and the notion of the Intellect "flowing" (*fayḍ*) from God.

75 See, for example, *Rasāʾil*, vol. 4, 486. For this Ismāʿīlī concept, see Daftary, *The Ismāʿīlīs*, 137–138.

76 See, for example, *Rasāʾil*, vol. 4, 486–488.

77 See Chapter 6 for a comparison of the terminological and doctrinal contents of the *Epistles* and the *Comprehensive Epistle* as regards the Ismāʿīlī question.

78 See, in particular, the positive appraisal of Abū Bakr, ʿUmar, and ʿUthmān in *Rasāʾil*, vol. 4, 74–75, 269; *On God and the World*, 283.

itual guide of humankind.[79] Saif has recently argued that the notion of imāmate in the *Epistles* corresponds better to Zaydite and Ibāḍite views than to Ismāʿīlīsm; this would also explain the apparent contradiction of the authors' positive appraisal of the first caliphs.[80] It is also true that the *Epistles* criticize the "seveners" (*subāʿiyya*) for placing excessive emphasis on the number seven, which could be a reference to the Ismāʿīlīs with their heptadic cycles.[81] Perhaps the greatest doctrinal divergence of the *Epistles* with regard to Ismāʿīlī theology, however, concerns the concept of God: for Ismāʿīlī authors, God is absolutely transcendent to the extent that even His description by negation is illicit.[82]

Some of these discrepancies could be explained away by the authors' deliberate ambiguity, necessitated by the hostilities between the Būyid rulers of Iraq and the Ismāʿīlī missionaries. Thus, if we take the authors to be of Ismāʿīlī affiliation, the brotherhood they advocate can be understood as a reference to the imāmate and the guiding intellect to the imām as an incarnation of the cosmic intellect.[83] But this still leaves the other problematic aspects which make it difficult to view the Brethren of Purity as orthodox Ismāʿīlīs. What seems clear, however, is that Ismāʿīlī thought exercised an influence on the *Epistles* regardless of whether their authors were in fact formally affiliated to some Ismāʿīlī denomination, or whether Ismāʿīlī ideas rather inspired them to construct their own freewheeling interpretation of Islam. The aim of this work is not to solve the question of authorship or the author's religious identity. The Ismāʿīlī question will, nevertheless, be of importance in the second part of this book with regard to the doctrinal relation between the *Epistles* and its two summaries.

---

79 Netton, "Brotherhood."

80 Liana Saif, "Ikhwān al-Ṣafā''s Religious Reform and Magic: Beyond the Ismaʿili Hypothesis," *Journal of Islamic Studies* 30, no. 1 (2019). The Zaydite hypothesis would also fit with the fact that the Būyid rulers were inclined toward Zaydism. On the other hand, the deliberate ambiguity of the *Epistles* could be interpreted against the background of Zaydite hostility toward the Ismāʿīlīs.

81 *On Music*, 111; *Sciences of the Soul and Intellect, Part 1: An Arabic Critical Edition and English Translation of Epistles 32–36*, ed. and trans. Paul E. Walker et al. (Oxford: Oxford University Press and the Institute of Ismaili Studies, 2016), 8.

82 For the conception of God in the *Epistles*, see Marquet, *La philosophie des Iḫwān al-Ṣafāʾ*, 49–63; for an overview of Ismāʿīlī theology, see Daniel De Smet, "Ismāʿīlī Theology," in *The Oxford Handbook of Islamic Theology*, ed. Sabine Schmidtke (Oxford: Oxford University Press, 2018), 313–324. The conception of God and creation in the *Epistles* versus the *Comprehensive Epistle* will be compared in Chapter 6.

83 For prophets and imāms as incarnations of the Intellect in al-Sijistānī and al-Kirmānī, see De Smet, *La quiétude de l'intellect*, 360–368; Walker, *Early Philosophical Shiism*, 114–123.

3 Dating

The question of when the *Epistles* were composed is largely intertwined with the problem of authorship. Thus, it is not surprising that the work's dating has also been contested in scholarship. Many scholars have based their dating on an assumption of the authors' Ismāʿīlī affiliation; from this basis, they have identified the "major event" predicted in the *Epistles* with either the founding of the Fāṭimid caliphate in 909 or the Fāṭimid conquest of Cairo in 969. Al-Tawḥīdī's account is, moreover, again of central importance in that he refers to the *Epistles* in 983–985 and describes the Brethren of Purity as a group that was active at that time. Al-Tawḥīdī, then, gives us the latest possible date for when the work must have existed. While it is in theory possible that the redaction of the *Epistles* would have continued beyond this, there is no strong external or internal evidence that points to a later date.[84] Internal evidence for the earliest possible date includes citations of poems, in particular by al-Mutanabbī (d. 965) and al-Bustī (d. 1010), a reference to the Ashʿarite theological school, whose eponymous founder died in 931, and the alleged influence of al-Fārābī (d. 950) on the list of qualifications of the just ruler provided in the *Epistles*.

In consequence, since Dieterici in the nineteenth century,[85] the most common view has been to date the *Epistles* in the period between the 960s and the 980s. However, this evidence alone by no means resolves the question because of (1) the possibility of a long period of redaction and (2) the problematic nature of the manuscript tradition. Surveying manuscripts, in fact, suggests that the few citations of relatively late poets and the single reference to Ashʿarites may be interpolations introduced to the manuscript tradition well after the work's completion.[86] Al-Fārābī's alleged influence can be explained by a common source.[87] Many scholars have therefore suggested a consider-

84 For a later dating, see Paul Casanova, "Une date astronomique dans les épîtres des Ikhwān al-Ṣafāʾ," *Journal Asiatique* 5 (1915). Based on calculations of astronomical conjunctions referred to in the *Epistles*, Casanova argued that the *Epistles* should be dated at 1047–1051. The argument is highly speculative, however.

85 Friedrich Dieterici, *Die Philosophie der araber im IX. un X. Jahrhundert n. Chr. aus der Theologie des Aristoteles, den Abhandlungen Alfārābīs und den Schriften der Lautern Brüder: Erstes Buch: Einleitung und Mikrokosmos (die Welt)* (Leipzig: J.C. Hinrichs, 1876), 141–142.

86 See Abbas Hamdani, "The Arrangement of the *Rasāʾil Ikhwān al-Ṣafāʾ* and the Problem of Interpolations," in *The Ikhwān al-Ṣafāʾ and Their Rasāʾil: An Introduction*, ed. Nader El-Bizri (Oxford: Oxford University Press and the Institute of Ismaili Studies, 2008), 83–100, at 92–97.

87 Ibid, 97, 199–201.

ably earlier date for the redaction of the *Epistles*. Both Marquet and Hamdani relate the redaction of the *Epistles* to Ismāʿīlī chronology. Marquet accepts al-Tawḥīdī's account as the end point but assumes a long period of redaction; the first parts of the *Epistles* would have been written before 909.[88] Hamdani rejects al-Tawḥīdī's testimony altogether and argues that the entirety of the *Epistles* was composed between 873 and 909.[89]

The postulation of veiled references to Ismāʿīlī history is a highly speculative basis for dating the *Epistles*. But there is considerable reason independent of any hypothesis of authorship to suggest an earlier date of composition. First, as stated above, the twin Andalusian works of the *Degree of the Sage* and *Aim of the Sage* are probably dated 950–953 and 954–960, respectively, which is likely to make them the earliest works that explicitly refer to and cite from the *Epistles*. This means that the work's Iraqi origin must predate at least the year 950; it is probably significantly earlier since it must have taken some time for the work to travel to al-Andalus.[90] De Callataÿ has further argued that the *Epistles* also influenced the Andalusian mystic Ibn Masarra (d. 931), which might necessitate an even earlier date.[91] Second, the *Epistles* contain evidence that the work underwent successive revisions. This makes it plausible to assume that the *Epistles* took their final form over a lengthy period, perhaps even during many generations. Marquet assumed a long period of redaction; both Baffioni and de Callataÿ have recently inclined toward the theory of the *Epistles* as a temporally stratified work.[92] This means that an earlier date does not necessarily contradict al-Tawḥīdī's account; it might instead represent the end point of a process of redaction, the origins of which could lie as early as in the second half of the ninth century.

---

88 For a more detailed account of Marquet's chronology, see pp. 40–45.

89 Abbas Hamdani, "Shades of Shiʿism in the Tracts of the Brethren of Purity," in *Traditions in Contact and Change: Selected Proceedings of the XIVth Congress of the International Association for the History of Religions*, ed. Peter Slater and Donald Wiebe (Waterloo: Wilfrid Laurier University Press, 1983), 447–460; Idem, "Brethren of Purity;" Idem, "The Arrangement of the *Rasāʾil Ikhwān al-Ṣafāʾ*."

90 For al-Qurṭubī's eastern journey in the 930s, during which he possibly became acquainted with the *Epistles*, see Godefroid de Callataÿ and Sébastien Moureau, "A Milestone in the History of Andalusī Bāṭinism: Maslama b. Qāsim al-Qurṭubī's *Riḥla* in the East," *Intellectual History of the Islamicate World* 5 (2017).

91 Godefroid de Callataÿ, "Philosophy and Bāṭinism in al-Andalus: Ibn Masarra's *Risālat al-Iʿtibār* and the *Rasāʾil Ikhwān al-Ṣafāʾ*," *Jerusalem Studies in Arabic and Islam* 41 (2014). For Ibn Masarra's father visiting Baṣra, see Sarah Stroumsa and Sara Sviri, "The Beginnings of Mystical Philosophy in al-Andalus: Ibn Masarra and His *Epistle on Contemplation*," *Jerusalem Studies in Arabic and Islam* 36 (2001): 201.

92 See, for example, Baffioni, "Ikhwân al-Safâ';" Godefroid de Callataÿ, "Brethren of Purity," in *Encyclopaedia of Islam Three*, ed. Kate Fleet et al. (Leiden: Brill, 2013), 85.

Third, there are philological and doctrinal arguments for an earlier date. The language and contents of the *Epistles* agree better with the Kindīan period of Arabic philosophy in the ninth century than the more consolidated Aristotelian tradition of tenth century Baghdād. Marquet believes that linguistic and doctrinal similarities suggest that the *Epistles* were composed around the same time as the Jābirean corpus, which Paul Kraus has dated between the second half of the ninth and the first half of the tenth centuries.[93] Hamdani has argued that the doctrinal contents of the *Epistles* imply that the work was composed during the period after al-Kindī (d. ca. 870), the first Muslim philosopher, and before al-Fārābī (d. 950), the most famous representative of the Baghdād school.[94] The *Epistles* do not cite contemporary philosophical or scientific authors by name, with the exception of the famous astrological authority Abū Maʿshar al-Balkhī (d. 866),[95] who was al-Kindī's student. Al-Kindī's treatise of philosophical definitions, *On the Definitions and Descriptions of Things* (*Risāla fī ḥudūd al-ashyāʾ wa-rusūmihā*), possibly later redacted by al-Kindī's students, seems to constitute a source for Epistle 41 (*Definitions and Descriptions*).[96] There are also various doctrinal parallels between the *Epistles* and the views held by al-Kindī and his circle of students.[97] Both accord high prominence to astrology;[98]

---

93 Marquet, *La philosophie des alchimistes*, 130–135.

94 Abbas Hamdani, "The Ikhwān al-Ṣafāʾ: Between al-Kindī and al-Fārābī," in *Fortresses of the Intellect: Ismaili and Other Islamic Studies in Honour of Farhad Daftary*, ed. Omar Alí-de-Unzaga (London and New York: I.B. Tauris and the Institute of Ismaili Studies, 2011), 189–212.

95 See Ibid, 196.

96 See Poonawala's introduction to this epistle in Ikhwān al-Ṣafāʾ, *Sciences of the Soul and Intellect, Part III: An Arabic Critical Edition and English Translation of Epistles 39–41*, ed. and trans. Carmela Baffioni and Ismail K. Poonawala (Oxford: Oxford University Press and the Institute of Ismaili Studies, 2017), 277–284.

97 For the general features of al-Kindī's school, see Peter Adamson, "The Kindian Tradition: The Structure of Philosophy in Arabic Neoplatonism," in *The Libraries of the Neoplatonists*, ed. Cristina D'Ancona (Leiden: Brill, 2007), 351–370. For some parallels between al-Kindī and the *Epistles*, see Alfred L. Ivry, *Al-Kindī's Metaphysics: A Translation of Yaʿqūb Ibn Isḥāq al-Kindī's Treatise "On First Philosophy"* (*fī al-Falsafah al-Ūlā*) (Albany: State University of New York Press, 1974), 20, 21, 170, 174. For the compatibility of the *Epistles* with the Kindīan school, see also de Vaulx d'Arcy, *Les Épîtres*, 22–60. The author develops this into the more specific argument that the *Epistles* should be attributed to al-Kindī's student al-Sarakhsī (d. 899).

98 For the primacy of astrology in al-Kindī, see Peter Adamson, "Abū Maʿšar, al-Kindī and the Philosophical Defense of Astrology," *Recherches de théologie et philosophie médievales* 69 (2002); Idem, *Al-Kindī* (Oxford: Oxford University Press, 2007), 191–206. The interest in astrology also extended to al-Kindī's students, al-Balkhī being the most influential astrologer in early Islamic history; a treatise by the title *On the Main Principles of Philosophy and the Establishment of Astrology* (*Kitāb Arkān al-falsafa wa-tathbīt aḥkām al-nujūm*)

and both attest to the doctrine of cosmic harmony in the form of microcosmic relations prevailing between parts of the cosmos and of the human body and soul.[99] Both are influenced by Neopythagorean and Platonic ideas concerning the primacy of mathematics;[100] and both authors show considerable interest in the Sabian religion of Ḥarrān.[101] In contrast, nothing in the *Epistles* seems to require postulating the influence of al-Fārābī or any other member of the Baghdād school.

If the beginnings of the *Epistles* went back to al-Kindī's school, a plausible date for the beginning of the work's composition would fall at the end of the ninth or the beginning of the tenth century. None of the above of course proves that the *Epistles* must be dated at so early a date; nor is it the purpose of this book to establish such dating. What this book aims to show instead is that the *Epistles* contain abundant evidence that the work that we now possess is the end result of a long and complex process of redaction. If we accept al-Tawḥīdī's account of a society of the Brethren of Purity during the second half of the tenth century, it may well be that the origins of that society go back to the times of al-Kindī's students.

4 The *Comprehensive* and *Super-Comprehensive Epistles*

Besides the *Epistles*, the Ikhwānian corpus consists of two works that summarize the *Epistles*: (1) the *Comprehensive Epistle* (*al-Risāla al-jāmiʿa*) proceeds from the first epistle to the last and thereby constitutes a systematic commentary of the entire *Epistles*; (2) the *Super-Comprehensive Epistle* (*Risālat Jāmiʿat al-jāmiʿa*) does not follow the epistle order but instead aims to recapitulate the essential doctrinal contents of the *Epistles*. Both works present themselves as pertaining to the authors of the original *Epistles* and employ the

---

is also attributed to al-Sarakhsī. For this treatise, see Franz Rosenthal, *Aḥmad b. aṭ-Ṭayyib as-Saraḫsî* (New Haven: American Oriental Society, 1943), 120–121.

99 For microcosmism in al-Kindī, see Adamson, *Al-Kindī*, 172–180; for the *Epistles*, see Inka Nokso-Koivisto, "Microcosm-Macrocosm Analogy in *Rasāʾil Ikhwān aṣ-Ṣafāʾ* and Certain Related Texts" (doctoral thesis, University of Helsinki, 2014).

100 For the idea of mathematics in the school of al-Kindī and the *Epistles*, see Gerhard Endress, "Mathematics and Philosophy in Medieval Islam," in *The Enterprise of Science in Islam: New Perspectives*, ed. Jan P. Hogendijk and Abdelhamid I. Sabra (Cambridge: MIT Press, 2003), 121–176, at 127–135; for Neopythagorean ideas in al-Kindī, see also Adamson, *Al-Kindī*, 27–28, 173–180. For both authors, the Neopythagorean ideas draw, in particular, on Nicomachus of Gerasa's (second cent. CE) *Introduction to Arithmetic* (*Arithmetikē eisagōgē*).

101 See Mattila, "Sabians."

formulaic expressions that characterize their style. There have been no systematic studies of either work; the precise relation between these two works and the *Epistles* has therefore remained unclear.

The *Comprehensive Epistle* claims to be more than a mere summary or commentary. The work presents itself as the culmination of the *Epistles* in the sense that it explains the hidden meanings of the ideas that were only alluded to in the original epistles. Thus, it supposedly forms an integral part of the *Epistles* as its final part which is indispensable for correctly understanding the meanings of the previous epistles. Most scholars have accepted, either implicitly or explicitly, that the two works were written by the same authors.[102] The theory of common authorship is supported by (1) the fact that the *Epistles* refer to the *Comprehensive Epistle* and (2) the relative terminological, stylistic, and doctrinal uniformity of the two works. Many scholars have also noted that the *Comprehensive Epistle* is more esoteric and explicitly Ismāʿīlī in character.[103] This could be taken to confirm the thesis that the *Comprehensive Epistle*'s aim is to unveil the hidden allusions of the *Epistles*, perhaps in the sense of revealing their Ismāʿīlī meanings. The first part of Kacimi's doctoral dissertation, the most extensive study of the work so far, seems to accept this in stating that the work's aim is not to summarize the entire *Epistles* but to reveal the "finality" of each epistle.[104] This assumes that the two works have shared objectives, as opposed to the *Comprehensive Epistle*'s author inserting novel readings that were not present in the original epistles. Since Kacimi also accepts that certain Ismāʿīlī ideas are more prevalent in the *Comprehensive Epistle*, he suggests that the work was written by the person who gave the *Epistles* its final, unified, form;

---

102 For statements supporting common authorship, see, for example, Carmela Baffioni, "Uso e rielaborazione degli autori classici nella *Risāla al-Ǧāmiʿa*," in *La diffusione dell'eredità classica nell'età tardoantica e medievale: Filologia, storia, dottrina: Atti del Seminario nazionale di studio (Napoli-Sorrento, 29–31 ottobre 1998)*, ed. Carmela Baffioni (Alessandria: Edizioni dell'Orso, 2000), 1–10, at 6; Hamdani, "The Arrangement of the *Rasāʾil Ikhwān al-Ṣafāʾ*," 90–91; Mourad Kacimi, "Reflexiones sobre la relación de la *Risālat ǧāmiʿat al-ǧāmiʿa* con los Iḫwān al-Ṣafāʾ," *eHumanista/IVITRA* 8 (2015): 329–330; Ehud Krinis, "*Al-Risāla al-Jāmiʿa* and Its Judeo-Arabic Manuscript," in *Islam: Identité et altérité: Hommage à Guy Monnot, O.P.*, ed. Mohammad Ali Amir-Moezzi (Turnhout: Brepols, 2013), 311–329; Netton, *Muslim Neoplatonists*, 2.

103 See, for example, Baffioni, "Ibdāʿ;" Corbin, *Histoire*, 193; de Callataÿ, "Brethren," 304–306; Diwald, *Arabische Philosophie*, 26–27; Ebstein, *Mysticism*, 45–46; Krinis, "*Al-Risāla al-Jāmiʿa*;" Shlomo Pines, "Shīʿite Terms and Conceptions in Judah Halevi's Kuzari," *Jerusalem Studies in Arabic and Islam* 2 (1980): 171–172, 74–77, 229.

104 Mourad Kacimi, "Edición crítica árabe y estudio de la *Risālat al-ŷāmiʿa ḏāt al-fawāʾid al-nāfiʿa, tāŷ Rasāʾil Ijwān al-Ṣafāʾ* (finales del s. IV H./X e.C., principios del s. V H./XI e.C.): Tomo I" (doctoral thesis, University of Alicante, 2015), 72.

this also explains the existence of references from the *Epistles* to the *Comprehensive Epistle*.[105]

There are also arguments against the theory of common authorship; some scholars, such as Carusi and De Smet, are convinced that the *Comprehensive Epistle* is a later work.[106] Besides the work's more esoteric nature, the manuscript tradition may provide a further argument. Apparently, no manuscripts exist that carry the *Epistles* and the *Comprehensive Epistle* as a single work.[107] This contradicts the assumption that the two works form a single whole, within which the *Comprehensive Epistle* represents the final part to be studied after the fifty-two epistles. Perhaps this could be explained by that the latter work, due to its esoteric nature, is aimed for a more restricted audience. But it is also the case that the transmission history of the *Comprehensive Epistle* is more intimately connected with the Ismāʿīlīs than that of the *Epistles*.[108] Indeed, most manuscripts attribute the work to either al-Majrīṭī or imām Aḥmad. If the two works were composed by the same authors, even in the qualified sense that their final redactor was the same person, one might expect this to be reflected in a shared manuscript tradition.

The *Super-Comprehensive Epistle* has received even less scholarly attention.[109] The title of the work suggests that it is a summary of the *Compre-*

105 Kacimi, "Edición crítica: Tomo I," 76–82; Idem, "Reflexiones," 329–330.

106 Paola Carusi, "Alchimia islamica e felicità nella Risāla Ĝāmiʿa: Inalterabilità delle sostanze e pace dell'anima," in *Le Felicità nel Medioevo: Atti del Convegno della Società italiana per lo studio del pensiero medievale (S.I.S.P.M.), Milano, 12–13 settembre 2003*, ed. Maria Bettetini and Francesco D. Paparella (Louvain-la-Neuve: Fédération internationale des instituts d'études médiévales, 2005), 277–296, at 277–278; Daniel De Smet, "La Risāla al-Jāmiʿa attribuée aux Ikhwān al-Ṣafāʾ: un précurseur de l'ismaélisme ṭayyibite?" in *Labor limae: Atti in onore di Carmela Baffioni*, ed. Antonella Straface, Carlo De Angelo, and Andrea Manzo (Naples: Centro di studi maġrebini, 2014–2015), 269–298, at 295–297.

107 A partial exception is the Tübingen manuscript, which contains a summary of the *Epistles* and the latter part of the *Comprehensive Epistle*. But even there, the combination seems to be accidental since the manuscript contains the entirety of neither the *Epistles* nor the *Comprehensive Epistle*.

108 For the employment of the *Comprehensive Epistle* by the Ṭayyibites of Yemen, and the possible implications this has for the work's authorship, see De Smet, "La Risāla al-Ĝāmiʿa;" Idem, "L'auteur des *Rasāʾil*," 155; Idem, "Henry Corbin et les études ismaéliennes," in *Henry Corbin: Philosophies et sagesses des religions du livre: Actes du colloque "Henry Corbin", Sorbonne, les 6–8 novembre 2003*, ed. Mohammad Ali Amir-Moezzi, Christian Jambet, and Pierre Lory (Turnhout: Brepols, 2005), 105–118, at 114–115; Hamdani, "*Rasāʾil*," 284; Ivanow, *A Guide to Ismaili Literature*, 20.

109 For studies of some of the doctrinal content of the work, see Carmela Baffioni, "Eléments platoniciens et néo-platoniciens de la représentation d'Adam dans la *Jāmiʿa al-Jāmiʿa*," in *Adorare caelestia, gubernare terrena: atti del colloquio internazionale in onore di Paolo*

hensive Epistle, rather than of the *Epistles*. Like the *Comprehensive Epistle*, it is more explicitly Ismāʿīlī in doctrine than the *Epistles*. Moreover, all known manuscripts are late and proceed from the Ismāʿīlī community of Salamiyya in Syria.[110] Since it is also the case that neither the *Epistles* nor the *Comprehensive Epistle* refers to this work, it seems likely at the outset that the work is a later Ismāʿīlī summary, and that it should not be attributed to the authors of the *Epistles*. Even if this turns out to be the case, however, it is still an important testimony on the evolution of the Ikhwānian corpus.

5 Sources

Until recently, all research on the *Epistles* was based on three uncritical editions printed in Bombay, Cairo, and Beirut in 1888, 1928, and 1957, respectively.[111] None of the three editions reveals its manuscript sources. In fact, all three constitute a single edition, insofar as they are copied from each other with minor correction of errors. Thus, the origin of the Beirut edition, which has in effect been the standard edition in recent decades, goes back to the Bombay edition made by Ṭayyibite Ismāʿīlīs, which, in turn, is probably based on a single manuscript of Ismāʿīlī origin.[112] At the time of writing, the project of Oxford University Press and the Institute of Ismaili Studies to produce critical editions for all the epistles is approaching its conclusion. These editions employ a subset of nineteen manuscripts that the editorial committee considered the most representative among the more than hundred manuscripts of the *Epistles* in the libraries of the world.[113] These include the oldest known manuscript of Atıf

---

Lucentini (*Napoli, 6–7 novembre 2007*), ed. Pasquale Arfé, Irene Caiazzo, and Antonella Sannino (Turnhout: Brepols, 2011), 99–112; Idem, "La figure d'Adam dans la *Jāmiʿa al-jāmiʿa*," in *Adam, le premier homme*, ed. Agostino Paravicini Bagliani (Florence: SISMEL edizioni del Galluzzo, 2012), 115–134.

110 See Kacimi, "Reflexiones".

111 For these three editions, see Poonawala, "Why We Need an Arabic Critical Edition," 47–56; Tibawi, "Ikhwān al-Ṣafā'," 32–33, 36–37.

112 See Omar Alí-de-Unzaga, "The Missing Link: MS 1040: An Important Copy of the *Rasāʾil Ikhwān al-Ṣafāʾ* in the Collection of the Institute of Ismaili Studies," in *Texts, Scribes and Transmission: Manuscript Cultures of the Ismaili Communities and Beyond*, ed. Wafi A. Momin (London: I.B. Tauris, 2022), 81–136; De Smet, "La Risāla al-Ǧāmiʿa," 296–297.

113 For the editorial plan of the critical editions, and the manuscripts employed, see Nader El-Bizri, "Prologue," in *The Ikhwān al-Ṣafāʾ and Their Rasāʾil: An Introduction*, ed. Nader El-Bizri (Oxford: Oxford University Press and the Institute of Ismaili Studies, 2008), 1–32, at 20–23.

Efendi 1681, dated 1182 CE, which most editors have used as their base manuscript, as well as most of the other most ancient manuscripts.

The Oxford edition does not provide a thorough survey of all existing manuscripts, nor have the editors of individual epistles been able to establish a *stemma codicum*. This suggests that the textual tradition of the work is complex, and the manuscript tradition perhaps diverged into different variants at an early date. It is also true that the oldest manuscript is at least 200, and possibly even 300, years posterior to the redaction of the *Epistles*. Thus, there remains a considerable time gap for which we have no direct manuscript evidence. All of this means that answers to many questions concerning the genesis of the *Epistles* may be unreachable through the manuscript evidence that we possess.

Despite the shortcomings, the new editions represent a huge improvement on the Beirut edition.[114] In fact, the Oxford editions have transformed our view of the *Epistles* as a uniform and stable text. The most important general revelation is the highly fluid nature of the textual tradition that the considerable variation between manuscripts manifests.[115] All epistles include textual vari-

114 For a detailed rebuttal of the Oxford edition, see Guillaume de Vaulx d'Arcy, "The *Epistles of the Brethren of Purity* Edited by the Institute of Ismaili Studies: When the Re-Edition of a Text Can Be Its Destruction," *Mélanges de l'Institut dominicain d'études orientales* 34 (2019). The author's major conclusions are: (1) the edition fails to offer a coherent edition and translation of the *Epistles* as a whole; and (2) the Beirut edition presents a more trustworthy text as compared not only to the new edition but also to MS Atıf Efendi 1681, which most editors have chosen as their base manuscript. As regards the first conclusion, many of the shortcomings listed by the author are real. Since the editors of the individual epistles rely on different base manuscripts, subsets of manuscripts, and editorial methods, the edition of the *Epistles* does not represent a uniform view on the work as a whole. This is the case even though in the manuscript tradition the work is mostly transmitted as a single work. Since there was no effort to harmonize the translations, identical terms and phrases, including the stock expressions running throughout the *Epistles*, are translated differently from one epistle to another. As for the second conclusion, it does not follow that the Beirut edition is superior to the Oxford edition, even if the Beirut edition, or any given manuscript, obviously contains a more uniform text. On the contrary, while the manuscript basis of the Beirut edition is unknown, the new edition shows the discrepancies within the manuscript tradition. While it is undoubtedly true that the Beirut edition should be considered one variant along with the manuscripts, the author fails to demonstrate his claim that the Beirut edition is based on the most ancient testimonies. It seems extremely unlikely that the manuscript on which the Bombay editor based his edition just happened to possess the authentic text. As this study will also show, the Beirut edition must in fact represent a late version of the text.

115 The footnotes in Table 1 present a summary of the manuscript variation for each individual epistle, based on the technical introductions of the editors complemented by an incomplete review of manuscripts.

ation that goes beyond what is habitual for a well-established work during this period; most epistles also involve lengthy additions or omissions in some manuscripts. Particularly interesting are additions with Ismāʿīlī characteristics, as they reflect the Ismāʿīlī inclinations of some later copyists.[116] Interestingly, some of these added passages also appear in the *Comprehensive* and *Super-Comprehensive Epistles*, which are of clearly Ismāʿīlī nature. Beyond this, for some epistles, the manuscripts carry entirely different versions of a particular epistle. The most striking case is Epistle 52 (*Magic*), for which the manuscripts provide three variants with no textual relation to each other.

The new critical editions are indispensable for the present book since many of its arguments are based on manuscript variation. Besides the evidence given by the editions, manuscripts are also employed directly. This is necessary, first, because, at the time of writing, critical editions for all epistles are not yet available. Second, even for the epistles for which critical editions exist, the editors have not always noted all the variant readings that are relevant for the purposes of this book. Third, the book also makes use of manuscripts that are not included in the critical editions. This will hopefully make the manuscript evidence more conclusive, even if a comprehensive survey of all existing manuscripts is beyond the scope of this study.

I have consulted a total of twenty manuscripts of the *Epistles* for this book. These include nine manuscripts used in the Oxford editions, to which I will refer by the Arabic letters employed in that edition:

1) Istanbul, Atıf Efendi 1681 (dated 578 AH/1182 CE) [ع]. This is the oldest known manuscript for the *Epistles*. Most editors of the Oxford editions have chosen it as their base manuscript, not only because of its antiquity but also because of its reliable readings.
2) Paris, Bibliothèque nationale de France, 6647–6648 (675/1275) [د]
3) Istanbul, Esad Efendi 3638 (686/1287) [أ]
4) Istanbul, Esad Efendi 3637 (ca. seventh/thirteenth cent.) [ن]
5) Istanbul, Feyzullah Efendi 2130–2131 (704/1304) [ف, ق]. These two manuscripts contain the first and second, and the third and fourth, parts of the *Epistles*, respectively.
6) Istanbul, Köprülü 871 (820/1417) [ل]
7) Istanbul, Köprülü 870 (ca. 857–886/1453–1481) [ك]
8) Paris, Bibliothèque nationale 2303 (1020/1611) [ر]
9) Paris, Bibliothèque nationale 2304 (1065/1654) [ز]

116 These additions, studied in a series of articles by Carmela Baffioni, will be discussed in more detail in Chapter 6.

In addition, I have employed the following manuscripts of the *Epistles* not used by the Oxford editions, to which I will refer with Latin letters:

10) A private manuscript sold at a Sotheby auction on April 26th, 2017 (683/1284) [A]. The manuscript contains the second half of the third section and the entire fourth section of the *Epistles*.
11) Istanbul, Yeni Camı 1199 (887/1482) [B]
12) Princeton University Library, Garrett Collection 4263 (956/1549) [C]
13) Paris, Bibliothèque nationale 2305 (1153/1740) [D]
14) Istanbul, Damat Ibrahim 808 (undated) [E]
15) Istanbul, Raghıp Paşa 839 (undated) [F]
16) Istanbul, Raghıp Paşa 840 (undated) [G]

I have also consulted the following incomplete manuscripts of the *Epistles*:

17) Universitätsbibliothek Tübingen, MaVI86 (612/1215) [H]. The first half of the manuscript (fols. 1–41) contains a summary of the first twenty-two epistles; the second part carries the latter half of the *Comprehensive Epistle*.
18) Istanbul, Amcazade Hüseyn Paşa 452 (1090/1680) [I]. The manuscript contains the first part of the *Epistles* followed by Epistles 33, 25, and 41.
19) Istanbul, Carullah 982 (undated) [J]. The manuscript contains a selection of twenty-two epistles.
20) Paris, Bibliothèque nationale 6000 (undated?) [K]. The manuscript contains the first eight epistles.

Besides the consulted manuscripts and critical editions, the Beirut edition will also be an important testimony. Its text is clearly based on a manuscript, which, in many cases, is at variance with most other manuscripts. It is also of major significance because most research, including some misconceptions, was until recently based on its text.

Regarding the two summaries, the arguments of this book are not based on manuscript variation but on differences between the *Epistles* and these summaries. For the *Comprehensive Epistle*, Kacimi's critical edition, based on sixteen manuscripts,[117] will provide the primary source, which is systematically compared with the Ghālib edition. Based on the critical edition, the text of the *Comprehensive Epistle* is much more stable than that of the *Epistles*. Concerning the *Super-Comprehensive Epistle*, not many manuscripts are known in any case. Thus, the final chapter of this book will employ the edition based on the three known manuscripts.[118]

117 For a survey of the manuscripts of the *Comprehensive Epistle*, see Kacimi, "Edición crítica: Tomo I," 161–205.

118 For the known manuscripts and the Tāmir edition, see Kacimi, "Reflexiones," 331–332.

PART 1

Redaction History of the Epistles

CHAPTER 1

Gradual Redaction

The *Epistles* of the Brethren of Purity, as we have them in the Beirut and Oxford editions, consist of fifty-two tractates; each tractate is devoted to a single science or subject matter. The fifty-two epistles are grouped into four main sections: (1) mathematical-logical, (2) physical, (3) psychical-intelligible, and (4) nomic-theological; the four sections are preceded by an introductory *Fihrist*, or table of contents, which introduces the contents and aims of both the *Epistles* as a whole and each epistle individually.[1] The epistles proceed gradually along the curricular order that medieval Arabic authors inherited from Late Antiquity. Each epistle functions well enough as an independent tractate, but together they constitute a relatively coherent unity of an encyclopedia of philosophical and religious sciences.

However, there is considerable evidence that the *Epistles* did not always possess the form that they have now. First, as regards the total number of epistles, the text of the *Epistles* consistently refers to fifty-one, rather than fifty-two, epistles.[2] Among contemporary witnesses, al-Tawḥīdī speaks of fifty epistles,[3] and al-Qurṭubī alternately of fifty or fifty-one epistles.[4] Moreover, in all the most ancient manuscripts, there are in fact only fifty-one epistles.[5] Second, when viewed against the late ancient and early Islamic schemes of classification of sciences, many epistles seem to occupy an arbitrary, or even erroneous, position within the whole. Third, the *Epistles* include passages on the order of either

1 For the arrangement of the *Epistles*, see Table 1.

2 *Rasāʾil*, vol. 3, 29, 408, 538; vol. 4, 64, 173, 186; *On Magic: An Arabic Critical Edition and English Translation of Epistle 52, Part 1*, ed. and trans. Godefroid de Callataÿ and Bruno Halflants (Oxford: Oxford University Press and the Institute of Ismaili Studies, 2011), 6, 8, 104; *On the Natural Sciences*, 413; *On Composition*, 115; *On God and the World*, 353; *On Life, Death, and Languages: An Arabic Critical Edition and English Translation of Epistles 29–31*, ed. and trans. Eric Ormsby (Oxford: Oxford University Press and the Institute of Ismaili Studies, 2021), 39; *On Ethics*, 83, 171. The Beirut edition refers to fifty-two epistles three times in the *Fihrist* (*Rasāʾil*, vol. 1, 21, 43) and a single time in Epistle 1 (Ibid, 77). In most manuscripts, the number is fifty-one also in these cases, or non-existent in the case of the last reference in the *Fihrist*.

3 Al-Tawḥīdī, *Kitāb al-Imtāʿ*, vol. 2, 5. See also al-Sijistānī, *The Muntakhab*, § 309, 161. Al-Sijistānī mentions fifty-one epistles. But we only possess a later redaction of the work; the redactor could have updated the number of epistles.

4 The relevant passages of the *Degree of the Sage* (*Rutbat al-ḥakīm*) are cited in de Callataÿ, "Magia," 317, 325.

5 See Table 1.

the sciences or of the epistles which contradict the current configuration of the work. Fourth, the *Epistles* and the manuscripts that carry them contain various further indications that attest to successive redactions. For many epistles, manuscript variation is substantial enough to warrant the conclusion that the contents of the epistle in question evolved in time. Differences concerning the positions and titles of epistles suggest similar conclusions for the arrangement of the work as a whole; this is supported further by contradictions between epistle transitions and the current order of the epistles. Finally, an analysis of cross-references between the epistles implies that the epistles were not composed in the order in which they now appear. When all this evidence is taken together, it will be possible to formulate a hypothesis on what the redaction history of the *Epistles* was like.

The problem of the *Epistles*' redaction has not aroused as much scholarly interest as questions concerning the work's authorship or absolute dating. Nevertheless, a few scholars have formulated a theory of redaction.[6] Previous theories have been founded primarily on a strong hypothesis of authorship, and only secondarily on the internal evidence provided by the *Epistles*. For Marquet, the *Epistles* present the official doctrine of the Fāṭimids;[7] thus, the establishment of the Fāṭimid caliphate in North Africa in 909 is a key date for his chronology. According to Marquet, then, the fourth section of the *Epistles*, apart from Epistles 42 (*Opinions and Doctrines*), 51 (*Constitution of the World*), and 52 (*Magic*), represents the work's initial core.[8] The remaining nine epistles deal with religious subjects, including philosophical theology and cosmology, and represent the authors' original doctrinal views that conform to the early Fāṭimid doctrine. Passages in Epistles 48 (*The Call to God*) and 50 (*Governance*) presage the coming age of the virtuous and therefore predate 909;[9] Epistle 48, in effect, constitutes a manual for Fāṭimid propaganda, and is thus the very earliest epistle.[10] Epistle 44 (*Beliefs of the Brethren of Purity*) rejoices over a

6 For an overview of the chronologies proposed by Marquet, Hamdani, and Madelung, see Table 2. In Marquet's case, the table is based on remarks scattered over publications spanning several decades.

7 See p. 21.

8 Marquet, *La philosophie des Iḫwān al-Ṣafāʾ*, 8–15; Idem, *Les "Frères de la pureté"*, 5–6, 9, 31–32, 46, 52–53, 58, 68, 81–83, 102, 110, 157–158, 185, 186, 188, 226, 232–236, 237, 240, 250, 253, 270–271, 273, 274, 293, 305, 313, 322–323, 343, 369–372. The latter work (9) suggests that Epistle 52 is composed of old material but was added to the *Epistles* at a late stage.

9 For Epistle 48, the first passage is present only in the Beirut edition (*Rasāʾil*, vol. 4, 146) but not in the most ancient manuscripts; Epistle 48's second passage and the passage in Epistle 50 occur in *On God and the World*, 241–246, 282–286; *The Call to God*, 184–190.

10 Marquet, "Ikhwān al-Ṣafāʾ," 1072–1073; Idem, "Iḫwān al-Ṣafāʾ," 234. For the suggestion that

recent victory, and therefore can be dated 910;[11] a poem in Epistle 47 (*Divine Law*) contains a veiled reference to Fāṭimid caliph al-Qā'im (r. 934–946), and should therefore be dated during his reign.[12] Marquet also suggests that Epistle 43 (*The Way to God*) at one point formed an introduction to the fourth section of the *Epistles*;[13] this would presumably make it relatively early, although later than the very first epistles.

Like many scholars today, Marquet proposed that the work we have now is a result of a lengthy period of redaction. During a time-span of eighty years or so, the *Epistles* were gradually expanded around the inner core of nine epistles of the fourth section; the technical epistles, that is, most epistles in the first and second sections, were the last to be written.[14] Among these, however, Epistle 3 (*Astronomy*) was composed early, and thus it only slightly postdates the epistles of the fourth section.[15] The final redaction took place during the two decades between the Fāṭimid conquest of Cairo in 961 and the time of al-Tawḥīdī's report in 980.[16]

Marquet's main thesis relies on the Fāṭimid hypothesis. But his conclusions about the gradual expansion of the work are also based on internal evidence. Thus, according to Marquet, the first section of the work initially only contained five epistles, including a single epistle on logic; Epistles 5 (*Music*) and 6 (*Arithmetical and Geometrical Proportions*) similarly formed one epistle; Epistle 4 (*Geography*) was added later, and Epistles 7 (*Theoretical Sciences*) and 8 (*Practical sciences*) were moved to the first part from the third section and Epistle 9 (*Ethics*) from the second section of the *Epistles*.[17] The second section initially contained eight epistles but was later expanded by the insertion of additional epistles; the text of existing epistles was also expanded, and some epistles were divided into multiple epistles.[18] Thus, for example, the famous

---

Epistle 45 was written immediately after Epistle 48, see also Idem, *Les "Frères de la pureté"*, 46–53.

11 Marquet, "Iḫwān al-Ṣafā';" Idem, "910 en Ifrīqiyā: une épître des Iḫwān al-Ṣafā'," *Bulletin d'études orientales* 30 (1978): 234. Later, Marquet (*Les "Frères de la pureté"*, 40) was no longer entirely convinced that this epistle refers to the year 910.

12 Yves Marquet, "Note annexe: à propos d'un poème ismaïlien dans les épîtres des Iḫwān aṣ-Ṣafā'," *Studia Islamica* 55 (1982).

13 Marquet, *Les "Frères de la pureté"*, 32.

14 Marquet, *La philosophie des alchimistes*, 131.

15 Marquet, *Les "Frères de la pureté"*, 32, 40. Specifically, Marquet argues that Epistle 3 is anterior to Epistle 1, and by extension Epistle 2, since Marquet considers the latter two epistles to be closely connected.

16 Marquet, "Ikhwān al-Ṣafā'," 1072; Idem, *La philosophie des Iḫwān al-Ṣafā'*, 8.

17 Marquet, "Ikhwān al-Ṣafā'," 1073; Idem, *La philosophie des Iḫwān al-Ṣafā'*, 11–13.

18 Marquet, "Ikhwān al-Ṣafā'," 1073; Idem, *La philosophie des Iḫwān al-Ṣafā'*, 12–13.

animal fable that currently fills most of Epistle 22 (*Animals*) was written as the very last part of the entire *Epistles*.[19] In this case, the rationale again relates to the Fāṭimid hypothesis of authorship: the ending of the story supposedly contains a reference to ruling imāms.[20] The passage with this reference is only present in the Beirut edition, however, but not in most manuscripts. Thus, the example highlights the problematic nature of Marquet's enterprise of dating epistles based on their supposed allusions to Fāṭimid history; many such allusions have clearly been introduced to the manuscripts at a much later stage.

Other sections of the *Epistles* also bear evidence of these methods of expanding and repositioning epistles. Thus, for example, Epistle 42 (*Opinions and Doctrines*) complements the discussion of doctrinal diversity in Epistle 31 (*Differences of Languages*); it was therefore added to the fourth section at a later date; Epistle 41 (*Definitions and Descriptions*) initially occupied a different position within the work.[21] Marquet was, of course, also aware that most of the internal references, even in the Beirut edition text he used, give the number of epistles as fifty-one, rather than fifty-two. Thus, he concluded that Epistle 51 (*Constitution of the World*) is spurious because it reproduces sections of Epistle 21 (*Plants*).[22]

In his formidable career of research on the *Epistles*, Marquet only had the Beirut edition text at his disposal. Even that text, despite its false impression of uniformity, contains internal contradictions. Many of Marquet's conclusions are based on these contradictions. These very same contradictions form a large part of this book's argument in the following two sections of this chapter and the second section of the next chapter. Thus, the conclusions concerning the gradual expansion, rearrangement, and division of epistles will be partly similar. Like Marquet, this book will, then, claim that the final form of the *Epistles* was not the initial plan but the result of a complex process of redaction. However, Marquet seems to have been unaware of the major variation that many epistles manifest in manuscripts. This manuscript variation reinforces Marquet's general thesis. Marquet's theory of the relative chronology of the epistles, in contrast, was mainly founded on the Fāṭimid hypothesis of authorship. In this question, this book will arrive at diametrically opposite results.

---

19 Marquet, "Iḫwān al-Ṣafāʾ," 234.

20 *Rasāʾil*, vol. 2, 377. The concluding phrase states: "The epistle of animals was completed by the help of the Creator of created beings and Muḥammad and his family, the rightly guided imāms."

21 Marquet, *La philosophie des alchimistes*, 131–132; Idem, *La philosophie des Iḫwān al-Ṣafāʾ*, 13–15.

22 Marquet, *La philosophie des Iḫwān al-Ṣafāʾ*, 11. The corresponding sections of Epistles 21 and 51 are *On the Natural Sciences*, 452–464; *On God and the World*, 336–351.

While Hamdani's starting point is also in Fāṭimid history, regarding the general thesis of the redaction process, his theory is almost the exact opposite of Marquet's.[23] Thus, Hamdani's assumptions about the changes that the *Epistles* underwent are much more conservative. For Hamdani, the entirety of the *Epistles* was composed between the occultation of the twelfth Shīʿite imām in 873 and the establishment of the Fāṭimid caliphate in 909. Epistles 48 (*The Call to God*) and 50 (*Governance*), predicting the imminent age of the virtuous, therefore pertain to the last epistles that were written. In general, the authors followed a carefully thought-out plan, and consequently composed the epistles in roughly the order and form in which they appear now. Epistles 51 (*Constitution of the World*) and 52 (*Magic*) later switched places, and thus represent a minor exception to this rule. As a result, Hamdani's relative chronology is also diametrically opposed to Marquet's; the epistles of the fourth section were the last to be composed. The introductory *Fihrist* was composed after the third section, and the writing of the *Comprehensive Epistle* began around the same time. The redaction of the Ikhwānian corpus was completed by the final editor's rushed copy-editing to update the *Fihrist* and some of the cross-references to correspond to the work's final form. Any internal contradictions that remain are presumably due to the unfinished nature of this final phase of redaction, as well as to later interpolations in the manuscripts.

Based on the considerable variation that the *Epistles* manifest in the manuscripts, Hamdani's thesis of a straightforward composition process seems untenable. The theory is ultimately founded on the Beirut edition text's appearance of a carefully planned work, which the new critical editions have shown to be false. The fact that Marquet and Hamdani, despite their shared premise of Fāṭimid authorship, arrived at precisely opposite results, also shows the arbitrariness of the method of binding the *Epistles* to alleged hidden references to Fāṭimid history. Still, as regards redaction chronology, this book's conclusions are closer to those of Hamdani than of Marquet.

More recently, Madelung has proposed a further theory of redaction chronology, which takes the manuscript variation into account.[24] Once again, the theory is tied to a strong hypothesis of authorship. For Madelung, al-Qurṭubī, the Andalusian scholar to whom the alchemical and magical works of *Degree of the Sage* and *Aim of the Sage* are now attributed, is also the final redactor of the *Epistles*.[25] Thus, Madelung relates the manuscript variants of some epistles

23 Hamdani, "The Arrangement of the *Rasāʾil Ikhwān al-Ṣafāʾ*," 88–91.
24 Madelung, "Maslama."
25 For this theory of authorship, see pp. 17–19.

to both Ismāʿīlī history and al-Qurṭubī's personal history. He assumes that the founding of the Brethren of Purity is related to the split of the Ismāʿīlī *daʿwa* into several factions in 899, and that the composition of the *Epistles* consequently began sometime after this. During his eastern travels in the 930s, al-Qurṭubī joined the society of the Brethren of Purity; he then took charge of the redaction of the *Epistles*, and gave the final touch to the *Epistles* after his return to al-Andalus around 937.

In consequence, Madelung postulates three, or even four, versions of the *Epistles*. Epistle 7 (*Theoretical Sciences*) presents a plan for the *Epistles*, which does not entirely correspond to their current configuration; thus, it pertains to the earliest epistles. The first redaction of the work, which predates al-Qurṭubī's arrival in Iraq, consists of fifty epistles that follow the four-division of sciences presented in Epistle 7.[26] The epistle variants of 32a, 49a, and 52c also pertain to this original redaction.[27] The second redaction was made by al-Qurṭubī in Baṣra, and consists of fifty-one epistles grouped under a new four-division, which corresponds to an earlier variant of the *Fihrist*.[28] It includes the epistle variants 32b, 42a, 50a, and 52a, as well as the variant of Epistle 12 (*On Interpretation*) in which Epistles 12 and 13 constitute a single epistle. The final redaction made by al-Qurṭubī after his return to al-Andalus corresponds to a later variant of the *Fihrist*, and includes the epistle variants 32c, 42b, 49b, 50b, and 52b, as well as Epistles 12 and 13 as separate epistles and the *Comprehensive Epistle*. The *Epistles* still underwent some changes after al-Qurṭubī's death in 964, including Epistles 4 (*Geography*) and 5 (*Music*) shifting places.

In sum, Madelung does not present so much a theory about the order in which the epistles were composed as about the chronology of the different epistle variants that the new critical editions have revealed to exist. Thus, like Marquet, but based on novel evidence, Madelung argues for a fairly long period of redaction. His theory is highly interesting and agrees with the current book's aim of employing manuscript evidence for the end of reconstructing a redaction history of the *Epistles*. But the relative dating of epistle variants based on al-Qurṭubī's supposed role as the master redactor is highly speculative. Madelung never properly argues for al-Qurṭubī's authorship; nor does he provide a clear rationale for his dating of the variants. It is also possible to approach

26 For the classification of sciences in Epistle 7, see Table 4.

27 For the different variants of each epistle, see Table 1.

28 While there is no critical edition for the *Fihrist*, Madelung is aware that the text is very different in the Beirut edition versus MS ع; thus, he postulates two variants for the text. In fact, as will be shown in the last section of this chapter, the manuscripts contain more than two variants of the *Fihrist*.

the problem of redaction chronology without taking any stance on the question of authorship. This is precisely the approach of this book. In the three chapters of the first part of this book, I will present seven kinds of internal evidence that support the thesis of gradual redaction. At the end of the third chapter, I will finally provide a hypothesis about the *Epistles*' redaction history based on that evidence.

1 Rationale for the Order of the Epistles

Most of the evidence presented in this book consists of internal inconsistencies in the *Epistles*. The first contradiction concerns the overall structure of the work as we currently have it. Set against the customary schemes of classifying philosophical sciences during this period, the order of the epistles makes only partial sense. That is, the authors subscribe to a specific rationale in their arrangement of the epistles, but the way the work is ordered in part contradicts that rationale. Moreover, the authors present several explicit curricular classifications, which, in most cases, also in part contradict the actual order of epistles. These inconsistencies suggest that the original plan for the *Epistles* was different, and thus provide evidence for the work's evolution. The current section compares the configuration of the *Epistles* with the standard classificatory schemes; the next section addresses the authors' explicit statements on the classification of sciences.

As is well-known, in the late ancient Platonic schools of Alexandria and Athens, Aristotle's works, designed to precede the study of Plato, were grouped into a set curricular order: (1) logic, consisting of eight works, including *Rhetoric* and *Poetics* in the expanded *Organon*, and preceded by Porphyry's (d. ca. 305 CE) *Eisagōgē*; (2) physics, consisting of eight works; and (3) *Metaphysics*.[29] The position of logic at the beginning is justified by its instrumental value for all philosophical studies; otherwise, the curricular progression from physics to metaphysics corresponds to a pedagogical one from lower to higher knowledge,

29 See, in particular, Dimitri Gutas, "Paul the Persian on the Classification of the Parts of Aristotle's Philosophy: A Milestone between Alexandria and Baġdâd," *Der Islam* 1 (1983); Idem, "The Greek and Persian Background;" Christel Hein, *Definition und Einteilung der Philosophie: von der spätantiken Einleitungsliteratur zur arabischen Enzyklopädie* (Frankfurt am Main: P. Lang, 1985); Leendert G. Westerink, "The Alexandrian Commentators and the Introductions to their Commentaries," in *Aristotle Transformed: The Ancient Commentators and Their Influence*, ed. Richard Sorabji (Ithaca: Cornell University Press, 1990), 325–348.

or from the existents of the material world toward their ultimate immaterial causes.[30] While Aristotle composed no mathematical works, the mathematical *quadrivium* of arithmetic, geometry, astronomy, and music either occupied the position of Platonic propaedeutics preceding the study of philosophy, or was placed after logic.[31] Practical philosophy followed logic in the curricular order, and was divided into (1) ethics, (2) economics, and (3) politics on quantitative grounds; that is, these sciences are concerned with the good of an individual, a household, and a polity, respectively.

Practically all philosophers of the Islamic world adopted the late ancient curricular views, although in slightly different forms. For the Brethren of Purity, the chronologically most relevant classificatory accounts are presented by al-Kindī and al-Fārābī.[32] Both authors emphasize the normative basis of classification of philosophy in the sense that the philosophical sciences should be studied in an order that corresponds to a gradual progression from lower to higher knowledge.[33] For al-Kindī, mathematics forms a necessary prerequisite for the study of philosophy.[34] Theoretical philosophy consists of four parts: (1) logic, (2) physics, (3) psychical existents, and (4) theology. The objects of study for the last three parts are distinguished by the ontological grounds that they are respectively (1) entirely material, (2) immaterial in essence but associated with matter, and (3) entirely immaterial.[35] Al-Kindī further specifies several Aristotelian works for each part, which constitute the individual sciences or fields of knowledge for that section.[36] As regards both the major and minor

30 See, for example, the excerpts translated in Richard Sorabji, *The Philosophy of the Commentators, 200–600 AD: A Sourcebook: Volume 1: Psychology (with Ethics and Religion)* (Berkeley, Los Angeles, and London: University of California Press, 2004), 319–321.

31 Hein, *Definition*, 170–225.

32 Al-Fārābī, *Iḥṣāʾ al-ʿulūm li-Abī Naṣr al-Fārābī*, ed. ʿUthmān Amīn (Cairo: Maṭbaʿat al-saʿāda, 1931); al-Kindī, "Risālat al-Kindī fī kammiyyat kutub Arisṭūṭālīs wa-mā yuḥtāj ilayhi fī taḥṣīl al-falsafa," in *Rasāʾil al-Kindī al-falsafiyya*, ed. Muḥammad Abū Rīda (Cairo: Dār al-fikr al-ʿarabī, 1950), 363–384. A comparative summary of these classifications is presented in Table 3.

33 For al-Kindī, see "Risālat al-Kindī," 365, 378. Al-Fārābī states this most explicitly in his second curricular text, the first part of the *Attainment of Happiness* (*Taḥṣīl al-saʿāda*), where the progression of philosophical sciences is portrayed as a gradual ascent from effects to causes. See, for example, al-Fārābī, *Kitāb Taḥṣīl al-saʿāda*, ed. Jaʿfar Āl Yāsīn (Beirut: Dār al-Andalus, 1981), 52–55.

34 Al-Kindī, "Risālat al-Kindī," 365, 369–370, 376–378.

35 Ibid, 364–365.

36 See Table 3. The corresponding works by Aristotle for each section are presented in Ibid, 365–368 (logic) and Ibid, 368 (physics, psychical existents, and theology). It may be noted that the logical section excludes Porphyry's *Eisagōgē*.

divisions, for the most part, al-Kindī's scheme of theoretical philosophy is faithful to the late ancient classification. The main exception is that al-Kindī elevates psychology into an independent section, as opposed to forming the final part of physics. Moreover, al-Kindī identifies the sphere of study for metaphysics exclusively with theology, even though only the last book of Aristotle's *Metaphysics* deals with theological subject matter. As for practical philosophy, al-Kindī, like all subsequent Arabic philosophers, transposes it to the very end of the philosophical curriculum. This is justified on the grounds that practical philosophy, which serves the practical end of becoming virtuous, represents the "fruit" (*thamara*) of theoretical sciences,[37] apparently in the sense that knowledge about virtue is in some sense grounded in theoretical knowledge.

Al-Fārābī's classificatory account is similarly based on late ancient precedents. Consequently, it follows the same general scheme where philosophy is divided into (1) logic, (2) physics, (3) metaphysics, and (4) practical philosophy. But it also differs from al-Kindī's classification in four main respects. First, al-Fārābī transfers mathematics to a position between logic and physics. The rationale is that logic is a necessary instrument for the study of philosophy, rather than an actual part of philosophy;[38] mathematics, in contrast, forms a proper first step for the study of existents insofar as its objects are in themselves abstract but also have extramental existence as instantiated in matter.[39] Second, al-Fārābī retains the late ancient classification where psychology forms a part of physics, as opposed to constituting an independent sphere of study. Third, he introduces linguistic sciences as a novel philosophical propaedeutics that prepares the student for the study of logic, in particular.[40] Fourth, he expands on the contents of some of the main sections. He adds three new sciences to the mathematical *quadrivium*, as well as dividing metaphysics into three subsections—all based on Aristotle's *Metaphysics*—that investigate existents as existents, the first principles of theoretical sciences, and immaterial existents, respectively.[41] Finally, as regards practical philosophy, al-Fārābī treats it as a single "political science" (*al-ʿilm al-madanī*) that comprises both ethics and political philosophy.[42] He then complements the political science with the

37 See Ibid, 369, 384. For Avicenna's similar statement that the final section of metaphysics, dealing also with the question of happiness and afterlife, is the "fruit" of physics and metaphysics, see Avicenna, "Aqsām al-ʿulūm al-ʿaqliyya," in *Tisʿ rasāʾil fī al-ḥikma wa-l-ṭabīʿiyyāt* (Cairo: Dār al-ʿarab li-l-bustānī, 1326 AH), 104–118, at 114–116.

38 Al-Fārābī, *Iḥṣāʾ*, 11–12.

39 Ibid, 35.

40 Ibid, 3–11.

41 Ibid, 60–63.

42 Ibid, 64–66.

Islamic sciences of jurisprudence (*fiqh*) and theology (*kalām*), for which he accords the ancillary functions of inference of particular moral and legal judgments, and of dialectical defense of religious beliefs, respectively.[43]

The classifications by al-Kindī and al-Fārābī differ in important respects. This shows that the classical paradigm could be adopted in slightly different forms, as well as expanded with novel additions to better fit the Islamic context. Yet, the division of theoretical philosophy itself invariably follows an ontological ascent that starts with purely corporeal existents, then proceeds through the ensouled existents of plants and animals up to the human being endowed with intellect, and finally arrives at the purely incorporeal beings of the celestial world, and ultimately God. The entirety of the curricular order is justified by either ontological or pedagogical grounds; no philosophical science or subject of study occupies an arbitrary position within the whole.

The authors of the *Epistles* share this general principle with al-Kindī and al-Fārābī. They emphasize that the epistles should be studied in exactly the order in which they appear, because they constitute a pedagogical program that progresses toward gradually higher objects of knowledge.[44] The *Comprehensive Epistle*, similarly, employs with reference to the epistles the term "ladder of salvation" (*sullam al-najāt*), suggesting that the conveyed knowledge consists of ascending steps that ultimately bring about the soul's salvation in the afterlife.[45] The position of mathematics at the beginning is justified by the authors' Pythagorean belief in the ontological primacy of numbers.[46] While the authors never clearly justify the division of the *Epistles* into precisely these four sections,[47] they explicitly state that each section leads a student toward a gradually higher form of knowledge:

> Know, O brother, may God support you and us with His spirit, that the goal of the wise monotheist philosophers (*al-falāsifa al-ḥukamāʾ al-muwaḥḥidūna*) in investigating the propaedeutic sciences (*al-naẓar fī al-ʿulūm al-riyāḍiyya*), and training their students in them, is to proceed and arrive from them to the physical sciences. As for their goal in investigating the

43 Ibid, 69–77.

44 See, for example, *Rasāʾil*, vol. 1, 46; *On Magic*, 6–7. The former reference occurs in the final part of the *Fihrist* which, as we will see in the last section of this chapter, is only present in its later variant.

45 *Al-Jāmiʿa*, eps. 1, 7; ed. Ghālib, 17, 108; ed. Kacimi, 5, 117.

46 *On Arithmetic & Geometry*, 8–9.

47 Passages in the *Fihrist* (*Rasāʾil*, vol. 1, 21) and Epistle 52a (*On Magic*, 7) refer to the four-division, but do not provide an explanation for its classificatory basis.

> physical sciences, it is to climb and ascend from them to the divine sciences (*al-ʿulūm al-ilāhiyya*), which is the utmost goal (*aqṣā ghāya*) of the sages and the end of ascent in knowledge of reality (*al-nihāya allatī ilayhā yurtaqā bi-l-maʿārif al-ḥaqīqiyya*).[48]

The authors of the *Epistles* also follow the central aspects of the late ancient classificatory scheme. When the arrangement of the *Epistles* is compared with the classification of sciences by al-Kindī and al-Fārābī, it is immediately obvious that it has much more in common with the former than with the latter. All al-Fārābī's elaborations on the classical scheme, such as the extended *quadrivium* or the introduction of linguistics, are absent in the *Epistles*.[49] This supports the view argued for in the introduction that the *Epistles* are doctrinally close to al-Kindī's school. However, it must also be noted that al-Fārābī's innovations were not necessarily adopted by his successors, such as Avicenna, either.[50] As regards the general division of the *Epistles*, the mathematical sciences are placed at the beginning due to the shared Neopythagorean inclinations of al-Kindī and the Brethren of Purity. The fact that al-Kindī makes psychology an independent part of philosophy, while the third part of the *Epistles* consists of psychical-intelligible sciences, appears at first glance to be an even more striking similarity, since it goes against the standard classification of Aristotelian philosophy.[51] But the similarity is, in fact, entirely superficial since, for al-Kindī, this section consists of Aristotle's psychological treatises, none of which are accorded tractates of their own in the *Epistles*; nor do the epistles of the third section discuss psychology in this sense.

---

48 *On Arithmetic & Geometry*, 64. See also *Sciences of the Soul and Intellect, Part III*, 72–73 [tr. Baffioni, 184–185, modified]: "The one who seeks to investigate the true natures of things (*al-naẓar fī ḥaqāʾiq al-ashyāʾ*), to search for their causes, and to ask about their reasons … should be exercised in the four rational mathematical sciences (*al-riyāḍiyyāt al-ʿaqliyya al-arbaʿa*), and in the investigation (*naẓar*) of the science of logic and of the natural sciences … then she should investigate this kind (*fann*) of science called the science of theology (*ʿilm al-ilāhiyyāt*), because this science is the ultimate goal at which the human being can arrive in knowledge (*al-ghāya al-quṣwā allatī yantahī ilayhā al-insān fī al-maʿārif*) that is close to the ranks of the angels."

49 For the view of grammar as an introductory science in Epistle 7, however, see Table 4. The subject matter of Epistle 31 (*Differences of Languages*) is also related to languages, but the epistle is neither concerned with grammar nor is it placed as an introduction to logic.

50 See Avicenna, "Aqsām."

51 The similarity is noted in de Vaulx d'Arcy, *Les Épîtres*, 43–44. The author employs this as an argument for situating the *Epistles* within the school of al-Kindī.

As regards the general four-division of the *Epistles*, at first glance it appears unexpected from the perspective of the late ancient and early Arabic classifications of philosophy. The first section consists of mainly mathematical and logical epistles, which are grouped together into a "preparatory" (*riyāḍiyya*) section. While this represents an innovation, it perhaps makes sense insofar as both mathematics and logic are considered necessary prerequisites for theoretical knowledge. The second part is devoted to the natural sciences and occupies the position within the whole that we would expect. The final part of theoretical philosophy, however, should be metaphysics, understood either in the Kindīan sense as philosophical theology or in the Fārābīan sense as a combination of first philosophy and theology; instead we have psychical-intelligible sciences. But since the subject matter of philosophical theology for both al-Kindī and al-Fārābī consists of the incorporeal psychical and intellectual beings of the celestial world as the secondary causes for the existents of the sublunar world, as well as of God as the primary cause, this is not so strange. The third section can therefore be understood as addressing metaphysics in the Kindīan theological sense, given that at least some of the epistles correspond to this designation. The final section of "nomic-divine" sciences should finally correspond to practical philosophy, and this is also largely the case in the extended sense, where practical philosophy also includes the Islamic religious sciences. In consequence, if we ignore the nominal four-division, the classificatory scheme of the *Epistles* follows the order: (1) mathematics, (2) logic, (3) physics, (4) metaphysics/theology, (5) practical philosophy/religion. This is entirely consistent with contemporary Islamic classifications of knowledge.

A closer inspection of the epistles contained in each section gives the surprising result that their order at times follows precisely the conventional curricular order, while at other times it involves unexpected changes, additions, and omissions. As regards the first section, Epistles 1–6 contain the mathematical *quadrivium* in the correct order, but also include two additional epistles: an epistle on geography (4) and an epistle on arithmetical and geometrical proportions (6). While both epistles are innovations with respect to the *quadrivium*, the latter obviously falls within the subject matter of mathematics; geography may also be considered mathematical in the sense of being an extension of astronomy in the sublunar world.[52] Epistles 10–14 follow the logical curricu-

52 This is suggested in Godefroid de Callataÿ, "The Classification of Knowledge in the *Rasāʾil*," in *The Ikhwān al-Ṣafāʾ and Their Rasāʾil: An Introduction*, ed. Nader El-Bizri (Oxford: Oxford University Press and the Institute of Ismaili Studies, 2008), 58–82, at 76. Moreover, like astronomy, Arabic geography was in part based on Ptolemy (d. ca. 170 CE). For the Ptolemaic and other influences in Epistle 4, see the translator's introduction in *On Geography:*

lum, but the last four works of the Arabic logical corpus are absent. This is perhaps due to the authors following a pre-Fārābīan tradition that restricted the *Organon* to its first books.[53] More conspicuously, the position of Epistles 7–9, on theoretical and practical sciences and ethics, seems out of place in the first section. A logical position for a classificatory account of sciences would be at the beginning of the *Epistles*, perhaps within the *Fihrist*. As for Epistle 9 (*Ethics*), since al-Kindī, ethics was presented as a follow-up on theoretical philosophy; the *Epistles* in fact conform to this scheme in devoting the fourth part to practical philosophy and religious sciences. Epistle 50 (*Governance*), in particular, is partly ethical in content.

As for the second section on natural philosophy, its first part (Epistles 15–24) follows the Aristotelian physical curriculum with two notable exceptions. First, an epistle corresponding to Aristotle's *De anima* is missing, even though it should form a prominent conclusion for the Aristotelian physical corpus, and the *Epistles* discuss psychology elsewhere. Second, Epistles 20 (*Essence of Nature*) and 23 (*Constitution of the Human Body*) do not correspond to any Aristotelian or pseudo-Aristotelian treatise. Epistle 20 is devoted to nature as a faculty of the Universal Soul and should therefore probably pertain to the third section; in any case, its position between epistles on minerals and plants is unjustified. Epistle 23, on the other hand, logically follows upon an epistle on animals, but is unexpectedly devoted to the human body, as opposed to the soul. As for the second part of the second section (Epistles 25–31), none of its treatises bear any relation to the Aristotelian curriculum. Epistles 25–30 are, however, plausibly physical in their subject matter, and might therefore be considered a legitimate extension of the curriculum of natural philosophy. Epistle 31 (*Differences of Languages*), in contrast, is entirely out of place within natural philosophy; a position in the first section would make more sense, given the connection between linguistics and logic.

In contrast to the first two sections, it is remarkable that none of the epistles in the third and fourth sections correspond directly to the subsections of the late ancient or early Arabic classifications. Nevertheless, insofar as the subject matter of the third section is philosophical theology in the sense of the supernal

---

An Arabic Edition and English Translation of Epistle 4, ed. and trans. Ignacio Sánchez and James Montgomery (Oxford: Oxford University Press and the Institute of Ismaili Studies, 2014), 27–46.

53 This is suggested in Baffioni's introduction to her edition and translation of the logical epistles in *On Logic: An Arabic Critical Edition and English Translation of Epistles 10–14*, ed. and trans. Carmela Baffioni (Oxford: Oxford University Press and the Institute of Ismaili Studies, 2010), 1–3.

causes of the terrestrial world, Epistles 32 and 33 (*Intellectual Principles according to the Pythagoreans/Brethren of Purity*) constitute a proper beginning. Epistles 34 (*The World is a Big Human Being*), 35 (*Intellect and the Intelligibles*), 36 (*Cycles and Revolutions*), 39 (*Movements*), and 40 (*Causes and Effects*), in contrast, should perhaps instead form part of natural philosophy in the second section. Since these epistles also address metaphysical themes, however, they are perhaps not entirely out of place in the third section. The inclusion of Epistle 37 (*Love*) may be justified on similar grounds as it treats love as a cosmic principle. Epistles 38 (*Resurrection*) and 41 (*Definitions and Descriptions*), on the other hand, seem to occupy an incorrect position; the former would be a better fit in the fourth section, and the latter in the first section. It is also surprising that no treatise is devoted to God, either in the third section or in the *Epistles* as a whole, even though a treatise corresponding to Book 12 of Aristotle's *Metaphysics*, or the pseudo-Aristotelian *Theology of Aristotle* for that matter, should presumably represent the culmination of philosophical theology.

If the subject matter of the fourth section is understood as practical philosophy complemented by Islamic religious sciences, the inclusion of Epistles 42–48 and 50 may be justified, even if their order seems arbitrary. Epistles 49 (*Spiritual Beings*) and 51 (*Constitution of the World*), however, again seem out of place, and should pertain either to the second or the third section. The position of Epistle 52 (*Magic*) is also arbitrary, especially since nothing in the other epistles suggests that the philosophical curriculum should be concluded by an epistle on magic.

The point of all this is not to claim that the authors of the *Epistles* should have faithfully followed the Aristotelian curriculum. Even self-acclaimed Aristotelians, such as al-Kindī and al-Fārābī, took liberties in developing that curriculum to better suit their overall conception of philosophy. The authors of the *Epistles* do not identify themselves as orthodox Aristotelians; their aim is rather to present a synthesis between different strands of classical philosophy and science on the one hand, and between classical sciences and Islam, on the other. Therefore, it should not be surprising that they show a greater degree of innovation with respect to the essentially Aristotelian classification of philosophical sciences. But even so, it is also true that they follow the late ancient curricular order for the most part, and explicitly commit to the pedagogic rationale on which it is based. In consequence, some epistles (7–9, 20, 31, 38, 41, 49, 51, 52) occupy a clearly incorrect position in the sense that they do not comply with the organizing principles to which the authors otherwise subscribe. This supports the claim that the current configuration of the *Epistles* is a result of gradual evolution rather than a conscious editorial decision. That this is indeed the case is confirmed by the many passages within the *Epistles* where the authors adhere to the standard philosophical curriculum but contradict the configuration of the *Epistles* as we know them.

2 Curricular Passages

The *Epistles* contain many passages that explicitly address the classification of either sciences in general, or of the epistles in particular. In the following, I will refer to these as the 'curricular' passages of the *Epistles*. The most important curricular passage is the remarkably elaborate scheme of philosophical and non-philosophical sciences presented in Epistle 7 (*Theoretical Sciences*).[54] Less complete classifications appear in Epistles 1 (*Arithmetic*), 2 (*Geometry*), 15 (*Matter and Form*), and 52a (*Magic*).[55] These show us the authors' normative views on how the sciences should be classified. In some cases, they refer to the actual order of the epistles, as opposed to a theoretical scheme of classification. In both instances, the passages almost invariably contradict the current configuration of the *Epistles*. Thus, they seem to bear witness to previous redactions of the work.[56] In this section, I will first analyze all the curricular passages, and then address their value as evidence for the redaction history of the *Epistles*.

Starting with Epistle 7, the tractate presents a complete classificatory scheme for knowledge in general, rivaling similar accounts by contemporary Arabic authors.[57] To this end, the epistle presents a three-tier classification

54 The classification of Epistle 7 is presented in Table 4.

55 The classifications of Epistles 1, 2, and 15 are summarized in Table 5.

56 This is suggested in de Callataÿ, "The Classification," 61 with regard to the discrepancies between Epistle 7 and the arrangement of the *Epistles*. As stated above, Madelung takes Epistle 7 to represent the structure of the first redaction of the *Epistles*.

57 *On Composition*, 72–96. For a detailed comparison of the actual order of the epistles and the classificatory account of Epistle 7, see de Callataÿ, "The Classification." Besides the philosophical classifications of al-Kindī and al-Fārābī discussed above, al-Khwārizmī's (d. ca. late tenth century) *Keys of the Sciences* (*Mafātīḥ al-ʿulūm*) and al-Tawḥīdī's (d. 1023) *Epistle on the Sciences* (*Risāla fī al-ʿulūm*), in particular, offer close parallels in that their classifications include both non-philosophical and philosophical sciences. For al-Khwārizmī, see C.E. Bosworth, "A Pioneer Arabic Encyclopedia of the Sciences: al-Khwārizmī's *Keys of the Sciences*," *Isis* 54, no. 1 (1963); for al-Tawḥīdī, see Marc Bergé, "Êpitre sur les sciences d'Abū Ḥayyān al-Tawḥīdī: Introduction, traduction, glossaire technique, manuscrit et édition critique," *Bulletin d'études orientales* 18 (1964). For these and other classifications, including Epistle 7, in the wider Islamic context, see Hinrich Biesterfeldt, "Arabisch-islamische Enzyklopädien: Formen und Funktionen," in *Die Enzyklopädie im Wandel vom Hochmittelalter bis zur frühen Neuzeit: Akten des Kolloquiums des Projekts D im Sonderforschungsbereich 231 (29.11.–1.12.1996)*, ed. Christel Meier (Munich: Wilhelm Fink,

of sciences divided into (1) propaedeutic (*riyāḍiyya*), (2) religious (*sharʿiyya waḍʿiyya*), and (3) philosophical (*falsafiyya ḥaqīqiyya*). These three main sections are clearly conceived as a progression toward gradually higher forms of knowledge, meaning that the propaedeutic and religious sciences are both conceived as introductory with respect to the philosophical sciences that concern the knowledge of reality itself. Since the *Epistles* are an encyclopedia of philosophical sciences, our main concern is with the third section. Some aspects of the first two sections are nevertheless also of interest.

The first section of propaedeutic sciences consists of mundane skills and crafts related to the aim of gaining livelihood in this world;[58] thus, they are situated at the low end of sciences in the authors' hierarchy of kinds of knowledge. It is conspicuous, however, that magic is included in this section, given that the very last treatise of the *Epistles* is also devoted to magic. There is, then, a blatant contradiction between magic considered to be a low science dealing with mundane matters, versus it constituting the culmination of philosophical knowledge.[59] This implies that the conception of the authors or later editors of the *Epistles* on the value of magic as a craft and object of study was upgraded at some point.

The second section consists of the traditional Islamic religious sciences, but also reflects the authors' Ṣūfī inclinations in including the sciences of asceticism (*zuhd*) and Sufism (*taṣawwuf*). It also reflects their Shīʿite leanings in ascribing the science of the interpretation of revelation (*taʾwīl*) to the imāms.[60] More importantly, the lower tier of religious sciences shows that, even if the *Epistles* pursue the goal of a synthesis between philosophy and religion, the authors nevertheless consider the traditional religious sciences as clearly distinct from, and lower in the hierarchy of sciences than, philosophical theology.

Our main interest lies in the third section of philosophical sciences; this presents the following classification: (1) mathematics (*riyāḍiyyāt*), (2) logic

---

2002), 43–83; Wolfhart Heinrichs, "The Classification of the Sciences and the Consolidation of Philology in Classical Islam," in *Centres of Learning: Learning and Location in Pre-Modern Europe and the Near East*, ed. Jan Willem Drijvers and Alasdair A. MacDonald (Leiden: Brill, 1995), 119–139.

58 *On Composition*, 72–73.

59 For an argument that knowledge about magic, and specifically Epistle 52b, in fact constitutes the intellectual culmination of the *Epistles*, see Saif, "Ikhwān al-Ṣafāʾ's Religious Reform." As I will argue in many sections of this book, the concluding position of a magical epistle is not supported by other parts of the *Epistles*.

60 *On Composition*, 73–74. The absence of *kalām* (rationalistic theology) among the religious sciences could perhaps be related to the authors' Ismāʿīlī inclinations. For the Ismāʿīlī rejection of Ashʿarite and Muʿtazilite *kalām*, see De Smet, "Ismāʿīlī Theology."

(*manṭiqiyyāt*), (3) physics (*ṭabīʿiyyāt*), and (4) theology (*ilāhiyyāt*).[61] As we have seen, this is one variant among the standard Islamic classifications of philosophical sciences; it also partially coincides with the actual order of the epistles. At the same time, it contradicts the division of the *Epistles* into its four main sections. This again raises the question of why the authors or editors decided to adopt this novel division, even though they attest to the customary division in their normative account of the classification of philosophy.

More questions arise when the contents of the four sections are compared with the contents of the *Epistles*. In general, the sections in Epistle 7 roughly follow the standard curriculum, with some interesting differences, while they exclude the additional epistles that complement that curriculum in the *Epistles*. Thus, the mathematical sciences consist of the classical *quadrivium*, without the two additional epistles;[62] the physical sciences are similarly composed only of the Aristotelian curriculum of natural philosophy.[63] The physical section of Epistle 7, however, omits Aristotle's *De anima*, which is also the case in the *Epistles*. The logical section is more complicated. As in the *Epistles*, there are five logical sciences, but the logical curriculum now consists of *Poetics*, *Rhetoric*, *Topics*, *Posterior Analytics*, and *Sophistics*.[64] This is not the traditional order of Aristotelian logical treatises. Nor is this the logical curriculum in the *Epistles*; for four of these parts of logic, there is no corresponding treatise in the *Epistles*. The sole convergence is the treatise on *Posterior Analytics*. Despite this, Epistle 7 also goes on to summarize the contents of *Eisagōgē*, *Categories*, *On Interpretation*, and *Prior Analytics*; the authors then state that they have written an epistle for each of them,[65] which is indeed the case in the *Epistles*. This means that the authors were familiar with the entire logical curriculum; thus, the discrepancy between the two logical curricula is even more mysterious.

The final theological part of philosophy in Epistle 7 consists of five sections: (1) Creator, (2) spiritual beings (*rūḥāniyyāt*), (3) psychical beings (*nafsiyya*), (4) governance (*siyāsa*), and (5) return (*maʿād*).[66] This division is almost entirely non-Aristotelian. Insofar as the first three sections deal with God and the intellectual and psychical existents of the supernal world, they correspond to metaphysics as philosophical theology in the Kindīan sense. But the third

61 Ibid, 74–96.
62 Ibid, 75–76.
63 Ibid, 82–88.
64 Ibid, 77.
65 Ibid, 78–82.
66 Ibid, 88–95.

section in fact addresses not only astral souls but also the sublunar psychical existents of plants and animals.[67] The governance in the fourth section is divided into five further sub-sections: (1) prophetic (*nabawiyya*), (2) kingly (*mulūkiyya*), (3) general (*ʿāmmiyya*), (4) particular (*khāṣṣiyya*), and (5) individual (*dhātiyya*).[68] The descriptions of these sub-sections make it clear that the last three correspond to the Aristotelian tridivision of practical philosophy into ethics, economics, and politics, while the first two deal with the religious subjects of prophecy, caliphate, and imāmate. The fifth sub-section is concerned with eschatology.[69]

Thus, the fourth part of philosophical sciences in Epistle 7 lumps together philosophical theology, cosmic and sublunar psychology, practical philosophy, and the religious themes of prophecy and eschatology. It thereby diverges from the classifications by al-Kindī and al-Fārābī. Yet, it does have a parallel in the order of presentation followed by al-Fārābī's *On the Principles of the Opinions of the Inhabitants of the Virtuous City* (*Fī mabādiʾ ārāʾ ahl al-madīna al-fāḍila*) or Avicenna's *Beginning and Return* (*al-Mabdaʾ wa-l-maʿād*).[70] The rationale for the arrangement of both works follows the Neoplatonic scheme of descent and reascent of existence, in the sense that they consist of one part devoted to the cosmological order from God down to the sublunar world, and of another part devoted to the "return," in the sense of the ethical, political, prophetic, and eschatological aspects of the human soul's ontological reascent toward its spiritual origin.[71] Rudolph has argued that al-Fārābī adopts the order of presentation of his treatise from the theological "principles of religion" (*uṣūl al-dīn*) genre, which typically addresses the themes of God, the cosmos, ethics and

67 Ibid, 90.

68 Ibid, 90–94.

69 Ibid, 95.

70 Al-Fārābī, *Al-Farabi on the Perfect State: Abū Naṣr al-Fārābī's Mabādiʾ ārāʾ ahl al-madīna al-fāḍila*, ed. and trans. Richard Walzer (Oxford: Clarendon Press, 1985); Avicenna, *al-Mabdaʾ wa-l-maʿād*, ed. ʿAbd Allāh Nūrānī (Tehran: Institute of Islamic Studies and Tehran University, 1984).

71 This also coincides with Avicenna's conception of the last part of metaphysics, where natural theology is followed by questions of the human soul's relation with supralunar celestial beings, and thus deals with the problems of happiness, the afterlife, and prophecy. Gutas famously designated this last part of Avicennian metaphysics the "metaphysics of the rational soul." See Dimitri Gutas, *Avicenna and the Aristotelian Tradition: Introduction to Reading Avicenna's Philosophical Works: Second, Revised and Enlarged Edition, Including an Inventory of Avicenna's Authentic Works* (Leiden and Boston: Brill, 2014), 288–296. For Avicenna's statement of the order in which metaphysics should be studied, see also Avicenna, *The Metaphysics of the Healing: A Parallel English-Arabic Text*, ed. and trans. Michael E. Marmura (Provo: Brigham Young University Press, 2004), I.5, 22.

politics, prophecy, and the afterlife.[72] A similar line of influence could also plausibly apply to the authors of Epistle 7. It should also be noted that the Ismāʿīlī theologian al-Kirmānī follows a similar structure in his *Rest of the Intellect* (*Rāḥat al-ʿaql*),[73] given the debate on the Brethren of Purity's Ismāʿīlī affiliation.

When the fourth section of the *Epistles* is compared with the classificatory scheme in Epistle 7, the result is that, in general terms, the former corresponds to the subject matter delineated for theology in the latter: the epistles of the fourth part deal with cosmology (49 and 51), ethical and prophetic governance (45, 47, 48, 50), eschatology (43, 44), and religious beliefs (42, 46). But the order of the epistles in no sense follows the rationale presented in Epistle 7; nor does the fourth part include an epistle on God, which in Epistle 7 constitutes the first part of theology. Moreover, many epistles with themes that should fall under theology (32–33) and eschatology (38) are included within the third section of the *Epistles*. It has been suggested that the authors split the theological section into two parts:[74] the third section of the *Epistles*, then, corresponds to cosmology and philosophical theology; the fourth part corresponds to practical philosophy and religion. Even then, some of the epistles (38, 49, 51) are currently placed in the wrong section. Moreover, the order of the third section still does not follow the scheme of Epistle 7, which should proceed from God downwards. To make the situation even more confusing, in Epistle 1 (*Arithmetic*), the authors seemingly contradict both the theoretical order of theology in Epistle 7 and the actual order in the *Epistles*. In continuation of the passage cited above, which stated that theological sciences constitute the ultimate end of study, the authors take a stand on the proper starting point of the theological sciences:

> The first degree (*daraja*) of study (*naẓar*) in the theological sciences (*al-ʿulūm al-ilāhiyya*) is knowledge of the substance of the soul (*jawhar al-nafs*), and investigation (*baḥth*) of its beginning and origin before its connection (*taʿalluq*) with bodies, and inquiry (*faḥṣ*) about its return (*maʿād*) and where it ends after its separation from the body, which is called death, and the quality (*kayfiyya*) of the reward for good people (*muḥsinīna*) in the world of spirits, and the punishment for bad people in the abode of

72 Ulrich Rudolph, "Reflections on al-Fārābī's *Mabādiʾ ārāʾ ahl al-madīna al-fāḍila*," in *In the Age of al-Fārābī: Arabic Philosophy in the Fourth/Tenth Century*, ed. Peter Adamson (London: The Warburg Institute, 2008), 1–14.

73 Al-Kirmānī, *Rāḥat al-ʿaql*, ed. Muṣṭafā Ghālib (Beirut: Dār al-Andalus, 1983).

74 de Callataÿ, "The Classification," 75.

> afterlife (*dār al-ākhira*). Another thing is also that when people are called to knowledge of their Lord, they have no way to that knowledge except through knowledge of themselves (*maʿrifat nafsihi*).[75]

Here it would seem to be the case that the first subject to be studied in theology is the substance and origin of the soul, as well as the reward or punishment it receives in the afterlife. It is only by means of knowledge of oneself as an embodied soul that one may acquire knowledge about God. Again, it is, then, surprising that the *Epistles* do not contain an epistle devoted to either psychology or God.

In sum, Epistle 7 shows that the authors have a clear conception of how the philosophical sciences should be ordered. But the actual order and content of the *Epistles* follows that conception only to some extent. The first three parts of the classification in Epistle 7 roughly agree with the late ancient scheme, but in the first two parts of the *Epistles* the corresponding sections are complemented by additional epistles. The epistles of the third and fourth sections of the *Epistles* coincide with the subject matter attributed to theology in Epistle 7 but not with the order and subjects of individual epistles. Why should this be the case? It could be that the authors in Epistle 7 present a theoretical classification of sciences, which they do not entirely apply in their composition of the encyclopedia. Toward the end of the epistle, however, the authors relate their classificatory scheme explicitly to the redaction of the *Epistles*:

> We have made an epistle for each branch (*faṣl*) of the above-mentioned sciences and mentioned in them some of those meanings (*ṭarafan min dhālika al-maʿnā*) to awaken the negligent and guide the beginners, excite the interests of students, and serve as a path for those who would learn. Be happy by means of that, my brother, and show this epistle to your brethren and friends; make them desirous of knowledge, urge them to renounce this world, and show them the way to the final abode.[76]

The passage gives the impression that the authors have arranged the *Epistles* based on the classificatory scheme of Epistle 7, and that they have already composed an epistle for each of the sciences included in that scheme. That this is not entirely the case implies an evolution in the authors' conception of the proper contents and arrangement of a philosophical encyclopedia. As a pre-

75 *On Arithmetic & Geometry*, 65.

76 *On Composition*, 95–96; tr. de Callataÿ, 119, modified.

liminary hypothesis, Epistle 7 could, then, represent the original plan of composition which the authors or compilers substantially revised at a later stage of the redaction process. Since for the most part the classification in Epistle 7 follows the standard classification, this would mean that the Aristotelian curriculum formed the starting point, while the authors later inserted additional epistles and revised the order of some epistles. Regarding the chronology of redaction, this would make all the technical epistles in the first two sections of the *Epistles* the earliest; the non-classical epistles in these parts, as well as the novel four-division of the *Epistles*, would be later. Epistle 7 itself would have to postdate the original core of the epistles and predate the non-classical epistles.

These conclusions are supported by two further curricular passages in Epistles 1 (*Arithmetic*) and 2 (*Geometry*).[77] Both passages reiterate the classification of philosophical sciences into (1) mathematics, (2) logic, (3) physics, and (4) theology. The mathematical sciences are identified with the *quadrivium*, while the parts of the last three sections are not specified. In the second passage, the authors identify the subject matter of physics with corporeal existents and of theology with immaterial existents (*al-ṣuwar al-mujarrada al-mufāraqa li-l-hayūlā*). They also state that they have composed an epistle on each "kind of these sciences" (*nawʿ min hādhihi al-ʿulūm*).[78] Thus, the conclusion here is that the *Epistles* originally followed the classical four-division of philosophy, within which the mathematical section consisted only of the *quadrivium*.

The curricular passage in Epistle 15 (*Matter and Form*) is of particular interest because the reference is not to a theoretical classification but to the treatises that the authors have already composed. It is therefore worth citing this lengthy passage in full:

> And know that the study of natural sciences is a part of the art (*ṣināʿa*) of our noble brethren … and that natural objects are bodies and the necessary and transitory accidents that inhere in them. On these sciences we composed (*ʿamilnā*) **seven epistles**, the first of which is this epistle where we explained (*dhakarnā*) [1] **matter, form, motion, place, and time**, because these five things embrace all bodies. And in the epistle [2] ***On Senses and Sensibles***, we explained concisely the things that inhere in bodies. This is followed by the epistle where we explained the [3] **heaven and the world**, and where we described the composition and number of

77 *On Arithmetic & Geometry*, 10–11, 73–76. These classifications are summarized in Table 5.
78 Ibid, 76.

the spheres, the extent of their diameters, the speed of their rotation, the magnitude of the planets (*kawākib*) and the kinds of their movements, and the characteristics and maps [*takhṭīṭ*] of the zodiacs. This is followed by the epistle where we explained [4] **generation and corruption** and the essence of the four elements that are under the sphere of the Moon, that is, fire, air, water, and earth, and where we described how they are transformed into each other and how the generated beings come from them. Then the **fourth epistle** follows, where we explained the [5] **meteorological phenomena** and the changes that occur in the air above our heads. Then the **fifth epistle** follows, where we explained the [6] **substances of minerals** and described how they are generated in the interior of the earth, in the cavities of mountains, and in the depths of the seas. Then the **sixth epistle** follows, where we explained [7] **plants** and described their genera, species, properties, and benefits and harm. Then the **seventh epistle** follows, where we explained concisely (*bi-qawl wajīz*) the [8] **genera of animals**, their species, and the diversity of their natures.

And we composed (*ʿamilnā*) five more epistles before these, namely, the **propaedeutic** (*al-riyāḍiyyāt*), the first of which is the epistle where we explained [1] **number** and its species and properties, and how it originates from the one that comes before the two. Then the epistle follows where we explained [2] **geometry**, the species of measures, and how they originate from the point which in the art of geometry is like the one in the art of number. Then the epistle follows where we explained [3] **stars**, described the spheres and planets, and explained that their relations with the Sun are like the relation of number with the one and [the relation] of geometrical measures with the point. Then the epistle follows where we explained the [4] **arithmetical, geometrical, and harmonic proportions**, and that the origin of them all from the relation of equality is like the origin of number from the one and like the origin of geometrical measures from the point. Then the epistle follows where we explained [5] **philosophical logic** (*al-manṭiq al-falsafī*) and described the ten categories, each of which is a *summum genus* (*jins al-ajnās*), and where we explained the quantity of their species and their properties—that one of them is substance and the other nine are accidents, and that their connection in existence with substance is like the connection of number with the one that comes before two.[79]

79 *On the Natural Sciences*, 51–54; tr. Baffioni, 126–127, modified.

This passage is highly informative because it both agrees and disagrees with the *Epistles* that we have. According to the passage, at the time of its writing, the first two sections of the *Epistles* consisted of a propaedeutic (*riyāḍiyyāt*) section, composed of mathematics and logic, and of a physical section, which is obviously also the case in the *Epistles*. However, in this passage, the first section is composed of the mathematical *quadrivium* and a single epistle on logic, as opposed to the fourteen epistles of the final configuration. Moreover, the title of the fourth treatise is not music, but "arithmetical, geometrical, and harmonious proportions," which coincides in part with the subject matter of Epistle 5 (*Music*), but also with the title and contents of Epistle 6 (*Arithmetical and Geometrical Proportions*). In addition, the contents of the epistle on logic are equated solely with the subject matter of the *Categories*. The second physical section consists of eight, as opposed to seventeen, epistles, and follows the order of the Aristotelian curriculum. The only exception is the treatise of *Senses and Sensibles*, which occupies a wrong position. However, it occupies an incorrect place also in the passage, as indicated by the otherwise correct numbering of the treatises. The inclusion of this treatise is also interesting because it is not included in the physical section of Epistle 7, while there is a corresponding treatise in the *Epistles*. It is also noteworthy that the passage employs the attribute "concise" of Epistle 22 (*Animals*), given that this is the longest epistle in the entire work. This suggests that the epistle at this point consisted only of the introduction, which is indeed concise, whereas the animal fable that comprises most the treatise was added later.

This passage supports the preliminary hypothesis about the redaction chronology in that the first two sections include none of the non-classical epistles. The import of the passage is that, with the exceptions noted above, the first two sections at some point consisted of the classical mathematical *quadrivium*, an epistle on logic, and epistles corresponding to the Aristotelian physical curriculum; the additional epistles were added later. However, the passage also contradicts the other curricular passages in two respects. First, there is only one epistle on logic, as opposed to the logical curricula contained either in the *Epistles* or in Epistle 7. Second, in contrast to the other curricular passages, this passage coincides with the *Epistles* in including mathematics and logic in a single propaedeutic section.

The final curricular passage occurs at the beginning of Epistle 52a (*Magic*):

> Know, may God support you, O brother, that in the fifty epistles that precede this epistle, we have explained the varieties of science and the marvels of wisdom. We ordered them (*rattabnāhā*) and gathered in them the many sciences, numerous objectives, and profound wisdoms. And

> we explained them (*bayyannāhā*) in accordance with what the levels of learners and the ranks of students making use of them require. However, just as it would not be appropriate for us to provide science to anyone to whom it does not pertain and who does not know its value, likewise it is neither permissible nor allowed for us to forbid it from anyone seeking guidance and knowledge, and one should not be ungenerous with it with regard to someone who deserves it. He among our noble brethren who has attained this epistle should hand over from it to all deserving people what is within the grasp of their understanding and known to be profitable to them and appropriate to their rank, gradually, according to the order (*tartīb*) that we have established in the epistle of the *Fihrist*. Every time that their soul ascends to some degree in science (*ʿilm*) and reaches some rank in knowledge (*maʿrifa*), it is raised up to what is after and carried on toward what follows it, until their soul reaches the limit of its perfection—this according to **four sections.** The first section is [I] **the preparatory/mathematical sciences** (*riyāḍiyya*) with which one begins. The second section is [II] **the corporeal and natural sciences** (*jismāniyya ṭabīʿiyya*), which follows the first. The third section is [III] **the sciences of the soul and the intellect** (*nafsāniyya ʿaqliyya*), which comes after the second. The fourth section is [IV] **the nomic and divine sciences** (*nāmūsiyya ilāhiyya*), which is the last of them. **This epistle is the last one of the epistles from the fourth section, and it is the fifty-first.** In this epistle we would like to explain the essence of magic and how talismans take effect. For these matters are like any known science or knowledge and like some applied wisdom. We shall provide testimony for them with what we have heard from the savants and what we have got to know from the books of the ancients before us who lived in the past.[80]

In contrast to the previous passages, this passage agrees perfectly with the current configuration of the *Epistles*. It correctly indicates not only the four sections but also the position of the current epistle as the final one of the fourth section, and thus of the entire work. Moreover, the passage relates the order of the epistles to that established in the *Fihrist*, or table of contents, that precedes the work. In the *Fihrist*, of course, the configuration of the epistles arranged in four sections agrees perfectly, or almost perfectly as it turns out, with the work's actual contents. It is noteworthy that this is the only curricular passage that mentions the *Fihrist*. As we will shortly see, the *Fihrist* almost certainly

80 *On Magic*, 6–8; tr. de Callataÿ, 88–89, modified.

either postdates the work, or at least pertains to the very last stage of its redaction. This, in consequence, suggests that Epistle 52a, or at least the passage in question, is also late. The view that Epistle 52a is of late origin is further supported by its exclusion from the scheme of philosophical sciences in Epistle 7 discussed above, as well as by the fact that the quoted passage is the only one among the curricular passages that conforms to the actual four-division of the *Epistles*. What is remarkable, then, is not that the *Epistles* include a single passage that identifies the four sections of the *Epistles* correctly, but that there are so many passages that do not. The conclusion is that at the time when Epistle 52a was written, the *Epistles* had assumed their current form; at the time of composition of the other curricular passages, the work's structure was very different.

Besides explicit references to the curricular order, there are more ambiguous references to the composition of the *Epistles*. Some of these also seem out of place in the current configuration of epistles. A prominent example is Epistle 43 (*The Way to God*), the subject of which is the Neoplatonic theme of soul's ascent to God by means of its liberation from matter. The authors, then, state that they must start with the subject of how the soul is purified from bad character traits (*al-akhlāq al-radīʾa*).[81] Hence, they will discuss it in "this book (*kitāb*)" in "some chapters" (*abwāban shattā*), and provide examples in each chapter; then, in "another book that follows this book" (*fī kitāb ākhar yatlū hādhā al-kitāb*), they will discuss the "straight way" (*al-ṭarīq al-mustaqīm*) to God. The reference to two books is repeated later when it is stated that the instructions for the purification of the soul are given in "these two books" (*hādhāyni al-kitābayni*).[82] The idea of two books that follow each other is strange; it could perhaps mean two epistles, but Epistle 44 (*Beliefs of the Brethren of Purity*) does not provide a thematic continuity with the epistle that precedes it. In two manuscript variants (MSS أ, ق), the first reference to "this book" is replaced by "our mathematical/preparatory epistles" (*rasāʾilinā al-riyāḍiyya*), which is equally strange since the epistle pertains to the fourth rather than the first section of the *Epistles*. A further possibility would be that the other book is the *Comprehensive Epistle*. The description does not seem to fit its contents, however, and in any case the idea is negated by what follows. The authors conclude that after these two "subjects/chapters" (*bābayni*), they will proceed to reveal the hidden mysteries that they have come to know through God's inspiration (*bi-ilhām Allāh*) or "commentaries on the books of saints and revelations of the prophets" (*tafāsīr*

81 *On Companionship and Belief*, 15.
82 Ibid, 24.

kutub awliyāʾihi wa-tanzīlāt anbiyāʾihi).[83] In some variants, the two chapters or subjects appear as "two books" (*kitābayni*). In either case, the reference makes no sense. This is one of the most concise epistles in the entire work, and there will be no further chapters discussing the promised subjects. Nor will the discussion follow in Epistle 45 (*Companionship of the Brethren of Purity*), which instead addresses friendship and the nature of the Brethren of Purity's society. In any case, talking of books and chapters with different subjects seems entirely out of place within the *Epistles* arranged in four sections.

Later, however, the epistle does refer to other epistles.[84] The authors urge people to follow gradual instruction (*taʿlīm*), as they described in the "two primary epistles" (*al-risālatayni al-awwaliyyatayni*) that they had "composed for students and aspirants" (*waḍaʿnāhumā li-l-mutaʿallimīna wa-l-murīdīna*); once their souls are purified, they may get answers by proofs (*bi-barāhīnihā*), as "we explained in the five epistles that we shaped in the human form (*bayannāhā fī al-rasāʾil al-khams allatī ṣawwarnāhā ʿalā ṣūrat al-insān*), and made their indications (*dalāʾil*) clear by examples that are in the human form (*bi-l-mithālāt allatī fī ṣūrat al-insān*)." These references make no sense either. Epistles 7 (*Theoretical Sciences*) and 8 (*Practical Sciences*) could perhaps be meant by the couplet of epistles referring to instruction, insofar as the former of them at least provides a curricular order. The referent of the group of five epistles is entirely unclear; there is no group within the *Epistles* that would in some sense refer to the human form. Moreover, as we will see later, in the *Epistles*, the references to epistles are usually unambiguous.

When summed up, the curricular passages give support to the view that the *Epistles* evolved considerably in time. If Epistle 7 represents an early plan for the composition of the *Epistles*, then even that plan must have changed. Epistle 7 presents the more or less classical four-division of the philosophical sciences, which is modified in the *Epistles*. As for the contents of the first two sections of the *Epistles*, apart from logic, the two schemes agree with respect to the sciences that correspond to the Aristotelian curriculum; in the *Epistles*, however, these are supplemented by many new treatises. For the third and fourth sections there is partial agreement in content, but this is not true with regard to the arrangement and order of the epistles. The most striking example is Epistle 52 (*Magic*), which according to Epistle 7 should have no place whatsoever within an encyclopedia of philosophical sciences, but currently holds the honor of concluding the entire work.

83 Ibid, 15–16.

84 Ibid, 20.

It is not clear, as suggested by Madelung, that a work corresponding to Epistle 7 ever existed. But all the other curricular passages also point to a version of the *Epistles* that was very different from the work that we now have. In particular, the first two sections of that earlier version would have followed more closely the late ancient curriculum of philosophical sciences, consisting only of the corresponding mathematical, logical, and physical treatises. In consequence, the differences between the normative curricular order and the current arrangement of the *Epistles* discussed in the previous section would, then, be due to changes that were made later during the redaction process. This would mean that all the 'non-classical' epistles within the first two sections were introduced later, either because they were (1) composed later, (2) split from an existing epistle, or (3) transposed to the first two sections from the other two sections. The passage in Epistle 15 suggests that the second option is true in the case of Epistles 5 (*Music*) and 6 (*Arithmetical and Geometrical Relations*), which seem to have initially formed a single treatise. The same passage further suggests that all the logical epistles (10–14) also at some point formed a single epistle devoted to the subject matter of *Categories*. As regards the third option, the fact that many epistles seem to occupy an incorrect position may be explained by the notion that the evolution of the general plan resulted in the repositioning of some of the epistles. Beyond this, the passage in Epistle 15 indicates that the animal fable was added later to Epistle 22 (*Animals*). This suggests a fourth method of redaction: expansion of the text of an existing epistle. The result of all this is a hypothesis of four parallel methods by which the gradual redaction of the *Epistles* took place: (1) introduction of entirely new epistles, (2) splitting of existing epistles into two or more new epistles, (3) transposition of epistles from one position to another, and (4) expansion of the contents of existing epistles. The curricular passages alone do not prove that any of these four methods were in fact employed. But, as we will see, there is more evidence for all four methods.

3 Large-Scale Manuscript Variation

In a sense, research on the *Epistles* was simpler a few years ago, when the existing editions gave the false impression of a uniform and timeless text. The still ongoing project for producing a critical edition for each epistle has complicated things by showing that, for many epistles, the variation between manuscripts is considerable. This clearly supports the hypothesis that the work underwent successive revisions. Much of the remaining evidence to be discussed in this first part of the book involves manuscript variation. Before proceeding to more specific kinds of evidence, an overview of the generic variation in manuscripts is pertinent.[85] On this general level, the manuscripts include at least the following kinds of variants that support the thesis of gradual evolution: (1) variant position of an epistle within the work; (2) two or more epistles form a single epistle or one epistle is split into two epistles; (3) lengthy additions or omissions within an epistle in one or more manuscripts; (4) different variants for the entire epistle with textual relation between the variants; and (5) different variants of an epistle with no textual relation between the variants.

Beginning with the first group, as regards the epistle positions in manuscripts, most of the variation concerns the first section of the work, and Epistles 4 (*Geography*), 5 (*Music*), and 6 (*Arithmetical and Geometrical Proportions*), in particular. When compared with the position of epistles in the modern editions, Epistles 4 and 5 are in the reverse order in the majority of the reviewed manuscripts; Epistle 6 is either the second or fourth epistle in three manuscripts. In addition, several manuscripts carry Epistles 7 (*Theoretical Sciences*) and 8 (*Practical Sciences*) in an inverse order. Interestingly, the variation, then, only concerns the epistles that are added to the classical curriculum; in contrast, the mathematical and logical epistles of the first section are always in the correct curricular order. At least for Epistles 4 and 6, this supports the implications of the curricular passages that these epistles were added later to complement the *quadrivium*, and that the authors or later editors were therefore indecisive about their correct position. For the other three sections, however, the manuscripts appear decidedly unanimous on what the order of the epistles should be like.[86]

85 These are summarized in the endnotes of Table 1. Most of the observations are based on the technical introductions of the critical editions of individual epistles, which are complemented by an incomplete survey of the consulted manuscripts.

86 In the case of Epistle 20 (*Essence of Nature*), however, a single manuscript gives it the incorrect number 13, although its actual position is correct. Based on the Kacimi edition, the manuscripts of the *Comprehensive Epistle* also include variant positions for some epistles. Most, but not all, manuscripts place Epistle 45 (*Companionship of the Brethren of Purity*) as the very last, but since the actual final Epistle 52 (*Magic*) appears to be of only minor interest to the work's author, this probably only reflects a differing view on which epistle is sufficiently important to occupy the position of the concluding epistle. While all other epistles follow the order of the *Epistles* in the modern editions, in some manuscripts, Epistles 18 (*Meteorology*), 19 (*Minerals*), 20 (*Essence of Nature*), and 26 (*The Human Being is a Microcosm*) of the second part also occupy variant positions. Since the *Comprehensive Epistle* is a summary of the *Epistles*, it is not clear whether this variation reflects the manuscripts of the *Epistles*, or whether it is rather due to an editorial decision to discuss the epistles in a different order.

As for the second group, the manuscripts include a few variants regarding the division of the text into epistles. In most manuscripts, Epistles 12 (*On Interpretation*) and 13 (*Prior Analytics*) form a single epistle. In two manuscripts (MSS ز and ك), it is both the case that Epistles 49 (*Spiritual Beings*) and 50 (*Governance*) form a single epistle and that epistle 48 (*The Call to God*) is divided into two epistles. Thus, in the great majority of the manuscripts the total number of epistles is fifty-one, which coincides with the references that both the *Fihrist* and the text of the *Epistles* make to the number of epistles. The case of the combination of Epistles 49 and 50 is particularly interesting, because it is as if the two manuscripts compensate for the deficit of one epistle in the total number by dividing another epistle into two in order to arrive at a preconceived total number of fifty-one epistles.[87]

Third, based on the critical editions, most epistles include additions or omissions of different lengths in one or more of the employed manuscripts; often, these additions occur at the end of the epistle. In many cases, the additions are lengthy enough to constitute several chapters, and in some cases, they are of a more Ismāʿīlī nature than the rest of the epistle.[88] Also, at times the additions reproduce material from other epistles, such as in the case where an added passage in a manuscript of Epistle 34 (*The World is a Big Human Being*) is identical with a passage in Epistle 21 (*Plants*).

Fourth, for many epistles, the manuscripts provide different versions for the entire epistle. In some of these cases, the manuscripts carry two or more variants of an epistle that are textually dependent on each other. Epistles 32 (*Intellectual Principles according to the Pythagoreans*) and 33 (*Intellectual Principles*

---

87 For the suggestion that the number 51, or 17 × 3, carries a special numerological significance for the Brethren of Purity, see Carusi, "Alchimia islamica," 277n3.

88 For manuscript additions with Ismāʿīlī characteristics, see Carmela Baffioni, "Esoteric Shīʿism in the Additions to Ancient Manuscripts of the *Rasāʾil Ikhwān al-Ṣafāʾ*," in *L'Ésotérisme shiʿite, ses racines et ses prolongements* (*Shiʿi Esotericism: Its Roots and Developments*), ed. Mohammad Ali Amir-Moezzi et al. (Turnhout and Bristol: Brepols, 2016), 281–326; Idem, "Embryology in an Ismāʿīlī Messianic Context: The Manuscript Tradition of the Iḫwān al-Ṣafāʾ," *Studia graeco-arabica* 8 (2018); Idem, "Ismaili Onto-Cosmological Doctrines in the Manuscript Tradition of the Ikhwān al-Ṣafāʾ," *Shii Studies Review* 3 (2019); Idem, "Onto-cosmology and Hierohistory in the Manuscript Tradition of the *Rasāʾil Ikhwān al-Ṣafāʾ*," in *Intellectual Interactions in the Islamic World: The Ismaili Thread*, ed. Orkhan Mir-Kasimov (London: I.B. Tauris and the Institute of Ismaili Studies, 2019), 193–218; Idem, "The Figure of Iblīs in the Ikhwān al-Ṣafāʾ-related Esoteric Literature," in *Raison et quête de la sagesse: Hommage à Christian Jambet*, ed. Mohammad Ali Amir-Moezzi (Turnhout: Brepols, 2020), 113–141; Idem, "The Esoteric Meaning of the Sin of Adam in the Manuscript Tradition of the *Rasāʾil* Iḫwān al-Ṣafāʾ," *Studi Magrebini* 20, no. 1 (2022).

according to the Brethren of Purity) appear as intertwined in the manuscripts, in the sense that all manuscripts share more or less the same sections of the text for the two epistles taken together, but they are arranged in a different order, which results in very different versions for the two epistles. For Epistles 41 (*Definitions and Descriptions*) and 48 (*The Call to God*), the variance between the order of the sections, as well as major additions and omissions in the manuscripts, also result in very different variants of the epistle. For the *Fihrist* and Epistle 49 (*Spiritual Beings*), the manuscripts carry shorter and longer variants of the epistle, where the latter is an elaboration of the former.

Fifth, for some epistles, the manuscripts provide completely independent variants of the text. For Epistle 10 (*Eisagōgē*), two manuscripts provide a short and entirely different, but more technical, version. For Epistle 50 (*Governance*), two manuscripts carry entirely different versions of the epistle, where one variant is textually related to Epistle 32 (*Intellectual Principles according to Pythagoreans*) in the Oxford edition text. A further variant appears to summarize all of the *Epistles*, while it also relates to the text of the former variant. For Epistle 51 (*Constitution of the World*), the manuscripts provide two entirely different versions, of which 51a draws on Epistle 32b and 51b on Epistle 34 (*The World is a Big Human Being*); variant 51c joins variants 51a and 51b together.

The most extraordinary case, however, is Epistle 52 (*Magic*), for which the manuscripts provide three long and completely different variants, 52a, 52b, and 52c, with no textual relation with each other, or, apparently, with any other epistle.[89] All three versions give at least the superficial appearance of being authored by the Brethren of Purity in the sense that they employ the authors' stock phrases and contain references to other epistles. All but two manuscripts employed in the Oxford critical editions, as well as the additional manuscripts consulted for this book, carry variant 52b; two manuscripts carry 52a and a single manuscript carries 52c as inserted after 52b. The Beirut edition and at least one manuscript, on the other hand, contain both 52a and 52b as a unified epistle. While the manuscript tradition, then, seems to suggest that 52b is the authentic variant, it is far from clear that this is the case. First, although the most ancient surviving manuscript (MS ع) contains 52b, the manuscripts with 52a and 52c are also early. Second, as argued by de Callataÿ, 52a seems to

89 Epistle 52a is edited in *On Magic*. See the technical introduction and introduction for an overview of the manuscript variants for Epistle 52 based on the information available at that time. The introduction speaks of a "short" (52a) and a "long" (52b) version; since then, a third variant (52c) has been uncovered of a roughly similar length as 52a. Both Epistle 52b, edited by Godefroid de Callataÿ, Sébastien Moureau, and Bruno Halflants, and Epistle 52c, edited by the current author, are presently in the stage of copyediting.

give the best appearance of genuinely forming part of the *Epistles*.[90] In particular, it is the only one of the three variants that contains a proper beginning and ending; its introductory passages bind it to the preceding epistles, while its final passages provide an appropriate conclusion for the work as a whole. None of this, however, proves that 52a is the authentic version of this epistle. If by authentic we mean a text that was not interpolated into manuscripts after the final redaction, it is even possible that two, or even all three, of the variants for Epistle 52 are authentic; that is, they pertain to different stages of the redaction process.[91] It is also possible that none of the three variants is authentic, and that a magical epistle came to form part of the manuscript tradition of the *Epistles* after the redaction of the work was finished. As seen in the previous section, this conclusion is supported by Epistle 7.

Beyond the manuscripts that contain the entire *Epistles*, some manuscripts instead convey a selection or a summary of the *Epistles*. Some of these appear to present a "greatest hits" selection of complete epistles; for example, MS Amcazade Hüseyn Paşa 452 (1680 CE) carries the entire first section, followed by Epistles 33, 25, and 41. The choice of epistles is likely to relate less to the original process of redaction than to a later copyist's preferences. Tübingen manuscript MaVI86 (dated 1215 CE) seems to represent a special case, however; its first forty-one folios consist of a summary of Epistles 1–23 of the *Epistles*. This is not the only summary of the *Epistles*, but Kacimi has suggested that this particular one might represent an original version of the *Epistles*.[92] When the manuscript's text is compared with the critical editions, however, it becomes clear that the editor's method is to reproduce large sections of the epistles, with only minor variations with respect to other manuscripts, and to cut the intervening parts, sometimes in mid-sentence. In consequence, it seems unlikely that this manuscript would provide us with the 'proto-*Epistles*;' rather, it probably represents one early manuscript variant among others. But while the Tübingen manuscript, then, mostly summarizes the twenty-three epistles in the exact same order as they appear in the critical editions, it also contains some interesting deviations. The summary omits Epistles 2–4 and 12–13 entirely, while Epistles 7 (*Theoretical Sciences*) and 8 (*Practical Sciences*) are reproduced in the inverse order. Moreover, it includes all the logical epistles within a single "chapter on the essence of speech/reason" (*faṣl fī bayān māhiyyat al-nuṭq*), even though, in general, although not always, the copyist devotes a separate

90 See de Callataÿ's introduction to Epistle 52a in *On Magic*, 5–10.

91 This is suggested by Madelung, for which see p. 44.

92 Kacimi, "Edición crítica: Tomo I," 144–145.

chapter to each epistle. On the whole, then, the summary gives additional evidence for a variant where Epistles 7 and 8 occupy inverse positions, as well as supports the case that the logical epistles initially constituted a single epistle. Since, as we will see, both of these cases are supported by further evidence, it seems probable that the Tübingen summary is indeed based on a relatively early variant of the text. It is not clear, however, whether the omission of Epistles 2–4 and 12–13 has any particular significance, or whether they were merely of no interest for the author of the summary.

All of the large-scale manuscript variation taken together shows that the text of the *Epistles* was far from stable even two or three centuries after the work's final redaction, from which time the most ancient manuscripts we have originate. Beyond this, the manuscript variation also supports at least three of the four methods of redaction suggested in the previous section. The first kind of variation shows that the current positions of the epistles are not cut in stone, and that some epistles at least were probably transposed from one position to another during the redaction process. The second kind provides similar evidence for the splitting of epistles in that it shows that not all current epistles form independent epistles in the existing manuscripts. The three final categories indicate that the contents of the epistles also evolved in time. Many of the additions that occur in only one or two manuscripts should in all likelihood be interpreted as later interpolations introduced by copyists with the aim of reinterpreting the work to comply with their personal doctrinal inclinations. But the different versions of the epistles seem to imply a rather heavy redaction of the contents of the epistles as well. In the cases where the two or more versions bear a textual relation with each other, the natural explanation is a chronology of an earlier and a later variant, where the latter is the result of an author or a later redactor either expanding on or re-editing the original epistle. The entirely different versions of an epistle could partly represent pseudepigraphs, but they also might be interpreted as an earlier and later version of an epistle, perhaps because the later redactors were unsatisfied with the previous variant of the epistle and decided to compose an altogether new version to replace it.

It is obviously also the case that neither the Oxford critical editions nor the consulted manuscripts provide evidence for so radically a different version of the *Epistles* as the curricular passages would suggest. But especially since all the manuscripts that we have postdate the final redaction of the work by two or three centuries, it seems likely that they tend to reflect the final form of the work. The variation in the manuscripts would, then, only be the tip of the iceberg with respect to the changes that the *Epistles* underwent during the redaction process. Thus, no instance of an original redaction of the work survives. What we have instead are hints of a prolonged redaction left within the manuscript tradition. In any case, the further evidence presented below will show that the changes that the *Epistles* underwent must have been much more substantial than the large-scale differences contained in existing manuscripts of the work.

4 The *Fihrist*

Before proceeding to the evidence that concerns the *Epistles* as a whole, it is necessary to first turn to the *Fihrist*.[93] In the Beirut edition and most manuscripts, the fifty-one or fifty-two epistles are preceded by an introductory section, the *Fihrist*. It consists of two main parts: (1) a table of contents that summarizes the contents and goal (*gharaḍ*) of each epistle and (2) a more general final section. The main part supposedly reveals what the authors perceived to be the core doctrine conveyed by each epistle; the final section provides an introduction for the work as a whole. Thus, the *Fihrist* represents a special case within the work in that it is not a separate epistle devoted to a specific science or subject matter but rather a preface that introduces both the work in general and each epistle, in particular. There is no critical edition for the text of the *Fihrist*, nor has it been the object of much scholarly attention. It is, nevertheless, of interest for the thesis of the gradual redaction of the *Epistles* for three reasons. First, scholars have tended to regard the *Fihrist* as an integral part of the work; by summarizing each epistle beforehand, it gives an impression of a carefully conceived plan by the work's authors. If it turns out, instead, that the *Fihrist* postdates the *Epistles*, its agreement with the current order of the *Epistles* cannot be employed as an argument for the stability of the work as a whole. Rather, it represents a later author's interpretation of the *Epistles*. Second, the manuscript variation of the *Fihrist* reveals a rather special case of how the *Epistles* evolved in time, regarding, in this case, the gradual expansion of the contents of a single epistle. Third, the conclusions on the status of the *Fihrist* here will be of major importance in the second part of the book, because, it turns out, there is a surprising textual relation between a specific *Fihrist* variant and the *Comprehensive Epistle*.

Starting with the question of the *Fihrist*'s status within the *Epistles*, it should be noted at the outset that not all manuscripts include it, even if most do.

93 For a detailed study of the *Fihrist*, on which this section draws, see Janne Mattila, "*Fihrist* of the *Rasāʾil Ikhwān al-Ṣafāʾ*: Textual Variants and Their Relation to *al-Risāla al-Jāmiʿa*," *Micrologus* 33 (2025).

Among the surveyed witnesses, the Beirut edition and sixteen manuscripts contain the *Fihrist* at the beginning; two manuscripts do not.[94] Since the earliest known witness (MS ع) also carries it, clearly the *Fihrist* has formed a part of the work for as long as we have manuscript evidence. Still, the two variant witnesses show that its inclusion is not universal. Moreover, the fact that most manuscripts carry it does not entail that the *Fihrist* formed a part of the original composition. It is perfectly plausible that the text was introduced at some point during the two or three centuries between the original composition and the first surviving manuscript. However, since al-Tawḥīdī mentions the *Fihrist*,[95] clearly some variant must have formed a part of the work at least since shortly after the work's final redaction.

That the *Fihrist* did form a part of the original composition is supported by the existence of two references in the text of the *Epistles*:

> Know, O brother, that we have finished with the epistle on definitions and descriptions, which is the last of the psychical-intelligible epistles, as we promised in the *Fihrist* at the beginning of this book of ours. Now we want to explain in this fourth part the discussion on theology (*al-kalām fī al-ilāhiyyāt*), which is the ultimate goal and utmost objective.[96]

> The one who has come into possession of these epistles among our noble brethren should present of them to every deserving one what is close to their understanding, and what he knows to suit and befit their grade, gradually and in the succession in which we ordered them (*ʿalā al-tartīb alladhī rattabnāhu*) in the *Fihrist*.[97]

The passages occur in Epistles 42 (*Opinions and Doctrines*) and 52a (*Magic*), respectively. Both references present the *Fihrist* as an integral part of the work in the sense that it establishes the general plan that the authors then follow in the actual epistles. The first reference occurs in a transitional passage at the beginning of Epistle 42; it is present in the Beirut edition but it is omitted in most manuscripts. Thus, this reference to the *Fihrist* is very likely to be a later interpolation. The second reference appears at the beginning of Epistle 52a, the shortest variant of the epistle on magic, which is of uncertain authenticity. Moreover, the passage in Epistle 52a is in fact in part a paraphrase of the final

94 This agrees with the results in Alí-de-Unzaga, "The Missing Link," 92. The vast majority of the manuscripts surveyed by the author contain the *Fihrist* at the beginning.

95 Al-Tawḥīdī, *Kitāb al-Imtāʿ*, vol. 2, 5.

96 *Rasāʾil*, vol. 3, 401.

97 *On Magic*, 7.

section of the *Fihrist*, which is only present in a later variant of the text.[98] This almost certainly means that Epistle 52a postdates the *Fihrist*. Since this is now the only reference to the *Fihrist* in the *Epistles*, it is more likely to provide additional support for the late date of Epistle 52a than to attest to the authenticity of the *Fihrist* as a part of the *Epistles*. That is, if the *Fihrist* provides indispensable guidelines that a student of the *Epistles* should follow, why does the only reference to these instructions occur in the very last epistle? Why do most references in the *Epistles* to its curricular order in fact contradict the curriculum of the *Fihrist*? Even two references would be a low number if the *Fihrist* indeed was an important part of the work from the beginning. But now we have one reference that was clearly inserted into the manuscript tradition at a much later date, and another reference in a possibly inauthentic treatise. The absence of clearly authentic references, then, supports the claim that the *Fihrist* was composed after the epistles. This does not necessarily mean that it is inauthentic, since it could have been the last part of the *Epistles* that the authors composed. But, as we will see in Chapter 3, the fact that many epistles refer to each other shows that cross-references were updated to some extent at the final stage of the redaction. That this was not done for the *Fihrist* suggests that it was introduced into the corpus after the final redaction.

This conclusion is further supported by evidence that indicates that the textual transmission of the *Fihrist* and the epistles is separate. Namely, as shown in the previous section, the manuscripts show variation with respect to the number, order, and configuration of epistles: (1) Epistles 12 (*On Interpretation*) and 13 (*Prior Analytics*), and (2) Epistles 49 (*Spiritual Beings*) and 50 (*Governance*), may constitute a single epistle; (3) Epistle 4 (*Geography*) is sometimes either the fifth or sixth epistle; (4) Epistle 6 (*Arithmetical and Geometrical Proportions*) is either the second or fourth epistle; and (5) the order of Epistles 7 (*Theoretical Sciences*) and 8 (*Practical Sciences*) is reversed. If the *Fihrist* were to correspond to the order and configuration of epistles in each manuscript, it would mean that the *Fihrist* was, in each case, edited to correspond to the actual configuration of epistles in that manuscript. This is not the case, however. Although the order and configuration of the epistles also varies in the *Fihrist*, this order often does not agree with the actual composition of the epistles. The variant position of Epistle 4 contradicts the *Fihrist* in the case of five manuscripts, and that of Epistle 6 in four cases. The combination of Epistles 12 and 13 contradicts the *Fihrist* in six cases, and the combination of Epistles 49 and 50 contradicts the *Fihrist* in two cases; the reverse order of Epistles 7 and 8

98 See Mattila, "*Fihrist*."

contradicts the *Fihrist* in four cases. While the *Fihrist* more often corresponds to the standard order and the actual configuration follows a variant order, it is also sometimes the case that the *Fihrist* diverges from the standard order while the actual epistles do not. The conclusion would appear to be that even if the *Fihrist* text is not stable either, it constitutes a separate unit from the actual epistles.

As for the text of the *Fihrist*, the variation between manuscripts is substantial enough to warrant the designation of various variants. The main part of the *Fihrist* provides a summary of each epistle following a fixed order of presentation: (1) the number of the epistle in one of the four sections (in most manuscripts), (2) a summary of its contents, and (3) a statement of the epistle's goal. For this main part of the *Fihrist*, the manuscripts convey three variants that are distinguished by progressively longer entries for each epistle. In the sample of seventeen testimonies (sixteen manuscripts and the Beirut edition) that contain a variant of the *Fihrist*, fifteen contain a short variant (*Fihrist a*), a single manuscript (MS D) includes an expanded variant (*Fihrist b*), and the Beirut edition has an even more elaborate variant (*Fihrist c*).

The variants bear a very specific textual relation to each other. This may be illustrated by means of the *Fihrist* entry on Epistle 3 (*Astronomy*).[99] The entry in *Fihrist a* is as follows:

> The third of them is an epistle on the stars which is like an introduction to the composition of the spheres, the attributes of the zodiac, and the trajectory of the planets. The goal is to arouse desire in souls to ascend to the world of the spheres and the layers of the heavens.[100]

Fihrist c gives a considerably longer entry:

> The third is an epistle on the stars which is like an introduction to the **knowledge of the** composition of the spheres, attributes of the zodiac, and the trajectory of the planets, **and to the knowledge of their influences on this world, the manner in which the elements and generated beings are affected by them in origination and decay, generation and corruption.** The goal is to arouse desire in **pure** souls to ascend to the world of the spheres and the layers of the heavens, **the degrees of the**

99 For a second example, and full Arabic citations of both sample passages, see Mattila, "*Fihrist*."

100 MS Atıf Efendi 1681, fol. 1^{b}.

> **spiritual beings and closest angels, highest abode, and supernal substances, and the arrival at the sanctity and trusted spirit.**[101]

The words in bold indicate the additions that the two expanded versions make to version *a*. The testimonies for *Fihrist b* and *c* contain the same text until the final phrase, which is included in the Beirut edition but not in MS D. Thus, *Fihrist b* represents an intermediate stage between *Fihrist a* and *c*, in the sense that it contains most, but not all, additions made to *Fihrist a* in *Fihrist c*; this relation is confirmed by other entries. The method employed by the editor of *Fihrist c* is one of expansion; the precise words of *Fihrist a* are retained while additions of words or phrases are inserted in the middle. That is, the three versions contain a shared core text that is gradually expanded.

As a result of this method of expansion, the *Fihrist* entries become quite different. In *Fihrist a*, the first section lists three categories of knowledge that Epistle 3 aims to convey: (1) the spheres, (2) the zodiac, and (3) the planets. The longest variant adds two additional categories: (4) astral influences on the sublunar world, and (5) the generation of sublunar bodies. In all variants, the second section relates the epistle's goal, or inner meaning, to arousing desire in the human soul to ascend to the supralunar world. In the longest variant, however, this is qualified by associating the ascent with proximity to spiritual beings or angels. The same relation prevails in the epistle entries for each of the fifty-one epistles; in each case, *Fihrist c* expands on the entry in *Fihrist a* with additions of varying length while retaining all or most of the wordings in *Fihrist a*. This clearly shows that the *Fihrist* was expanded gradually and that *Fihrist c* must postdate *Fihrist a*. This is important since, as we will see in Chapter 4, the *Comprehensive Epistle* depends textually on *Fihrist c*.

Curiously, a similar method of expansion is also employed for the *Fihrist*'s final section that follows the primary section listing the contents and goals of each epistle. Here, two manuscripts provide a very concise text:

> Abū Sulaymān said: The one who has come into possession of these epistles must only show them to the seekers of the sciences and lovers of wisdom. If he finds some who seek the right way and pursue guidance and salvation, he should present to each of them according to what he knows of their understanding and what befits them, and gradually and in proper order show them one [epistle] after another until they gain possession of the wisdom in themselves. After all that, they will seek all [of them] with

101 *Rasāʾil*, vol. 1, 22.

> eagerness and desire and follow them in the succession in which they are ordered in the *Fihrist*. They will receive an abundant reward for that. God guides whom He wills to the right path, but only those with knowledge understand it.[102]

This is the only variant in which the final section is attributed to a specific person. The person cited might be Abū Sulaymān al-Maqdisī in al-Tawḥīdī's account of the members of the Brethren of Purity. This does not necessarily confirm al-Tawḥīdī's testimony, since the attribution may have been introduced by a later author of the *Fihrist*, or even by the copyist. In any case, the passage very concisely addresses the proper use of the *Epistles*. They should not be divulged freely but only given to those who seek knowledge and wisdom, and they should be studied in the precise order specified in the *Fihrist*.

Most manuscripts convey a longer version, spanning a little less than one manuscript page. It provides essentially the same message but introduces it by means of a metaphor where the one in possession of the *Epistles* is compared to a man in charge of a flourishing garden. Again, most of the precise words of the short variant are included in the longer variant.[103] The third variant spans almost six edited pages in the Beirut edition; it includes considerably expanded versions of both the garden story and the instructions on the proper divulgation of the *Epistles*,[104] as well as a new textual block of cosmological subject matter.[105] The focus of the latter passage is on the hypostasis of the Intellect and the progression of further degrees of existence from it by the intermediacy of the Universal Soul. The passage has an Ismāʿīlī flavor, especially in its employment of the terms "Preceder" (*sābiq*) and "Follower" (*tālī*) for the two primary hypostases.

The textual history of the *Fihrist* is, then, surprisingly complex. It seems that the two parts of the *Fihrist*, the table of contents and the final section, are textually independent. Namely, the distribution of the different variants for the two parts is not at all aligned in the manuscripts. For the final section, two manuscripts (MSS ج and F) provide the shortest variant, most manuscripts the middle variant, and only the Beirut edition the longest version.

This textual status of the *Fihrist* provides two kinds of results for the thesis of the gradual redaction of the *Epistles*. First, it provides an instance of grad-

---

102 MS Atıf Efendi 1681, fol. 5a.

103 The parts of the passages of the two other variants that show the textual dependence are cited and translated in Mattila, "*Fihrist*."

104 *Rasāʾil*, vol. 1, 43–47.

105 Ibid, 41–42.

ual expansion of text within the *Epistles*. The method of editing by expanding on an existing text while retaining original wordings, even in cases where the original meaning is transformed, seems curious. It does not seem to have been applied in this precise manner for other epistles. Perhaps it is due to editors making additions in a manuscript's margins, which are later incorporated into the main text. But Epistle 49 (*Spiritual Beings*) provides an analogous case in the sense that its longer variant is an elaboration of a shorter variant. In any case, the example of the *Fihrist* supports the suggestion made above, based on curricular passages, that expanding the text of epistles constituted one method in the gradual redaction of the work.

Second, this shows that the concordance between the *Fihrist* and the rest of the *Epistles* cannot be used as an argument for the stability of the text of the *Epistles*. That is, it is not the case that the authors had a clear-cut plan, shown in the *Fihrist*, which they then realized in the rest of the *Epistles*. On the contrary, all the evidence suggests that the *Fihrist* was composed after the *Epistles*. Moreover, like many other epistles, the *Fihrist* itself also evolved in time in the form of progressive redactions. Tables 6 and 7 present a comparative summary of the contents and goals, respectively, attributed to each epistle in the two main variants.[106] In most cases, the additions that *Fihrist c* makes to *Fihrist a* seem innocuous; that is, they add little of substance. However, they often provide a different emphasis on an epistle's meaning, highlighting its esoteric aspects. Thus, in the case of Epistle 3, while *Fihrist a* provides for the subject of astronomy the inner goal of the soul's ascent to the spheres, in *Fihrist c*, this is related explicitly to spiritual beings and the holy spirit. The *Fihrist* does not, then, represent the original authors' views on the contents and goal of each epistle; rather, it presents the interpretations of the author of each *Fihrist* variant on the meaning of the epistles. As we will see in the next chapter, the precise configuration of epistles presented by the *Fihrist* was not the original form of the *Epistles*.

106 The tables also summarize the corresponding entries in the *Comprehensive Epistle*, to which we shall return in Chapter 4.

CHAPTER 2

Methods of Redaction

The previous chapter argued for the claim that the *Epistles* assumed their current form as the end result of a gradual process of redaction. The thesis was supported, first, by the contradiction between the actual configuration of epistles and references to that configuration in the *Epistles*, and, second, by actual variation of the configuration in the manuscripts. It was also shown that the *Fihrist* does not represent the original plan for the *Epistles* but rather postdates the final redaction. More specifically, the internal inconsistencies of the curricular passages suggested four distinct methods of redaction: (1) introduction of new epistles, (2) splitting of existing epistles, (3) expansion of existing epistles, and (4) reordering of epistles. The last three are also manifest in existing manuscripts of the *Epistles*: Epistles 12 (*On Interpretation*) and 13 (*Prior Analytics*), and Epistles 49 (*Spiritual Beings*) and 50 (*Governance*), appear as a single epistle; many epistles have additions as well as variants of different lengths; and Epistles 4 (*Geography*), 5 (*Music*), 6 (*Arithmetical and Geometrical Proportions*), 7 (*Theoretical Sciences*), and 8 (*Practical Sciences*) occupy variant positions.

The manuscripts also provide more subtle indications of redaction. Two kinds of such evidence will be discussed in this chapter. The first section addresses the variation of epistle titles; this provides further evidence for the second method, that is, the splitting of an epistle into two or more smaller epistles. The second section discusses discrepancies in the transitions between successive epistles; this supports the claim that many epistles initially occupied a different position within the work. Taken together, this evidence shows that there must once have been a redaction of the *Epistles* that was more radically different from the modern editions than any known manuscript.

1 Epistle Titles

In modern scholarship of the Brethren of Purity, one often gets the impression that each epistle has a clearly fixed title in the work. This impression is confirmed by the recent critical editions, where the editions and translations of the epistles are provided by such determinate titles. In fact, there is abundant variation regarding these titles in the *Epistles*. First, each epistle is provided with a title on the title page of that epistle. The manuscripts show great variation; for most epistles, the version selected for an edition is just one among many.

Second, within a single manuscript, the title of a given epistle in the *Fihrist* and the title page of that epistle may be different. Third, the titles employed in references from one epistle to another may differ from those employed on the title pages. The evidence that epistle titles provide for the gradual redaction thesis is based on two instances of this variation: differences between manuscripts regarding the epistle titles on title pages are summarized in Table 8; variation in cross-references is presented in Tables 9 and 10.

As regards the variation between manuscripts in title pages, for the most part, the differences are probably not particularly significant; an epistle's subject matter is merely presented in different words. A notable example is the employment of the transliteration of a Greek word versus an Arabic translation of its meaning for the mathematical and logical epistles. Thus, in some manuscripts, the title of Epistle 1 (*Arithmetic*) is given as *arithmāṭīqī*; in others, the title is "on the number" (*fī al-ʿadad*); many manuscripts also combine both the Greek original and its Arabic translation in their version of the title. In other cases, the concise version of the title is replaced by a longer variant. Again, there is considerable variation regarding the additional attributes, such as "on the number and its essence and quantity" (*fī al-ʿadad wa-māhiyyatihi wa-kammiyatihi*) versus "on the number and its special properties" (*fī al-ʿadad wa-khawāṣṣihi*). Due to the ever-growing list of attributes, sometimes the epistle titles become extremely long. Often, this is the case because the title is clearly taken from the corresponding epistle description in the *Fihrist*, perhaps by the copyist, as is evident in the fact that the title page retains the *Fihrist*'s title and goal format. All of this probably tells little of the redaction process, apart from the fact that the lack of agreement on precise epistle titles once again attests to the fluctuating nature of the textual tradition of the *Epistles*.

Of more significance for the gradual redaction thesis, however, are the cases where a variant title contradicts the actual contents of the epistle. More specifically, there are many instances where an epistle title overextends to encompass the subject matter of more than one epistle. Some of these instances occur in the title pages of manuscripts (Table 8). Thus, for both Epistles 1 (*Arithmetic*) and 2 (*Geometry*), a single manuscript complements the title with the additional element of either arithmetical or geometrical proportions, which coincides with the title of Epistle 6 (*On Arithmetical and Geometrical Proportions*); a single manuscript also contains the term "geometry" (*jūmuṭriyā*) for Epistle 6. Taken alone, these instances could merely represent a confusion by the copyist. But we have seen that the curricular passages suggest that an earlier redaction of the *Epistles* contained four, rather than six, mathematical epistles. If these variant titles are taken as vestiges of an earlier redaction, they give additional support for the hypothesis that Epistle 6 was added to the curriculum later, perhaps by splitting it from one of the original mathematical epistles.

There are many more such instances in the title pages of manuscripts. Based on at least a single manuscript variant, similar couples are formed by Epistles 9 (*Ethics*) and 31 (*Differences of Languages*); Epistles 23 (*Constitution of the Human Body*) and 26 (*The Human Being is a Microcosm*); Epistles 26 and 34 (*The World is a Big Human Being*); Epistles 29 (*Death and Life*) and 30 (*Pleasure and Pain*); Epistles 32 (*Intellectual Principles according to the Pythagoreans*) and 33 (*Intellectual Principles according the Brethren of Purity*); Epistles 32/33 (*Intellectual Principles*) and 51 (*Constitution of the World*); Epistles 47 (*Divine Law*) and 48 (*The Call to God*), perhaps combined further with Epistle 45 (*Companionship of the Brethren of Purity*); and Epistles 49 (*Spiritual Beings*) and 52 (*Magic*).

Not all of these combinations make as much immediate sense as the case of Epistle 6, but many of them do. The most evident case is that of Epistles 32 and 33; there is no obvious rationale why these should constitute two separate epistles. Moreover, as mentioned in the previous section, the contents of these two epistles are in fact intertwined in the manuscripts to the extent that it is not entirely clear how they should be reconstructed as two separate epistles in the critical edition.[1] A logical explanation for this confusion is that the initially single epistle on intellectual principles was at some point separated into two epistles. Most of the other instances of coupled epistles above have sufficiently shared subject matter to make the idea of a combined epistle plausible. For example, Epistle 30's discussion of pleasure and pain is frequently related to the wisdom of life and death in Epistle 29; the astrological basis of magic in Epistle 52 similarly forms a logical couple with Epistle 49 on spiritual beings. Also, there is no clear reason why the subject of microcosm should be divided between Epistles 26 and 34. For many other epistle couples, the connected subject matter is obvious in their titles.

The epistle titles also provide more subtle indications of the epistles' transformation. Various manuscripts attribute to Epistle 50 (*Governance*) the additional attributes of "commentary" (*sharḥ*) and "demonstration" (*bayān*/*burhān*), which make no sense with regard to its contents. But as we will see in Chapter 4, the genesis of the *Comprehensive Epistle*, which claims to be both a commentary and an ultimate demonstration of the *Epistles*, seems to be connected with Epistle 50. In some cases, the title speaks of a "chapter" (*faṣl*) rather than an "epistle"; thus, the title page of a single manuscript presents Epistle 36

1 See the editor's technical introduction in *Sciences of the Soul and Intellect, Part I*.

(*Cycles and Revolutions*) as a chapter. This suggests that the epistle in question might initially have formed a chapter within a larger epistle.

All these confusions concerning epistle titles in manuscripts could be dismissed as simple scribal errors. But we have two actual cases where two epistles form a single epistle. Epistles 12 (*On Interpretation*) and 13 (*Prior Analytics*) are combined in the majority of the reviewed manuscripts; only a few manuscripts make them into separate epistles, which is the way they appear in all modern editions. Similarly, Epistles 49 (*Spiritual Beings*) and 50 (*Governance*) are combined in two manuscripts. In both cases, the title pages of the manuscripts conform to the combined status of these epistles. Thus, for Epistle 12, the manuscripts include "analytics" (*anālūṭiqā*) in some form in the epistle's title; similarly, "governance" (*siyāsa*) is included in the title of Epistle 49. It makes, then, sense to extrapolate this to other epistle couples. At some point of the redaction process, they constituted a single epistle; the only vestige of their former status is preserved in the erroneous title pages of some manuscripts.

Moreover, the epistle titles employed in cross-references manifest even greater discrepancies (Table 9). Unlike the title pages in manuscripts, these cannot be due to the copyists; rather, they form a part of the body of the text. As we will see in Chapter 4, most references from the *Epistles* to the *Comprehensive Epistle* are clearly later interpolations. But the references between epistles manifest little variation in the manuscripts.

Here also, most variation concerns wordings, and thus merely shows that the epistle titles were not entirely fixed. Beyond this, many epistle references again treat two or more epistles as a single epistle. Some of these concern the same instances as in the title pages of manuscripts. Thus, Epistles 26 (*The Human Being is a Microcosm*) and 34 (*The World is a Big Human Being*), and Epistles 29 (*Death and Life*) and 30 (*Pleasure and Pain*),[2] are referred to as a single epistle. There are also new pairs between Epistles 3 (*Astronomy*) and 52 (*Magic*); Epistles 7 (*Theoretical Disciplines*) and 8 (*Practical Disciplines*); Epistles 21 (*Plants*) and 22 (*Animals*); and Epistles 25 (*Fall of the Sperm*) and 36 (*Cycles and Revolutions*); as well as between Epistles 32/33 (*Intellectual Principles*) and 49 (*Spiritual Beings*). As regards shared subject matter, most of these make immediate sense as a single epistle. Even Epistles 25 and 36, for which the connection may

2 For these two epistles, see the two references in Epistle 30: *On Life, Death, and Languages*, 34, 45. These refer the subject matter of pleasures to an "epistle on the aversion of animals to death" (*risālat karāhiyyat al-ḥayawān li-l-mawt*). The latter reference, as shown in the critical apparatus, only appears in some manuscripts.

not be intuitive, are connected by the fact that the formation of an embryo, in the former epistle, is mainly discussed in the astrological context of the latter epistle.

The most conclusive evidence of the epistle titles employed in cross-references, however, concerns two particular groups of epistles. The first regards the five logical epistles (10–14). The *Epistles* only contain a total of six references to these epistles individually: one reference to Epistle 10 (*Eisagōgē*), three references to Epistle 11 (*Categories*), and two references to Epistle 14 (*Posterior Analytics*); there are no references at all to Epistles 12 (*On Interpretation*) and 13 (*Prior Analytics*). Interestingly, the contents of these last two logical epistles are also missing in the summary of the Tübingen manuscript discussed in Chapter 1, which also presents the logical epistles as a single epistle. In contrast, the *Epistles* refer eleven times to an "epistle on logic" (*risālat al-manṭiq*). Clearly, this supports the evidence of the curricular passages, discussed in Chapter 1, that initially there was just one logical epistle. The fact that, among the individual logical epistles, the epistle on categories has most references, also agrees with Epistle 15 identifying the subject matter of the single epistle on logic with that of Aristotle's *Categories*.[3]

The second conclusive case concerns Epistles 32 (*Intellectual Principles according to the Pythagoreans*) and 33 (*Intellectual Principles according to the Brethren of Purity*). The *Epistles* do not contain a single reference to either of these epistles as an individual epistle. Instead, there are thirty-six references to an "epistle on the intellectual principles." Given all the previous evidence, in particular, the fact that the contents of these epistles are mixed in the manuscripts, it seems clear that they initially constituted a single epistle.

In addition, the cross-references in the *Epistles* contain a number of references to 'orphan' epistles (Table 10). These are references where neither the title nor the subject matter has any obvious counterpart within the existing epistles.[4] These include a "book on prosody" (*kitāb al-ʿurūḍ*) and an epistle on the "relations of letters" (*nisbat al-ḥurūf*); both mention the combination of letters as their subject matter, but the *Epistles* contain no such epistle. Some other titles, such as the "book of prophets" (*kitāb al-anbiyāʾ*), an "epistle of instruction/mathematics" (*risālat al-taʿālīm/taʿlīm*), an epistle on the "essence of the world soul" (*māhiyyat nafs al-ʿālam*), and an epistle on the "attributes and the subjects of attributes" (*risālat al-ṣifāt wa-l-mawṣūfāt*) point to subjects treated in the *Epistles* but not to any specific epistle. It is not clear whether this means

3 See pp. 60–61.

4 For these references, see the endnotes in Table 12.

that such epistles at one point existed. But the wandering references once again support the notion that the organization of the work into individual epistles evolved in time.

The results from the evidence of epistle titles may be divided into minimal and maximal conclusions. As a minimum, the evidence clearly shows that the structure of the *Epistles* evolved in time in the specific sense that some of the current epistles initially formed a part of a larger epistle. The manuscript tradition contains actual variants of the *Epistles* where two epistles form a single epistle. In the case of the logical epistles and Epistles 32 and 33, there is no other explanation for the fact that the internal references mostly or exclusively refer to them as a single epistle. In both cases, other evidence also points to them being initially combined. A maximal conclusion is that every instance of an epistle couple in the manuscripts represents an actual combination of epistles in an earlier redaction. In most cases, the suggested pairs of epistles seem plausible. But the fact that a single manuscript variant compounds the titles of two epistles is obviously not a sufficient proof for them to have previously formed a single epistle. Taken as a whole, what the evidence of epistle titles does prove beyond reasonable doubt, however, is that the method of splitting epistles into two or more smaller epistles was employed in the redaction process of the *Epistles*.

2 Epistle Transitions

The second kind of specific manuscript evidence concerns the transitions between epistles. Again, the evidence consists of internal contradictions; in this case, the contradictions indicate that many epistles initially occupied a different position within the whole. This gives additional support to the fourth method of redaction, that is, the repositioning of epistles.

Epistle transitions refer to the sections immediately at the beginning and end of each epistle. These introduce and conclude an epistle, respectively. Especially in the Beirut edition, the transitions contribute to the impression that each epistle constitutes a coherent unit, connected with the previous and following epistle, in accordance with a pre-conceived plan. These passages tend to manifest much variance in the manuscripts, however. As we saw in the previous chapter, it is often at the end of an epistle that some manuscripts introduce additional material. Also, epistles are often initiated and ended by invocatory formulas. Moreover, in most manuscripts, each epistle is ended by the formula: "The Epistle x is completed (*tammat al-risāla x*) ... and it is followed by Epistle y (*yatlūhā risālat y*)." None of these seem to be of any interest for the question of redaction, however. Rather, they just show the inclinations of each scribe. As regards the endings, they are absent in some manuscripts, while in those manuscripts in which they are present, they follow the same formulaic pattern throughout the manuscript, and invariably indicate the following epistle correctly. Thus, like the invocations, they are clearly mechanical additions introduced by the copyist, as opposed to forming an integral part of the text.

In contrast, what is of great interest for the thesis of gradual redaction are the beginnings of each epistle. Here, the standard form is the following:

> Now that we have finished with discussing x in epistle y (*lammā faraghnā min dhikr x fī risālat y*) we wish to discuss in this epistle … (*nurīd an nadhkur fī hādhihi al-risāla*).

Although these also manifest major variation in the manuscripts, they clearly originate from the textual tradition of the work, rather than from the pen of the scribe. In contrast to the epistle endings, then, which are uniformly present in each epistle, the epistle beginnings are randomly present for some epistles but absent for others; and when present, they often indicate the preceding epistle incorrectly. The distribution of the epistle *incipit*s is presented in Table 11.

The main evidence of the epistle beginnings consists of their correct or incorrect indication of the previous epistle. Beyond this, the *incipit*s also manifest manuscript variation, which takes one of two possible forms: either the passage containing a correct reference to the previous epistle is missing entirely; or the text of the passage containing an incorrect reference is expanded to include the correct reference. Moreover, while the Beirut edition also contains missing and incorrect transitions, in all the instances where there is variation, it is always the case that the Beirut edition variant contains the correct reference, often supported by one or two manuscripts, while the rest of the manuscripts either contain an incorrect reference or omit the entire passage. Also, in all those cases in which most manuscripts contain no transition, and only the Beirut edition, possibly supported by one or two manuscripts, contains a transition, the reference is always correct. The cases where the Beirut edition also has an incorrect reference are always shared by all manuscripts. The consistency of the pattern clearly shows that the Beirut edition must go back to a later version of the text, as was also apparent in the case of the *Fihrist*. The variation therefore attests to an effort to update the *incipit*s after the redaction process, although the end result still remains far from perfect.

Accordingly, only roughly a half of the epistles (24) contain any kind of initial transitory passage in most consulted manuscripts; most epistles contain one in the Beirut edition (41). The presence of initial transitions is, moreover, very unevenly divided over the *Epistles*. In the first section of the work, in most manuscripts, Epistles 6 (*Arithmetical and Geometrical Proportions*), 10 (*Eisagōgē*), 12 (*On Interpretation*), and 13 (*Prior Analytics*) lack any transition. In the Beirut edition, this is the case only for Epistles 10 and 13. In the second section, Epistles 15–22, 26 (*The Human Being is a Microcosm*), and 30 (*Pleasure and Pain*) contain an initial transition in the manuscripts; in the Beirut edition, all but Epistle 25 (*Fall of the Sperm*) contain a transition. In the third section, only Epistles 35 (*Intellect and the Intelligible*), 36 (*Cycles and Revolutions*), and 39 (*Movements*) carry a transition in most manuscripts; only Epistle 32 (*Intellectual Principles According to the Pythagoreans*) lacks one in the Beirut edition. In the fourth section, only Epistle 52a (*Magic*) has a transition in most manuscripts; in the Beirut edition, Epistles 42 (*Opinions and Doctrines*), 44 (*Beliefs of the Brethren of Purity*), and 51 (*Constitution of the World*) also have one.

The uneven distribution of *incipit*s raises further questions. First, one might be tempted to interpret the absence of an initial transition as evidence of a split epistle; that is, an epistle has no transitional phrase because it initially formed a chapter within an epistle rather than an independent epistle. The uneven distribution of initial transitions makes this unplausible; otherwise, most epistles in the fourth section would be split epistles. Yet, it is also true that in the first section of the work, where most epistles have an initial transition, all epistles missing one are epistles for which previous evidence suggested that they are split epistles. As for the larger picture, in the Beirut edition, most of the epistles that miss an *incipit* pertain to the fourth section. For most manuscripts, an even greater contrast emerges between roughly two halves of the *Epistles*. In the first section and the first half of the second section, that is, the technical mathematical, logical, and physical epistles mostly corresponding to the Aristotelian curriculum (1–24), only six epistles have no proper initial transition. In the second part of non-Aristotelian treatises (25–52), only six epistles have any kind of initial transition. It is not clear why this should be the case, but a plausible explanation is that the two halves constitute two different chronological layers. Perhaps this is because the practice of devising initial transitions was employed at some point of the redaction process but not at another. Moreover, the Beirut edition variant attests to a later effort to harmonize the rest of the epistles by providing them with up-to-date *incipit*s, although it seems mysterious why it should have been so unsuccessful with regard to the fourth section in particular.

The pattern of the Beirut edition variant correcting an older variant is reinforced further when the passages containing a correct versus incorrect reference in the two variants are compared. In most manuscripts, the *incipit* gives an incorrect reference to the previous epistle in eleven epistles; the Beirut edition text contains only six incorrect references; the remaining five epistle *incipits* give the correct reference against all or most manuscripts. Comparison of the Beirut edition text with manuscripts in these five cases illustrates a tendency to expand the epistle *incipit* to match the actual order of the epistles. The introductory passage of Epistle 35 (*Intellect and the Intelligibles*) provides an example:

> **Know, O brother, may God support you and us with His spirit, that we have now finished showing the statement of the sages that the world is a big human being, and we provided the resemblances (*mithālāt*), pointers (*ishārāt*), and similitudes (*tashbīhāt*) in accordance with the custom of our noble brethren. Before that we discussed the intellectual principles (*al-mabādiʾ al-ʿaqliyya*) and showed there the manner of the creation (*ikhtirāʿ*) of existents and generation (*takwīn*) of created beings. Also, we previously showed in the epistle on** [Now that we have finished discussing] the senses and sensibles [and shown] that all sensibles are corporeal accidents and that they are all [forms] in corporeal matter, and that the soul's perception of them takes place by means of the senses through its sensing faculty, and that all the senses are corporeal organs, and that sensation (*ḥiss*) is a change (*taghyīr*/*taghayyur*)[5] in the composition (*mizāj*) of those senses upon contact (*mubāshara*) of sensibles in it, and that sensation (*iḥsās*) is awareness (*shuʿūr*) by the sensing faculties of the changes of those compositions. And we wish now to discuss in this epistle, **called (*al-mulaqqaba bi-*)** the intellect and the intelligibles, and to show that intelligibles are all also spiritual forms which the soul sees in their essence.[6]

In the citation, the bolded section indicates the additions present in the Beirut edition variant only; the bracketed sections represent the additions and variants in the manuscripts but not in the Beirut edition; the plain text is shared by both main variants. In the current arrangement of the *Epistles*, Epistle 35

5 In the critical edition, the word is "changing" (*yughayyir*).

6 *Rasāʾil*, vol. 3, 231; *Sciences of the Soul and Intellect, Part I*, 105. The editor has not included the Beirut edition among the testimonies. The critical apparatus does not show any manuscript supporting the Beirut edition variant; nor does any of the consulted manuscripts not included in that edition agree with the Beirut edition.

is preceded by Epistle 34 (*The World is a Big Human Being*). In most manuscripts, this passage of transition starts with a reference to Epistle 24 (*Senses and Sensibles*) as the previous epistle, which is obviously incorrect. From the perspective of Aristotelian psychology, however, the two treatises form a natural sequence proceeding from sensation to intellection, as is the case in Aristotle's *De anima* or many Arabic philosophical works discussing psychological faculties. It seems therefore entirely plausible to presume that at some point these two epistles were positioned one after the other, and that both formed a part of the second section. The Beirut edition variant, however, inserts a lengthy addition at the beginning of the passage which correctly refers to Epistle 34 as the previous epistle, as well as to Epistles 32 and 33 on intellectual principles as the epistles preceding Epistle 34. At the same time, the longer passage retains the reference to Epistle 24, while editing the wording to suggest that it occupies an indeterminate position that precedes all the mentioned epistles, as opposed to being the immediately previous one.

This suggests that an existing *incipit* was later edited to correspond to the changed configuration of the *Epistles*. Epistles 7 (*Theoretical Sciences*), 26 (*The Human Being is a Microcosm*), 36 (*Cycles and Revolutions*), and 39 (*Movements*) show a similar pattern of correction. In most manuscripts, these refer incorrectly to Epistles 8 (*Practical Sciences*), 34 (*The World is a Big Human Being*), 20 (*Essence of Nature*), and the second or third section, respectively, as the previous epistle. In the Beirut edition, an additional passage in each case makes the correct reference to what is the previous epistle in the current configuration of epistles. For Epistles 7 and 8, this probably means that these epistles were initially in the inverse order; as we have seen, we have actual instances of manuscripts where Epistle 8 precedes Epistle 7. Epistles 26 and 34 both address the microcosmic parallels between the human being and the cosmos and could plausibly have been placed together.

Since more than half of the epistles lack an initial transition in most manuscripts, while the additional ones in the Beirut edition are clearly interpolations, the *incipit*s obviously only provide information about the redaction process for a part of the epistles. But for those epistles that in most manuscripts explicitly refer to a previous epistle, there are rather incredibly precisely as many correct references as there are incorrect ones (11/11). This points to a remarkable confusion concerning the order of the epistles. Unlike in the case of epistle titles, this confusion cannot be dismissed as scribal errors, since the *incipit*s pertain to the body of the text. Moreover, the tendency to correct the transitions in the Beirut edition shows that the incorrect references were later seen as a problem. We have actual manuscripts where some epistles occupy a different position. Based on the evidence of contradictory epistle transitions, it seems licit to extrapolate this to a more extensive method of redaction where some epistles were later repositioned within the whole.

Let us now look more closely at the remaining ten epistle transitions with incorrect references to the previous epistle. The first section contains such references for Epistles 5 (*Music*), 7 (*Theoretical Sciences*), 8 (*Practical Sciences*), and 9 (*Ethics*). All of these are epistles that seem out of place in their current position. For Epistle 5, the *incipit* appears to suggest Epistles 7 and 8 as the preceding epistles, which would place these two epistles in the middle of the *quadrivium*. These two epistles, however, also seem to have switched places. For Epistle 7, most manuscripts indicate Epistle 8 as the preceding epistle, while the *incipit* of Epistle 8 makes no reference to Epistle 7. As stated above, this is also their actual order in some manuscripts. The order of progression from practical to theoretical sciences appears logical, given that this is also their order in the overall scheme of sciences in Epistle 7. The *incipit* of Epistle 8, however, confuses the situation further:

> Now that we have finished discussing the corporeal substances (*al-jawāhir al-jismāniyya*), and described their matter, forms, composition (*tarkīb*), and the accidents that inhere in the composites (*wa-mā yaʿriḍ li-l-murakkab min al-aʿrāḍ*), and we showed also the manner of their perception by means of the senses that mediate their accidents (*yatawassaṭ aʿrāḍahā*) in our physical epistles (*fī rasāʾilinā al-ṭabīʿiyya*), we wish to discuss in our intellectual epistles (*fī rasāʾilinā al-ʿaqliyyāt*) the spiritual substances (*al-jawāhir al-rūḥāniyya*). Since all existents are either intelligible or sensible, substances, accidents, or their composites, form, matter, or their composites, corporeal, spiritual, or their composites, and since corporeal substances are all subject to being acted upon (*munfaʿila*) and perceivable by means of senses, while the spiritual substances act and are not perceivable by the senses, but only by the intellect and the intellectual acts that proceed from it (*wa-bi-mā yaṣdur ʿanhā min al-afʿāl al-ʿaqliyya*), and the practical disciplines/acts are after theoretical disciplines/acts in the corporeal substances, we have to discuss the theoretical disciplines in material things (*al-ṣanāʾiʿ al-ʿilmiyya fī al-hayūliyyāt*), their essences, quantities, and qualities, and the manner of bringing forth its discipline/act in the material things that are their substrates (*kayfiyyat iẓhār ṣināʿatihā fī al-hayūliyyāt al-mawḍūʿa lahā*), in order for it to indicate more clearly the establishment of the spiritual, active essences (*li-yakūn awḍaḥ fī al-dalīl ʿalā ithbāt al-dhawāt al-rūḥāniyya al-fāʿila*), and to be more manifest for the knowledge of their substances (*abyan li-maʿrifat jawārihihā*), their kinds of motions, wondrous powers (*ʿajāʾib quwwatihā*),

> strange knowledge (*gharāʾib ʿulūmihā*), novel actions/disciplines (*badāʾiʿ ṣanāʾiʿihā*), and different actions (*ikhtilāf afʿālihā*).[7]

Based on the order of epistles, this passage makes no sense whatsoever. The current position of Epistles 7 and 8 is immediately after the extended mathematical *quadrivium*, and before ethics and the logical epistles. Yet, the beginning of the passage makes no reference to mathematics. Instead, it states that Epistle 8, and consequently Epistle 7, follow upon the physical sciences, that is the entire second section of the *Epistles*. Epistle 8 is, then, situated at the beginning of the "intellectual epistles" (*al-rasāʾil al-ʿaqliyya*), which presumably means the third section. The latter part of the passage, moreover, justifies the position of these epistles at the junction of the second and third sections, although the rationale that the passage gives is somewhat hard to decipher. Perhaps the implication is that since the practical disciplines relate to knowledge and actions in the physical world, while the theoretical disciplines relate to intelligible knowledge, these two epistles describing the two kinds of knowledge provide the proper transition from physical to intelligible knowledge. In any case, the conclusion of the passage seems to be that both Epistles 7 and 8 at some point pertained to the beginning of the third section. This might not be an entirely logical place for these curricular epistles, but it is perhaps less arbitrary than their current position.

As for Epistle 9 (*Ethics*), it indicates that Epistle 25 (*Fall of the Sperm*) is the previous epistle:

> Now that we have finished discussing the vicissitudes (*taṣārīf*) of the states of the human being in the womb, from the day of the fall of the sperm to the day of the body's birth, and shown how the spiritual powers of the planets are attached to the constitution (*khilqa*) of the embryo, and how different character traits (*akhlāq*) fixed in nature (*al-markūza fī al-ṭabīʿa*) are imprinted in its disposition (*jibilla*) during nine months, one month after another, and this is the natural sojourn (*al-makth al-ṭabīʿī*) until the day of the child's birth, and the human being's initiation (*istiʾnāf*) of life in this world for 120 years, which is the natural age (*al-ʿumr al-ṭabīʿī*), we wish to discuss in this epistle the acquired character traits (*al-akhlāq al-muktasaba*) that are added to those natural dispositions (*al-ṭibāʿ al-markūza*) after birth by recurring habits and causes that incur their generation (*al-asbāb al-dāʿiya al-muwallida lahā*), whether exces-

7 *On Composition*, 99–101.

> sive or deficient, during the vicissitudes of the days of worldly life until the day of death.[8]

Again, this passage makes no sense for the current order of epistles; Epistle 9 should be preceded by Epistle 8 (*Practical Sciences*). The position suggested by the passage is instead in the middle of the physical epistles. For an ethical epistle, this conforms as little to the curricular order as the current position. But the *incipit* provides a rationale why this should be the case: Epistle 25 explains how astral influences affect the generation of character traits (*akhlāq*) in the womb; Epistle 9 explains how character traits are developed by habituation after birth. Since ethics is largely concerned with the reformation of psychical dispositions into virtues, the two epistles form a logical couple. It is perhaps for this reason that these epistles were at some point placed together within the physical section, as opposed to within the fourth section devoted to practical philosophy.

Besides the clearly incorrect references to a previous epistle, the *incipit* of Epistle 12 (*On Interpretation*) once again attests to the confusion that prevails in the logical epistles:

> Now that we have finished discussing the ten terms that the logical philosophers call the ten categories … and before that we discussed in another chapter (*faṣl ākhar*) [epistle] and explained the six terms that the philosophers used in their propositions, and in another chapter before that we explained that when the individual letters are combined they form words (*alfāẓ*), and when the meanings in the words are combined they become signs (*simāt*), and when the signs follow one another in succession (*tarādafat*), they become useful speech (*kalām mufīd*); in this chapter [epistle], we state that all speech is of three kinds.[9]

The *incipit* refers correctly to the *Categories* as the previous epistle, as well as to *Eisagōgē* as the epistle before that. But most manuscripts, including this time both the Beirut edition and MS ع, refer to both *Eisagōgē* and the current epistle of *On Interpretation* incorrectly as chapters; in some manuscripts, both references are corrected into epistles. In addition, the passage claims that Epistle 10 (*Eisagōgē*), which in the current arrangement is preceded by Epistle 9 (*Ethics*),

8 *Rasāʾil*, vol. 1, 296; *On Ethics*, 6. In the Beirut edition, the passage is preceded by a seemingly incomplete sentence: "Now that we have finished discussing the physical substances (*al-jawāhir al-jismāniyya*), and described their matter, form, and composition."

9 *On Logic*, 77–78.

is preceded by a further chapter discussing how letters are combined to form words and words to form speech. This description does not appear to correspond to the content of any existing chapter or epistle; it might, though, correspond to the 'orphan' epistle on letters mentioned in the previous section. All of this gives further support to the claim that the logical epistles initially constituted a single epistle, the chapters of which were at some point divided into five separate epistles. Perhaps initially a further epistle or chapter also existed which discussed the combination of letters.

The second section contains incorrect initial references for Epistles 19 (*Minerals*), 20 (*Essence of Nature*), 21 (*Plants*), and 26 (*The Human Being is a Microcosm*). Beginning with the easiest case, Epistle 26 gives Epistle 34 (*The World is a Big Human Being*) as the previous epistle. As stated above, this seems plausible given that the two epistles form a logical pair. Since Epistle 34 contains no transition in most manuscripts, this tells us nothing about the position of the two epistles within the four sections, however. As for the *incipit* of Epistle 19, here the reference is to various epistles, all of which are posterior to Epistle 19:

> Know, O brother, may God support you and us with His spirit, that we have shown in the epistle of opinions and doctrines that the world is originated, created, and devised, so that it came to exist after it did not exist, and that its Creator, Devisor, Originator, and Former is the Creator, who created it as He wished by saying "Be!" and so it came to be, as we showed in the epistle on the intellectual principles [epistle on the opinions, doctrines, and intellectual principles]. And we also showed the manner of the destruction of the world and the folding of the skies, the manner of the beginning of the afterlife, resurrection (*ḥashr*), judgment (*ḥisāb*), balance (*mīzān*), passing to the way (*al-jawāz ʿalā al-ṣirāṭ*), salvation from fire, and arrival and entrance to the gardens, and the manner of the proximity to the Merciful of sublimity and nobility in the epistle on resurrection (*baʿth*). And now we wish to discuss in this epistle something of the originated and created beings that are generated and corrupted under the sphere of the Moon.[10]

If this is taken as the standard transitory formula, the passage suggests that Epistle 19, which currently follows upon an epistle on meteorology, is preceded by Epistles 42 (*Opinions and Doctrines*), 32/33 (*Intellectual Principles*) and 38 (*Resurrection*). This is strange, because Epistle 19 currently occupies

10 *On the Natural Sciences*, 247–248.

the standard curricular position within the physical epistles, and the alternative position makes no sense. The *incipit* does provide the rationale that these epistles deal with the world's temporal creation and destruction, whereas Epistle 19 addresses minerals as the first degree of composed beings in our world. However, the *incipit* lacks the stock formula of "now that we have finished" (*faraghnā min*), which was present in all the previously discussed *incipits*. This may, then, be taken as a general reference to epistles with related subject matter, rather than to the immediately preceding epistles.

As suggested in Chapter 1, Epistle 20 (*Essence of Nature*) is not only an innovation within the Aristotelian physical curriculum, but its current position between epistles on minerals and plants also seems arbitrary. That the epistle in fact occupies an incorrect position within the whole is supported by the incorrect reference in its *incipit*, as well as by that of Epistle 21 (*Plants*), which refers to Epistle 19 as the previous epistle. The *incipit* of Epistle 20, on the other hand, refers to Epistle 8 (*Practical Sciences*) as the previous epistle:

> Now that we have finished discussing the human crafts (*al-ṣanāʾiʿ al-bashariyya*) in an epistle entitled the practical disciplines (*al-risāla al-mulaqqaba bi-l-ṣanāʾiʿ al-ʿamaliyya*), we wish to discuss in this epistle the physical crafts (*al-ṣanāʾiʿ al-ṭabīʿiyya*).[11]

The reference to Epistle 8 as the previous epistle is explicit, and the order seems to be justified by some kind of analogy between the human crafts and the 'crafts' of nature as a faculty of the Universal Soul. The *incipit* of Epistle 8, however, placed Epistles 7 and 8 at the beginning of the third section, with the seemingly contradictory justification that they provide a bridge from physical to spiritual existents. But if the subject of Epistle 20 is the Universal Soul and its effects on sublunar existents, and since the Universal Soul is a purely spiritual existent, both *incipits* do in fact make sense. This is supported further by the fact that all evidence suggests that Epistle 20 in any case currently occupies an arbitrary position, and that the third section has no epistle on the Universal Soul in the current configuration, despite the importance of the subject. Thus, Epistle 20 would occupy a position at the beginning of the third section corresponding to metaphysics in the sense of philosophical theology, preceded by Epistles 7 and 8 bridging the physical and metaphysical sections. This seems entirely plausible; it also concurs with the implication of the curricular passages that the first two sections initially consisted of the Aristotelian

11 Ibid, 355; tr. Baffioni, modified.

curriculum exclusively. That is, if the physical section consisted only of the Aristotelian physical treatises, it would make sense that a treatise on the principle of nature would follow it in the metaphysical section. Epistle 32/33 (*Intellectual Principles*) devoted to the hypostasis of Intellect would, then, be an entirely appropriate candidate to follow Epistle 20.

In the third section, Epistles 35 (*Intellect and the Intelligibles*), 36 (*Cycles and Revolutions*), and 39 (*Movements*) give an incorrect previous epistle in their transitional passage. The first of these instances was already discussed above. As for Epistle 36, it indicates that Epistle 20 (*Essence of Nature*) is the preceding epistle. The given rationale is that while the latter epistle explains the influences of the spheres and spherical souls on the sublunar world, the former epistle discusses the movement of the spheres. This supports the above conclusion that Epistle 20 should pertain to the third metaphysical section, since this is the current context of Epistle 36. The consequent progression at the beginning of the third section would, then, be the following: Epistles 7 and 8 provide a transition between physics and metaphysics; these are followed by Epistle 20 on the Universal Soul, Epistle 36 on the movements of the spheres, that is, a manifestation of the Universal Soul in the supralunar world, and Epistle 32/33 on the Intellect.

For Epistle 39, the *incipit* provides an ambiguous reference:

> Now that we have finished discussing bodies, and shown their essences, the quantity of their species, the quality of their states, and their specific attributes, and we have also shown that bodies are not free of movement and rest, and that what makes bodies move and rest are souls, in our physical and [other (*ghayrihā*)] theological epistles (*fī rasāʾilinā al-ṭabīʿiyyāt wa-l-ilāhiyyāt*) in a clear manner, we wish now to show in this epistle the essence of movements.[12]

In the current arrangement, the immediately preceding epistle is Epistle 38 (*Resurrection*), to which the *incipit* makes no reference, except in the corrected variant of the Beirut edition. Instead, the passage makes an ambiguous reference to a discussion of bodies and their psychical movers, which is attributed to physical and theological (*ilāhiyya*) epistles. The reference contradicts the current configuration of epistles; while bodies are discussed in the second section of physical epistles, the term 'metaphysical' or 'theological' (*ilāhiyya*) pertains to the fourth section, which mainly addresses subjects related to practical phi-

12 *Sciences of the Soul and Intellect, Part III*, 5.

losophy and religion. The reference is, then, clearly to the third section; thus, the passage supports the implication of the curricular passages that, in the initial four-division, the third part consisted of metaphysics in the sense of philosophical theology. The result is that the ambiguous *incipit* concurs with the current position of Epistle 39 in the third section. But it also supports the view that Epistle 38, which should be the previous epistle, currently occupies an incorrect place.

None of the epistles in the fourth section contain a transition referring to a previous epistle in most manuscripts; thus, there are no incorrect references, either. The only epistle with any kind of transition is Epistle 52a (*Magic*):

> Know, may God support you O brother, that in the fifty epistles that precede this epistle, we have discussed the varieties of sciences and marvels of wisdom. We have classified them and gathered together in them many sciences, various objectives, and thoughtful wisdom. And we have clearly expounded them in accordance with what the levels of the apprentices and the ranks of the students making use of them require.[13]

The passage contains no reference to the previous epistle but correctly concludes the *Epistles* by referring to fifty preceding epistles. Once again, then, this epistle differs from most other epistles by its more up-to-date information about the work's contents. As discussed previously, there are two competing candidates for the position of final epistle; neither of the other candidates contains a transitory passage. It seems, then, likely that there were in fact fifty preceding epistles at the time when Epistle 52a was composed because it was added to the work at a late date. Thus, the *incipit*s provide us with no information at all about the initial positions of epistles in the fourth section. However, both the beginning and end of Epistle 50 (*Governance*) are interesting.[14] The passage at the beginning introduces a chapter (*faṣl*), rather than an epistle; both the beginning and end of the epistle refer to themselves as pertaining to the "comprehensive chapter" (*al-faṣl al-jāmiʿ*). This does not suggest a variant position for the epistles; but it supports the view that Epistle 50 was not always an independent epistle. As we have seen, in two manuscripts, Epistles 49 (*Spiritual Beings*) and 50 are in fact combined.

---

13 *On Magic*, 6–7; tr. de Callataÿ, 88, modified.

14 *On God and the World*, 237–238, 288–289. Both passages are cited and translated in full in Janne Mattila, "The Authorship of *al-Risāla l-Ǧāmiʿa* Re-Examined," *Arabica* 68, no. 5–6 (2021): 478–479, 481–482.

Taken together, the evidence of epistle transitions points to substantial redaction of the *Epistles* in the specific sense that, during the redaction process, many epistles were shifted from one position to another. In part, erroneous references may also be due to the gradual addition of new epistles between existing epistles. The transition at the beginning of each epistle follows a standard format, which binds the subjects of the current and previous epistle together. Thus, their presence gives the impression of a well-conceived philosophical encyclopedia where each epistle occupies its proper place. The fact that the Beirut edition has such passages at the beginning of most epistles, and that they are more often correct than in most manuscripts, has undoubtedly contributed to the impression of a stable text in the scholarship of the Brethren of Purity. But the fact that less than half of the epistles have a passage of transition in most manuscripts, and that they are precisely as often incorrect as they are correct in their reference to the previous epistle, clearly shows that the impression of a stable textual tradition is false.

Regarding the more specific information concerning the positions of individual epistles, the evidence seems mixed. Epistle 19 (*Minerals*) currently occupies an entirely logical position within the curriculum; an alternative position in the third or fourth section makes no sense. However, Epistle 19, in contrast to all other passages of transition, makes no explicit reference to a previous epistle; thus, the reference is probably to epistles with related subject matter in general. In any case, the reference to Epistle 19 in the *incipit* of Epistle 21 (*Plants*) confirms its current position in the second section. Similarly, Epistle 5 (*Music*) currently seems to occupy the correct position. Yet, its *incipit* suggests explicitly that it is preceded by Epistles 7 (*Theoretical Sciences*) and 8 (*Practical Sciences*); the beginning of Epistle 8, in turn, indicates that these two epistles occupied a position between the second and third sections. One explanation is the complex nature of the redaction process; new epistles were added to the whole, other epistles were split into two or more epistles, and yet other epistles were repositioned within the whole. The combination of all these methods might result in cases where it is difficult to trace the original position of a particular epistle.

Yet, most of the evidence from epistle transitions in fact fits with the hypothesis that the composition of the *Epistles* started with the epistles corresponding to the classical curriculum and was later expanded with the introduction of non-classical epistles. First, the imbalance of initial passages of transition supports the view that the classical (1–24) and non-classical (25–52) halves of the work correspond to two distinct chronological layers. Second, apart from the exceptions mentioned above, the *incipit*s only provide incorrect references for epistles that currently seem to occupy an incorrect position. This is particularly clear for the first section, where the non-classical Epistles 7, 8, and 9 have an incorrect reference; moreover, Epistles 4 (*Geography*) and 6 (*Arithmetical and Geometrical Proportions*) make no reference to the previous epistle. The implication is that all five of the non-classical epistles of the first section are the result of either repositioned or split epistles. Similarly, within the Aristotelian part of the second section (15–24), the *incipit*s imply an incorrect position for Epistle 20 (*Essence of Nature*). In contrast, apart from Epistle 19 (*Minerals*), the rest of the epistles have either a correct transition or no transition at all. Finally, the *incipit*s of Epistles 20, 36 (*Cycles and Revolutions*), and 39 (*Movements*) all support the view that they initially pertained to a third section devoted to metaphysics or philosophical theology.

Most of the conclusions concerning the precise position of particular epistles are admittedly speculative. But at a minimum, the contradictions in the passages of transition show that many epistles were at some point preceded by a different epistle than is the case now. Again, we have actual examples of manuscripts where the order of epistles is different. The evidence of these passages merely shows that there was a time when the configuration of epistles was radically different from any known manuscript.

The fact that many epistles have changed their position within the whole, and, it seems, often from one section to another, raises the question of why this is the case. Why should the authors or later redactors deem it necessary to shuffle the positions of epistles? Arguably, this is due to the authors' evolving conception of the correct arrangement of an encyclopedia of philosophical sciences. If we accept the hypothesis that the origin of the *Epistles* of the Brethren of Purity lies in the context of al-Kindī's school, then the initial configuration of epistles conformed to the standard late ancient and early Islamic classification of philosophical sciences. This hypothesis is supported by the curricular passages, which invariably conform to the standard classificatory scheme. Yet, as was also the case within the school of al-Kindī, the authors of the *Epistles* are far from orthodox Aristotelians even in their Aristotelian treatises but are also influenced by Neoplatonic, Neopythagorean, and Hermetic strands of thought. Moreover, if we accept al-Tawḥīdī's account, the authors aimed for a synthesis of Greek philosophy and Islamic religion. Perhaps once the authors proceeded beyond the classical curriculum, they became less satisfied with the idea that the standard classification could fully incorporate their philosophical and religious ideas. The dissatisfaction extended to the overall division of philosophical sciences, which, based on the contradiction between the four-fold divisions in the curricular passages and the actual *Epistles*, clearly evolved in time. Thus, the novel four-division and the gradual addition of new, non-classical, epistles made it necessary to reposition some epistles. Moreover, perhaps they were uncertain about the correct position of some epistles because their subject matter was ambiguous, as in the case of Epistle 35 (*Intellect and the Intelligibles*), which is in effect a hybrid of psychology and metaphysics.[15] All of this is speculative; we may never know the precise forms of the *Epistles*' evolution. It is clear, however, that the evidence of both the epistle titles and passages of transition demonstrate that a version of the *Epistles* must once have existed that was very different from the one we currently possess.

---

15 See Walker's discussion of the seemingly confused aim of the epistle in *Sciences of the Soul and Intellect, Part I*, 111–113.

CHAPTER 3

Relative Chronology

The most important question about the redaction of the *Epistles* concerns the chronology of its composition. The first part of this book was initiated by a discussion of the theories of Marquet, Hamdani, and Madelung about the order in which the epistles were written. Although Marquet and Madelung both made use of internal evidence, all three based their theories on a strong hypothesis of authorship. It is not necessary to make strong assumptions about the Brethren of Purity's identity to construct a theory of the redaction history of the *Epistles*. The evidence discussed so far relates to the order of composition to some extent; in particular, the curricular passages suggest that the epistles corresponding to the classical curriculum were written before the non-classical epistles. But the *Epistles* also contain a particular kind of evidence that seems designed for investigating the relative chronology of the epistles, namely references from one epistle to another. It is surprising that these cross-references have been of so little interest in previous scholarship.

1 Cross-References

The final kind of technical evidence for the redaction history of the *Epistles*, then, consists of the cross-references between the fifty-two epistles (Table 12).[1] The *Epistles* contain more than 500 such references, which together form a complex network between the epistles. These provide by far the most important indications that we have for constructing an internal chronology of the *Epistles*. Like the initial transitory passages, the references give the impression that the redaction of the work followed a carefully conceived plan; each epistle occupies its logical position within the whole with thematic relations to other epistles. Thus, somewhat like a modern encyclopedia, every epistle encourages

1 The references were found by an electronic search of the Beirut edition text. The results were checked against the text of the critical editions, and, in the absence of a critical edition, against MS ع. Given this method, possible references that are absent in the Beirut edition but present in one or more manuscripts, are missing in the tables. The variance between witnesses in the case of cross-references is surprisingly minor, however. Thus, there should not be many such references. The manuscript variation and ambiguous cases are indicated in the endnotes of Table 12.

the reader to consult other epistles for more information about a particular subject.

The most common format of a cross-reference in the *Epistles* is one where the mention of a specific doctrine or claim is supported by the stock phrase "as we have shown (*bayyannā*) in Epistle x." The reference from Epistle 5 (*Music*) to Epistle 24 (*Senses and the Sensibles*) provides an example:

> The sound is an impact (*qarʿ*) occurring in the air from the collision of bodies against each other, as we have shown (*kamā bayyannā*) in the epistle on the senses and the sensibles. But it is necessary for us to explain (*nadhkur*) some of it in this epistle too.[2]

The reference is correct, since Epistle 24 does contain a chapter devoted to the faculty of hearing and the physical nature of sound.[3] In fact, the cross-references are rarely erroneous, although, as we have seen, there are a few references to non-existing epistles.

In contrast to the *incipits*, the cross-references show little variation between manuscripts, even in the case of the Beirut edition variant. Occasionally, a reference is absent in one or more manuscripts in a passage which is otherwise identical. In particular, the Tübingen manuscript shows a higher tendency to omit a reference in mid-sentence, although this might be because it is a summary of the work and the reference was consciously left out.[4] The occasional variation, together with the fact that the references between two epistles often proceed in both directions, indicates that cross-references were only gradually added to the *Epistles* during the course of its redaction. The minor variation also suggests, however, that the cross-references became an integral part of the text at an early stage, perhaps toward the end of the redaction process, and that there was no similar later effort to update them as in the case of the *incipits*.

It is not entirely clear how the cross-references should be read, given their sheer number, as well as the quantity of possible combinations resulting from references between fifty-two epistles. It is evident at the outset that they do not reveal the relative chronology of the epistles in any simple manner. First, the epistle references regularly proceed in both directions. That is, for example, Epistle 1 (*Arithmetic*) refers once to Epistle 32/33 (*Intellectual Principles*), while

2 *On Music*, 22; tr. Wright, 85, modified.
3 *Rasāʾil*, vol. 2, 407–408.
4 For this manuscript, see pp. 20*n*50, 69–70.

Epistles 32 and 33 refer three times to Epistle 1. This means that cross-references were inserted at different stages of the redaction process. Second, a given epistle obviously tends to refer to another epistle that shares its subject matter, so that the physical epistles, for example, often refer to each other. Also, given the authors' inclination to discuss their favorite themes in every epistle, there should be more references to those epistles whose subject matter is central to the *Epistles*. Thus, for example, since the Neopythagorean notion of correspondence between numbers and levels of reality is a fundamental idea for the authors, it is logical that references to Epistle 1 (*Arithmetic*) are distributed fairly evenly across the epistles of all four sections. Third, the epistles are of very different lengths, and the number of references an epistle makes should therefore be related to its length. Fourth, since some epistles were split from a larger epistle, an epistle might be both early and late at the same time. That is, it might have initially formed part of one of the earliest epistles, but it could still be one of the latest epistles as an independent epistle, and therefore there would be few references to it in other epistles. Thus, as we saw in Chapter 2, there are no references at all to Epistles 32 and 33 as separate epistles; but there are thirty-six references to the two epistles taken together. This means that the fact that a given epistle contains many references to other epistles, while it is itself referred to only a few times or not at all, for example, does not automatically mean that it must be a late epistle.

When we look at the general distribution of references, it is also clear that they are distributed in a more or less balanced manner between the epistles. This is particularly the case at the level of each of the four sections (Table 13). The epistles of the second section both make the most references and receive the most referents. This is at least partly the case because this section contains the highest number of epistles, which, moreover, refer extensively to each other. Otherwise, the rate of references that each of the sections makes is fairly even. The ratio between references and referents is also balanced for the first three sections, both in total and with respect to each of the three other sections. Only the epistles of the fourth section make almost double the number of references to other epistles as compared to the references they receive themselves. The evenness of the references seems somewhat counter-intuitive, insofar as the curricular progression of the *Epistles* is supposed to be cumulative, in that later epistles should build on the previous ones, and therefore refer to them more, rather than vice versa. It is, however, clear that the cross-references in the *Epistles* are not built this way, as epistles on the contrary regularly resort to evidence provided by a much later epistle. This gives the impression that the references were evened out across the work after all the epistles were finished. Since, despite all this, the fourth section is the only one that refers to other sections much more frequently than it is referred to by them, it seems plausible to assume that it was the latest to be written.

A second consideration concerns the number of references that each section makes to itself. Given the supposed thematic unity of each of the four sections, one could expect that the epistles in each section refer most frequently to other epistles in that section. The second section makes in fact overwhelmingly, and the first section slightly, more references to itself than to the other three sections. But the third section makes clearly less references to itself than to the first and second sections, while the fourth section makes much less references to itself than to the second section. Since the sections do not constitute entirely coherent units as some epistles were shifted from one section to another, and it is far from self-evident that the *Epistles* were composed one section at a time, this does not provide conclusive evidence. Yet, statistically at least, the implication is that the third and fourth sections are last in the chronology of redaction.

At the level of individual epistles, the first thing of note are the 'isolated' epistles that either never refer to other epistles or receive no references from other epistles. Both cases appear to require an explanation. There are only four epistles that contain no references at all to other epistles: Epistles 4 (*Geography*), 12 (*On Interpretation*), 13 (*Prior Analytics*), and 27 (*Spread of Particular Souls into Physical Bodies*). Among these, Epistles 12 and 13 are entirely isolated in that they also receive no references from other epistles. The latter fact is explained by the probability that these two epistles initially formed part of a larger logical epistle.[5] The former could be partly due to these being the two shortest epistles in the entire work. But it is still the case that they are set apart from all other logical epistles, which contain a high number of references to other epistles. Does this, then, mean that these two epistles were among the first to be written, and that in any case they precede the other logical epistles, or is the lack of references perhaps related to their subject matter? As for Epistle 4, it is not conspicuously short, and the lack of references therefore suggests that it is an early epistle. It is itself, however, referred to only five times, and exclusively by epistles of the second section, and never by those in its own section. It is particularly striking that Epistle 4 never refers to or is referred to by Epistle 3 (*Astronomy*) to which it is thematically related. This supports the possibility that Epistles 3 and 4 initially formed a single epistle, especially since Epistle 3 contains few references to other epistles and is repeatedly referred to by epistles of the second section. Finally, Epistle 27 is itself also referred to only three times, and only by other epistles in the second section, despite being occupied

5 See Chapters 1 and 2.

with subject matter of fairly central importance for the authors. This would suggest that it is late as an independent epistle, while the lack of references would be explained by it being split from another epistle, a hypothesis that is further supported by its shortness. Perhaps, then, it initially formed part of a larger epistle devoted to the Universal Soul, the previous existence of which is supported by our previous discussion, as well as the title of one of the 'stray' epistles.[6]

Among the five or six epistles that are never referred to by other epistles, for Epistles 12, 13, 32, and 33, this is clearly the case because they were not initially independent epistles; references to the larger epistles of which they formed part are plentiful. As for Epistle 37 (*Love*), it is only referred to by Epistle 52c (*Magic*), which is almost certainly a pseudepigraph. Since its subject matter should warrant references from other epistles, and it itself makes a high number of references to other epistles, the conclusion is that it is a late epistle. Epistle 28 (*Human Capability for Knowledge*) constitutes an uncertain case. This specific title is never employed in a cross-reference; however, there are three references to an "epistle of knowledge" (*risālat al-maʿārif*). It is not entirely clear whether these references designate Epistle 28. In any case, the fact that the epistle itself only makes three references to other epistles, and that the epistle is fairly short, suggest that it was split from a larger epistle. The *Epistles* in fact include two epistles, 24 (*Senses and Sensibles*) and 35 (*Intellect and the Intelligibles*), closely related to epistemology, as well as a stray epistle with the title "epistle of instructions" (*risālat al-taʿlīm/taʿālīm*).

While the former represent the simplest cases, all fifty-two epistles can be classified in an entirely mechanical way according to the number of references they make versus the number of times they are referred to. If none of the previously mentioned qualifications existed, an early epistle should have few references and many referents; a late epistle should have many references but few referents. Some epistles certainly stand out in this respect. The most conspicuous case is Epistle 32/33 (*Intellectual Principles*), which is referred to thirty-six times by other epistles, while Epistles 32 and 33 combined contain only five references to other epistles. Clearly, it is the case that these twin epistles are referred to so frequently because they convey the core metaphysical doctrine of the authors concerning the intelligible origin of reality. But the low number of references to other epistles is not explained by any of the qualifications discussed above. A symmetrically opposite case is Epistle 30 (*Pleasure and Pain*), which makes twenty-seven references to other epistles, but is itself referred

6 See first section of Chapter 2.

to only six times. This epistle was possibly initially combined with Epistle 29 (*Death and Life*), which has a similar relation between references and referents. The conclusion, then, would seem to be that Epistle 30, or Epistles 29 and 30 combined, is a late epistle.

However, the number of references alone is misleading, without taking into account the length of each epistle. Proceeding with the mechanical approach, Table 14 presents the number of references in each epistle in proportion to its length, as well as the relation of the reference ratio to the number of referents. The resulting number would ideally be directly proportional to the lateness of an epistle; that is, the lower the number the earlier the epistle. In Table 15, all fifty-two epistles are classified into categories of precedence based on this ratio alone. For the reasons mentioned above, it is highly unlikely that the chronology of the epistles could in fact be determined with precision by purely mathematical calculations. But it is, nevertheless, probable that at least many epistles fall into a roughly correct category in the table. Therefore, we may take this classification as a starting point for a more refined analysis. That analysis will be based on the assumption that epistles with related subject matter should refer to each other; this assumption will be applied to the internal and mutual relations between distinct curricular groups of the *Epistles*. By analyzing the reference ratios between distinct groups one by one, a hypothetical chronology of all epistles may be eventually reached.

Looking at the table, it does not seem to be the case that the epistles were redacted one section at a time because epistles in all four sections figure among both early and late epistles. The table nevertheless shows some regularities. First, the ratio of references suggests that Epistles 1 (*Arithmetic*) and 32/33 (*Intellectual Principles*) are both very early epistles; the former only contains two references to other epistles. Given the centrality of Neopythagoreanism for the *Epistles*, it seems entirely plausible to suggest that these two epistles pertain to the work's original core, or even that these were the very first epistles to be composed. Apart from Epistle 6 (*Arithmetical and Geometrical Proportions*), the rest of the mathematical epistles (2–5) are also situated among early epistles in the table. When the references within the mathematical epistles are compared, Epistle 1 only refers once to Epistle 2 (*Geometry*), while Epistles 2, 3 (*Astronomy*), and 5 (*Music*) each contain one or more references to Epistle 1, as well as a gradually increasing number of references to other epistles. Moreover, while Epistles 1 and 2 refer exclusively to other mathematical or Pythagorean epistles, with the exception of a single reference to Epistle 52 (*Magic*), Epistle 3 also contains two and Epistle 5 seven references to physical epistles, while Epistle 5 also refers twice to the logical epistles. The conclusion is that the authors expanded the mathematical *quadrivium* around the original Neopythagorean core gradually, so that Epistle 5, in particular, was composed either simultaneously with or after the physical epistles. As for Epistle 6, it only refers to Epistle 5, and while evidence suggests that it became an independent epistle at a later time, it was probably composed along with the mathematical epistles.

Second, we saw in Chapter 1 that the curricular passage of Epistle 15 (*Matter and Form*) identifies the physical sciences exclusively with the Aristotelian curriculum. This suggests that, within natural philosophy, the non-Aristotelian epistles are later additions. The ratios of references make it possible to test this conclusion. The physical epistles of the second section may be divided into two parts based on their Aristotelian or non-Aristotelian provenance. All the Aristotelian physical treatises (15–19, 21–22, 24) are situated among either early or middle epistles within the table. These epistles contain extensive references to other epistles in the second section, but they also refer a combined eleven times to the mathematical epistles, while they only contain a single reference to the logical epistles. As a conclusion, it seems that the Aristotelian physical treatises were composed either after or in part simultaneously with the mathematical epistles, but before the logical epistles. As for the non-Aristotelian epistles in the second section (20, 23, 25–31), most of them figure among middle or late epistles within the table, which would support the hypothesis of their later provenance.

A closer look at the mutual references between these two groups of epistles seems to confirm this conclusion. The non-Aristotelian physical epistles contain a combined thirty-eight references to the Aristotelian physical treatises, spread evenly across the epistles. In contrast, within the Aristotelian group, Epistles 17 (*Generation and Corruption*), 21 (*Plants*), and 24 (*Senses and Sensibles*) carry no references at all to the non-Aristotelian epistles; Epistles 15 (*Matter and Form*), 16 (*Heaven and World*), 18 (*Meteorology*), 19 (*Minerals*), and 22 (*Animals*) contain a combined two references to Epistle 23 (*Constitution of the Human Body*), three references to Epistle 20 (*Essence of Nature*), and three references to Epistle 25 (*Fall of the Sperm*). Given the tendency of the physical epistles to refer to each other, the imbalance is striking. It is particularly noteworthy that Epistles 20, 26 (*The Human Being is a Microcosm*), and 27 (*Spread of Particular Souls into Physical Bodies*), which expand the Aristotelian physics into a Neoplatonic framework, are either mostly or completely ignored by the Aristotelian treatises.

This leads to the general conclusion that the authors started the second section with the Aristotelian physical curriculum, which they later complemented with treatises that incorporate Neoplatonic cosmology into their natural philosophy. As for Epistle 23, its reference ratio, and the fact that it is referred to by two Aristotelian treatises, suggest that it is contemporary with the Aristotelian epistles. This in fact makes sense since it occupies the position of *De anima* in the Aristotelian curriculum. In the following, I will consequently classify the physical epistles of the second section into early (15–19, 21–24) and late (20, 25–27), while I will return to the rest of the epistles of this section later.

As a next step, it may be observed that, besides Epistle 32/33, the early physical treatises of the second section make a combined twelve references to six other epistles of the third section: three references to Epistles 35 (*Intellect and the Intelligibles*) and 40 (*Causes and Effects*) each, two references to Epistles 36 (*Cycles and Revolutions*) and 39 (*Movements*) each, and single references to Epistles 34 (*The World is a Big Human Being*) and 38 (*Resurrection*). Apart from Epistle 38, these constitute a group of epistles with largely physical subject matter within the third section, all of which are, moreover, situated among either middle or late epistles in the table. In comparison, the five physical epistles of the third section, despite their lower number, make a combined forty references to the early physical epistles of the second section, which clearly suggests that the former postdate the latter. As for the relation of these five epistles with the late physical epistles of the second section, the latter make ten references to the former, as opposed to four references in the reverse direction, which indicates that the former group predates the latter group. In consequence, we arrive at the following chronology of the epistles with physical subject matter: (1) early epistles in the second section, (2) epistles in the third section, and (3) late epistles in the second section.

Proceeding to the logical epistles (10–14), the refences to referents ratio locates them among late epistles in the table, both when they are viewed as separate epistles and when taken as a single epistle. A closer look at the references that the logical epistles make to the epistles of the three other sections shows that there are twelve references to the early physical epistles of the second section, three references to the physical epistles of the third section, three references to Epistle 32/33, two references to Epistle 38, and a single reference to the late physical epistles of the second section. In addition, there are single references to Epistles 31 (*Differences of Languages*) and 41 (*Definitions and Descriptions*), while there are no references to the fourth section. In comparison, there are three references to the logical epistles in the early and late physical epistles of the second section combined, three references in the physical epistles of the third section, two references in Epistle 42 (*Opinions and Doctrines*), as well as a single reference in Epistles 29 (*Death and Life*), 46 (*Belief*), and 48 (*The Call to God*) each. The evidence clearly indicates that the logical epistles postdate the early physical epistles, and less clearly that they might also postdate the physical epistles of the third section. Thus, they might be roughly contemporary with the late physical epistles of the second section. Consequently, we have arrived at the following categories of epistles in terms of their order of redaction: (1) mathematical-Pythagorean epistles (1–5, 32–33), (2) early physical epistles (15–19, 21–24), (3) middle physical epistles (34–36, 38–40), and (4) late physical epistles and logical epistles (10–14, 20, 25–27).

The rest of the epistles mostly do not pertain to the classical curriculum, and do not constitute clearly defined groups. Insofar as the final part of philosophy should consist of practical philosophy, including religious sciences, the epistles that deal with these subjects (9, 43–48, 50) may be postulated as a group. Among them, Epistles 9 (*Ethics*), 46 (*Belief*), and 48 (*The Call to God*) stand apart in that they contain a high number of references to other epistles of all four sections. Their combined ratio of references to referents with respect to the four chronological groups postulated so far is the following: (1) zero to one, (2) nine to one, (3) seven to one, and (4) nine to zero. The fact that these epistles refer fairly evenly to the last three groups, while all four groups refer a little or not at all to these epistles, suggests that they are late epistles that postdate all four groups. Apart from Epistle 47 (*Divine Law*), the rest of the epistles within this group, 43 (*The Way to God*), 44 (*Beliefs of the Brethren of Purity*), 45 (*Companionship of the Brethren of Purity*), and 50 (*Governance*), contain almost no references to epistles. But they also have few referents in other epistles: four in Epistle 46 (*Belief*), three in Epistle 48 (*The Call to God*), two in Epistle 9 (*Ethics*), and a single reference in each of Epistles 30 (*Pleasure and Pain*), 49 (*Spiritual Beings*), and 52 (*Magic*). Most of the references, then, come from Epistles 46 and 9, while none of the four posited chronological groups refer to these epistles. This suggests that these epistles also postdate the four postulated chronological groups, but that they are early with respect to other epistles in their group, and that they predate Epistle 46, in particular. In consequence, all epistles with ethical or religious subject matter (9, 43–46, 48, 50) postdate all of the mathematical, logical, and physical epistles, and will be postulated as the fifth chronological category. As for Epistle 47 (*Divine Law*), it contains no references to any identifiable epistle, while it is itself referred to five times by Epistle 46 (*Belief*), three times by Epistle 42 (*Opinions and Doctrines*), twice by Epistle 30 (*Pleasure and Pain*), and once by Epistles 8 (*Practical Sciences*), 9 (*Ethics*), 35 (*Intellect and the Intelligibles*), and 48 (*The Call to God*). The lack of references suggests that it is an early epistle, at least within the religious epistles, to which it should presumably contain references. Most of the referents, however, are from the practical-religious epistles, while two references are from Epistles (8 and 35) that probably predate this group. The tentative conclusion is that Epistle 47 pertains to the very earliest epistles within the ethical-religious section.

The references in all of the remaining epistles may finally be related to the five postulated chronological groups, while bearing in mind that, for any epistle, its rate of references with respect to each group is also conditional on the extent to which they share at least some subject matter. Thus, in the case of Epistles 7 (*Theoretical Sciences*) and 8 (*Practical Sciences*), since their subject is knowledge in general and the classification of sciences, in particular, it would make sense for them to refer evenly to all epistles. On the other hand, as we have seen,[7] the fact that the classification of sciences in Epistle 7 coincides only in part with the actual order of epistles suggests that the epistle predates the final redaction of the work. The combined ratio of references to referents of these two epistles with respect to each of the five categories is: (1) four to two, (2) nine to one, (3) three to five, (4) five to nine, and (5) zero to one. The upshot of the references is, then, that the two epistles refer to the first four groups but not to the fifth group, which would make them contemporary or posterior to the fourth group. In contrast, the consequence of the referents is that the first two groups contain almost no references to the two epistles, while the last three do, which would make them contemporary or posterior to the third group. Given that the curricular passages in Epistle 7 indicate that it precedes the actual configuration of the logical epistles, the two epistles will be placed between groups three and four.

As for Epistles 29 (*Death and Life*) and 30 (*Pleasure and Pain*), previous evidence suggests that they initially formed a single epistle; moreover, their combined ratio of references to referents would make them late epistles. The ratio of references to referents of these two epistles with respect to the five postulated categories is the following: (1) three to one, (2) twelve to zero, (3) eight to two, (4) seven to one, and (5) five to three. The partly physical subject matter of these epistles explains their numerous references to the second, third, and fourth groups; the paucity of references in the reverse direction suggests that these epistles postdate them. It should be noted also that these two epistles make five references to the particularly relevant Epistle 38 (*Resurrection*), while the latter only makes a single reference to them. The two epistles are also partly eschatological in content, which relates them to the epistles in group five; among this group, they refer once to Epistles 9 (*Ethics*) and 44 (*Beliefs of the Brethren of Purity*), twice to Epistle 47 (*Divine Law*), and three times to Epistle 46 (*Belief*); they are referred back twice by Epistle 9 and once by Epistle 46. The conclusion is that Epistles 29 and 30 postdate all or most of the physical epistles and were probably composed simultaneously with the later ethical-religious epistles.

---

7 See pp. 53–59.

Within the fourth section of the *Epistles*, a mixed compound of three epistles still remains. Epistle 42 (*Opinions and Doctrines*) is one of the longest treatises in the entire work. Its subject is philosophical and religious doctrinal divergence. The ratio of references to referents with respect to the five chronological groups for this epistle is the following: (1) seven to one, (2) eleven to two, (3) eight to two, (4) three to zero, and (5) four to one. Given that the epistle discusses a wide variety of philosophical and religious doctrines, it should refer to epistles in all or most sections; the reverse is not necessarily the case, and the very high ratio of references to referents does not therefore necessarily mean that it is a late epistle. But the fact that it does refer to epistles of all five groups nevertheless suggests that it was composed along with the ethical-religious epistles.

As for Epistles 49 (*Spiritual Beings*) and 51 (*Constitution of the World*), their subject matter is cosmology and metaphysics. For the former epistle, the ratio of references to referents is: (1) one to zero, (2) five to one, (3) zero to two, (4) three to two, and (5) one to zero. Among other epistles in the fourth section, the epistle refers three times to Epistle 52 (*Magic*) and once to Epistle 50 (*Governance*); while it is referred to once by Epistle 51, as well as various times by all three versions of Epistle 52. The epistle, then, appears to have a special connection with Epistle 52, although the references between the two epistles proceed in both directions. As for Epistle 51, it contains two references to the first group, seven references to the second group, and a single reference to both the third group and Epistle 49 (*Spiritual beings*), while it is itself only referred to once by Epistle 40 (*Causes and Effects*). The tentative conclusion is that both epistles postdate all physical epistles; they might precede the ethical-religious epistles to which they never refer, despite being positioned in the fourth section.

The remaining epistles both contain and attract few references and are thus hard to classify. Epistle 28 (*Human Capability for Knowledge*) is referred to three times by later epistles, if the "epistle of knowledge" (*risālat al-maʿārif*) indeed designates this epistle. It only itself refers to early mathematical, Pythagorean, and Aristotelian physical epistles. It therefore perhaps follows the Aristotelian physical treatises in chronology; this would also coincide with its subject matter. The subject matter of Epistle 31 (*Differences of Languages*) in part relates to Epistle 42 in that both address the question of doctrinal divergence; yet, the two epistles never refer to each other. Although the epistle is only referred to twice, by Epistles 10 (*Eisagōgē*) and 35 (*Intellect and the Intelligibles*), it itself only refers to early mathematical and physical epistles; this suggests it is posterior to the Aristotelian physical epistles. Epistle 37 (*Love*) is only referred to by Epistle 52c; it refers to mathematical and physical epistles, as well as twice to Epistle 9 (*Ethics*) and once to Epistle 49 (*Spiritual Beings*). This indicates that it postdates the practical-religious epistles. Epistle 41 (*Definitions and Descriptions*) is only referred to once by Epistle 14 (*Posterior Analytics*), while it contains two references to Epistles 1 (*Arithmetic*) and 18 (*Meteorology*) each. Given its subject matter, this suggests that it goes together with the logical epistles.

Finally, Epistle 52 (*Magic*) presents a problem of its own. As stated previously, the manuscripts provide three entirely different and mutually independent versions for this epistle. It is not clear which one, if any, pertained to the original work.[8] The cross-references throw some light on this question. Other epistles refer six times to an epistle on magic: Epistle 49 (*Spiritual Beings*) three times, and Epistles 2 (*Geometry*), 18 (*Meteorology*), and 42 (*Opinions and Doctrines*) a single time each. The low number of references contradicts the epistle's presumed importance as the culmination of the entire work, but obviously does not prove its late date of composition, especially since two of the referring epistles are early. However, all six references give a different title for the epistle: "epistle on talismans and incantations" (*risālat al-ṭilismāt wa-l-ʿazāʾim*) in Epistle 2; "epistle on auguries and physiognomy" (*risālat al-zajar wa-l-firāsa*) in Epistle 18; "epistle on auguries" (*risālat al-zajar*) in Epistle 42; and "epistle on magic and incantations" (*risālat al-siḥr wa-l-ʿazāʾim*), "epistle on magic and talismans" (*risālat al-siḥr wa-l-ṭilismāt*), and "epistle on astronomy, magic, and talismans" (*risālat ʿilm al-nujūm wa-l-siḥr wa-l-ṭilismāt*) in Epistle 49. All three references in Epistle 49, then, include the word 'magic' (*siḥr*), while the three other references do not. This is significant because, in all manuscripts, the first term of the epistle title is "magic," "essence of magic" (*māhiyyat al-siḥr*), or "the true nature of magic" (*ḥaqīqat al-siḥr*); variants 52a and 52b both start with a discussion of the term. Equally interesting is the context of the references. In Epistle 2, the context is magic squares in which the reader is referred to a discussion of their benefits; in Epistle 18, the context is the color of a rainbow as an augury of future events; and in Epistle 42, the Indian versus Greek views on whether the human imaginative faculty is capable of bringing about changes in other beings.[9] None of the three variants discusses magic squares. Variants 52a and 52b discuss auguries of the natural world (*zajar*) incidentally, while 52c does so systematically; none of them mentions rainbows, however. Both 52a and 52c attribute the employment of the imaginative faculty in magic to the Indians, in particular; Epistle 52a also contains the precise phrase of the reference in Epistle 42.[10] In consequence, the references in Epistles 2 and 18

---

8 See pp. 68–69.

9 *Rasāʾil*, vol. 3, 418; *On Arithmetic & Geometry*, 142; *On the Natural Sciences*, 226. The reference in Epistle 42 appears as *risālat al-zajar wa-l-ruqā* in MS ج (fol. 402[a]).

10 *On Magic*, 97; MS Esad Efendi 3638, fol. 315[a]. The Beirut edition variant of Epistle 52a states

correspond to none of the three variants, while the reference in Epistle 42 coincides precisely with 52a, and more generally with 52c.

In Epistle 49, the general context of all three references relates to the spiritual beings (*rūḥāniyyūn*) of the spheres, to which subject the epistle is largely devoted.[11] The general context of the first reference, in variant 49a, is the soul's afterlife in terms of its ascent to the spheres. The specific reference states that the proof (*dalīl*) for the existence of spiritual beings (*rūḥāniyyūn*) and their classes (*aṣnāf*) is established in both Epistles 49 and 52. In general, this agrees with all three variants of Epistle 52, since the subject of spiritual beings is prominent in each variant, but only Epistle 52a properly discusses the subject of Platonic afterlife. In 52b, the subject is only addressed tangentially and in 52c not at all. In the second reference of Epistle 49a, the context is again the eschatological one of the descent and ascent of particular souls, with specific reference to Tartarus as an abode where some souls are disciplined before their further reincarnation. The reference is unequivocally to Epistle 52a, which addresses the same subject in similar terms, and includes a reference to Tartarus, although as attributed to the Sabians.[12] In variant 49b, the context is the spiritual beings of the sphere of the Moon and their descent to the sublunar world, and the specific reference is to their employment in magical operations. Given the emphasis on the astrological basis of magic in at least parts of all three variants, the reference could apply to any of them, although Epistle 52b is clearly the most astrological of the three variants. All three references in Epistle 49, then, on a general level, relate to the subject matter treated in all three variants of Epistle 52; in their specifics, the first two references point to variant 52a.

---

that people in India "bring about effects of very wondrous things in others by their imaginations, which most people deny" (*yu'aththirūna fī ghayrihim bi-awhāmihim ashyā' 'ajība jiddan yunkiruhā akthar al-nās*); most manuscripts instead give "see in others by their imaginations" (*yarawna min ghayrihim bi-awhāmihim*). The formulation of the Beirut edition variant coincides exactly with the reference in Epistle 42: "It is told about the diviners of India that they bring about effects of wondrous things in others by their imaginations, which most people reject (*yu'aththirūna fī ghayrihim bi-awhāmihim ashyā' 'ajība yunkiruhā akthar al-nās*)".

11 *On God and the World*, 42, 43–44, 157–158.

12 *On Magic*, 50–54. In both the Beirut edition and the two manuscripts employed by the editors, the text in fact speaks of the "sea (*baḥr*) of *Ṭaws*/*Ṭāwus*," instead of Tartarus; in Epistle 49, the manuscripts have either *ṭūs*, *ṭarṭadīn*, or *ṭarṭāūs*. As suggested by the translator of Epistle 49a, Cyril V. Uy II (64*n*23 in the translation), it seems prudent to read the passages in both epistles as referring to Tartarus, which in Plato's *Phaedo*, *Gorgias*, and *Republic* occupies a similar eschatological position for the particular souls.

The references in the reverse direction reinforce the connection between Epistles 52 and 49; all three variants of Epistle 52 refer to this epistle. In Epistle 52a, the single reference is to "spiritual beings" (*al-rūḥāniyyūn*) in general;[13] in the two references in 52b, the reference is to the relation between the macrocosm and the microcosm, and the human form as God's vice-regent on Earth;[14] and in the five references of 52c, the reference is to astral souls as the origin of foreknowledge, the seven kinds of astral souls related to the seven spheres, demons 'eavesdropping' foreknowledge from the angels, the two angels related to divination, and scholarly disagreement on malevolent powers.[15] In contrast to the two other variants, however, 52c refers to Epistle 49 as one on "angels" (*malāʾika*) or "spirits of tranquility" (*sakīnāt*): these terms are not attributed to the epistle title of Epistle 49 in any other reference. As for references to other epistles, Epistle 52a refers twice to Epistle 3 (*Astronomy*) and once to the *Fihrist* and Epistle 42 (*Opinions and Doctrines*) each; 52b refers four times to the *Comprehensive Epistle* and once to Epistles 25 (*Fall of the Sperm*), 30 (*Pleasure and Pain*), 40 (*Causes and Effects*), and 44 (*Beliefs of the Brethren of Purity*) each; and Epistle 52c refers three times to Epistle 37 (*Love*). In sum, besides Epistle 49, the three variants share no other references. Moreover, all variants refer mostly to the late epistles; most importantly, variant 52a refers to the *Fihrist* and 52b to the *Comprehensive Epistle*, both of which are likely to postdate the *Epistles*.

All of this means, first, that the variation in the references to Epistle 52 indicates considerable ambiguity with respect to its subject matter. This might mean that they refer to different magical epistles. Second, two references are to subjects that are not discussed in any of the three existing variants of the epistle. Does this mean that there was a fourth version of the epistle? Third, both the references and referents attest to a strong connection between Epistles 49 and 52, indicating that Epistle 52 might initially have been attached to Epistle 49 in some sense. Fourth, all the identifiable references in other epistles refer to Epistle 52a; there are no identifiable references to 52b or 52c. This implies that 52a is perhaps the earliest of the three versions, and supports the arguments discussed in Chapter 1 for it being the likeliest candidate for being an authentic part of the *Epistles*. The following hypothesis may, then, be suggested. Astral magic was perhaps initially discussed in Epistle 49, while a magical epistle might also have existed that addressed magic squares and auguries, among other things. Epistle 52a predates the other two versions and was composed

---

13 *On Magic*, 11.

14 *Rasāʾil*, vol. 4, 367, 378.

15 MS Esad Efendi 3638, fols. 304ᵃ, 304ᵇ, 305ᵃ, 308ᵃ.

either at the end of the redaction process or after it; the two other versions almost certainly postdate the entire work. 52a and 52b were perhaps expanded from Epistle 49 by later editors who accorded greater prominence to magic; 52c, in contrast, appears more detached from the *Epistles* in both style and content.

The overall result of this analysis of cross-references is a hypothesis about the redaction chronology of the *Epistles*. This is presented in Table 16. In the table, each of the fifty-two epistles is placed in one of ten chronological groups, based on the preceding analysis of the reference ratio set against the relations between curricular groups. In some cases, the evidence appears strong; for example, it seems highly likely that Epistle 32/33 (*Intellectual Principles*) is an early epistle. In other cases, the interpretation of cross-references is sufficiently complicated to make the hypothesis as a whole speculative. Still, the evidence provided by cross-references both largely agrees with the evidence of curricular passages and further refines the order of redaction. The first composed treatises were the (1) mathematical epistles, as well as the Pythagorean Epistle 32/33 (*Intellectual Principles*); these were followed by the (2) Aristotelian physical treatises, the (3) non-Aristotelian physical epistles of the third section, (4) curricular epistles (7 and 8), (5) logical epistles, and the (6) non-Aristotelian physical epistles in the second section. All the mathematical and physical epistles of the Aristotelian curriculum would, then, have been composed before the introduction of the non-curricular epistles; the initially single epistle on logic was introduced at a slightly later stage. The cosmological, ethical, and religious epistles of the third and fourth sections were the latest to be composed. All versions of Epistle 52 (*Magic*), as well as all variants of the *Fihrist*, were added to the Ikhwānian corpus either at the very end of or after the redaction of the *Epistles*.

2 Hypothesis of Redaction History

The main goal of this first part was to present internal evidence for the redaction history of the *Epistles*. Based on this evidence, it is possible to construct a hypothesis about the evolution of the work without taking a stand on the seemingly unresolvable questions of the authors' identity and doctrinal affiliation. The three chapters have addressed seven kinds of evidence. Taken together, these are certainly conclusive for the general thesis that the work that we now have is the result of a complex process of redaction. First, the current arrangement of epistles, on the one hand, agrees with the classical curricular order but, on the other hand, contradicts it in arbitrary ways. This suggests that some epistles occupy an incorrect position within the whole. Second, the curricular passages in the *Epistles* also partly contradict the current configuration of epistles. In particular, they suggest that the work initially followed the classical four-division of philosophy, and that the work's first two sections consisted only of the mathematical and physical epistles within the classical curriculum, as well as of a single epistle on logic. Third, large-scale manuscript variation shows that some epistles occupy an ambiguous position within the work, that the division of the text into one versus two epistles is similarly ambiguous for some epistles, and that most epistles manifest considerable variation regarding the contents of the text. Moreover, the variation commonly includes major additions or omissions in at least one manuscript; in many cases, it also involves entirely different variants for an epistle, including many epistles for which the variants bear no textual relation with each other. These three considerations taken together suggest that four methods were employed in the gradual redaction of the *Epistles* to its final form: (1) the introduction of new epistles, (2) the splitting up of existing epistles, (3) the repositioning of epistles, and (4) the expansion of existing epistles.

Other kinds of evidence give further support for these methods of redaction. Fourth, the analysis of the *Fihrist* shows that it is likely to postdate the *Epistles*, and thus represents a later interpretation of the epistles rather than the authors' original plan. The *Fihrist* also provides a very particular example of how text was gradually expanded, and thus of the living nature of the manuscript tradition. Fifth, the ambiguity of many epistle titles, both at the head of epistles and as employed in cross-references, supports the view that many more of the current epistles were split from a larger epistle. The evidence proves that Epistles 10–14 (logic) and 32–33 (*Intellectual Principles*) in any case initially formed single epistles. Sixth, the presence of the formulaic epistle transitions at the beginning of some epistles but not others supports the contrast between the technical (1–24) and non-technical (25–52) halves of the work suggested by the curricular passages; the contradictions with the current order clearly prove that many epistles were moved from one position to another.

The seventh class of evidence, cross-references between epistles, provides a basis for constructing a hypothesis of relative chronology for the fifty-two epistles. In the big picture, the results agree with those of the curricular passages. That is, the analysis of cross-references appears to fit with the view that the epistles of the work's first two sections initially corresponded to the Aristotelian curriculum of philosophical sciences. The non-curricular treatises were, then, introduced later, whether because they were split from a larger epistle, transposed from another section, or composed after the classical epistles. Moreover, analysis of the references between the four sections supports the view that the last two sections were written after the first two, which means that the religious epistles of the fourth section, in particular, were also composed after the technical epistles.

As for the chronology of the *Epistles* as a whole, the discussion ended in a hypothesis that posits the following chronological groups: (1) mathematical and Pythagorean epistles (1–6, 32–33); (2) Aristotelian physical epistles (15–19, 21–24) of the second section; (3) non-Aristotelian physical epistles of the third section (28, 31, 34–36, 38–40); (4) curricular epistles (7–8); (5) logical epistles (10–14, 41); (6) non-Aristotelian physical epistles of the second section (20, 25–27); (7) cosmological and metaphysical epistles of the fourth section (49, 51); (8) earlier ethical and religious epistles of the fourth section (43–45, 47, 50); (9) later ethical and religious epistles (9, 29–30, 37, 42, 46, 48); and (10) very late epistles, which were introduced either at the very end of the redaction process or more probably after it (52a–c, *Fihrist a-c*). The evidence provided by the cross-references is complex and difficult to read. Thus, I do not claim that the cross-references prove that the epistles were written in this precise order. What I claim instead is that, based on the cross-references and curricular passages taken together, this is a plausible hypothesis of the general order of redaction.

But can the preceding be incorporated into a larger narrative that explains how and why the work came to gradually attain the shape that it has now? This question can be approached from the perspective of the four postulated methods of redaction, as well as the hypothetical chronology of epistles. First, why were new epistles introduced later to at least some of the four sections? Second, why were so many epistles split into two or more epistles? Third, why were some epistles repositioned from one section to another? Fourth, why were the existing epistles expanded, and sometimes even replaced by an entirely different version? Fifth, does the proposed chronology suggest a narrative that would explain why the epistles should have been composed in this order?

Starting with the last question, the chronology of redaction makes sense when viewed in the context of the multiple projects pursued by the authors of the *Epistles*. As claimed in the introduction, the work is not exclusively an encyclopedia of philosophical sciences nor a work with religious-messianic aspirations, but it combines both objectives. Even in their philosophical aspirations, the *Epistles* fuse the Aristotelian tradition with Neoplatonic, Neopythagorean, and Hermetic strands of thought. The postulated chronology suggests that the first epistles to be composed were the mathematical and Pythagorean epistles (group 1 in Table 16). The initial core of the work would, then, reflect the authors' Neopythagorean convictions about the numerical nature of ultimate reality, a view which, moreover, permeates the *Epistles* as a whole. Thus, it makes sense to locate the beginnings of the work in the late ninth century, given the similar Neopythagorean inclinations in the school of al-Kindī. From the perspective of the Kindīan tradition, it also makes sense that the authors subscribed to the Aristotelian curriculum of philosophical sciences. After the mathematical-Pythagorean epistles, then, the authors proceeded to write the physical epistles corresponding to the Aristotelian curriculum (group 2). The logical epistles (group 5) pose a problem for this interpretation, given that they appear to postdate both the Aristotelian and some of the non-Aristotelian physical epistles; in any case, the evidence suggests that initially there was just one logical epistle, and even the *Epistles* of modern editions do not incorporate the full logical curriculum. But perhaps the authors did not initially see Aristotelian logic, at least in its more technical aspects, to be particularly relevant for their philosophical-religious project. As a result, they saw it fit to complete the curriculum with logic only at a later stage.

Even if the authors initially subscribed to the Aristotelian tradition in their classification of philosophical sciences, their philosophical and religious views did not entirely accord with it. The chronological groups 3 and 6 consist of epistles with physical subject matter that have no place within the late ancient and early Islamic Aristotelian curriculum. It seems that the authors, then, at some point came to view the Aristotelian physical curriculum as insufficient for fully conveying their views about the human being and the cosmos. Therefore, they decided to compose additional epistles that complement the physical corpus with the non-Aristotelian subjects they wanted to address. This seems clearly to be the case for the additional physical epistles of the second section with largely Neoplatonic subject matter (group 6); the physical epistles of the third section (group 3) would have been introduced before them. Moreover, it also seems to be the case that the authors composed the curricular epistles (group 4), depicting the overall view of their project of a philosophical encyclopedia, before the additional physical epistles of the second section, because Epistle 7 completely ignores the existence of the latter epistles.

Finally, the chronology suggests that the epistles corresponding to practical philosophy and religion were composed at a later stage of the redaction process. This further supports the notion that the authors' initial motivation was to write an encyclopedia of philosophical sciences in a classical sense, even if their eschatological interests, in particular, are apparent even in the epistles of the first section. Later, they decided to elaborate the work to address religious themes more directly in various further epistles. The curricular passages support this narrative of the project's gradual evolution; they manifest complete ignorance about the non-classical epistles of the first two sections, and the classification of Epistle 7 concurs to only a very small extent with the actual configuration of epistles with religious or ethical subject matter.

As for the third question, the repositioning of epistles from one section to another also fits with this narrative. Once the authors expanded the original project to incorporate a large number of epistles that formed no part of the classical scheme, it would seem natural that they saw it necessary to review the overall arrangement of the epistles. The curricular passages suggest that during the first stage of the project they followed the Aristotelian four-division of philosophical sciences; at some stage they adopted the current four-division, instead. Once the authors abandoned the classical plan with respect to both the division and configuration of epistles, it also became incumbent to rearrange the position of some epistles within the sections. The first and fourth questions may also be answered by the gradual evolution of the project. The authors introduced new epistles into each of the four sections, as well as expanded or introduced variant versions for existing epistles, because their conception about the scope of their encyclopedia had evolved from what it had been initially.

The motivation for splitting epistles into two or more epistles seems more difficult to explain. For the logical epistles there is a natural explanation in that a single epistle was split into the five that correspond to five of the nine parts of logic within the Aristotelian logical curriculum. Even this, however, appears to have proceeded gradually, since many of the existing manuscripts carry only four logical epistles. But what reason was there to make Epistle 32/33 (*Intellectual Principles*) into two epistles, or to separate Epistle 6 (*Arithmetical and Geometrical Proportions*) from one of the mathematical epistles? In both cases, neither the length nor the subject matter warrants the splitting; the resulting epistles are short, and Epistles 32 and 33 are of shared subject matter. As for Epistle 6, it is unclear why the subject merits an independent epistle, since its themes could have been addressed within one or more of the mathematical epistles. Perhaps, then, the authors were aiming for a pre-conceived number of epistles, such as fifty-one, which possessed particular significance for them. In any case, it is conspicuous that, despite the otherwise considerable variation, most manuscripts consistently refer to the total number of epistles as fifty-one, and that most manuscripts in fact have fifty-one epistles, even in cases where the division of the text into epistles is different.

Admittedly, much of the proposed narrative of redaction is speculative. Some aspects of the narrative follow directly from the evidence; for example, the idea of an initially classical philosophical encyclopedia is supported by both the curricular passages and cross-references. Other aspects, such as the precise relative date of particular epistles, is not determined beyond doubt by the cross-references. What the internal evidence shows beyond reasonable doubt, however, is that the redaction history of the *Epistles* was extremely complex. Our evidence comes from manuscripts that postdate the beginning of that history by two or three centuries. It is therefore probably impossible to reconstruct precisely the different stages through which the work evolved, let alone postulate an original redaction of the *Epistles*. Yet, clearly, there must have once existed a version of the *Epistles* that was radically different from the work that we now possess. Initially, there was an encyclopedia of philosophical sciences that followed the Aristotelian curriculum more closely; later, this was expanded to better incorporate both the philosophical and religious perspectives of either the original or later authors. Many epistles were more concise, and, in some cases, even their contents were different from the current epistles. Moreover, there were initially much fewer epistles than the current fifty-one or fifty-two epistles, perhaps only twenty or thirty. The authors gradually expanded their number by introducing new tractates and splitting existing epistles into two or more epistles.

PART 2

The Comprehensive Epistle

CHAPTER 4

The Relation between the *Comprehensive Epistle* and the *Epistles*

The major scholarly dilemma concerning the *Comprehensive Epistle* (*al-Risāla al-jāmiʿa*) is its relationship with the *Epistles*. The treatise is a systematic, epistle-by-epistle, summary or commentary of the entire *Epistles*. Some known manuscripts summarize the *Epistles* in the sense of reproducing a selection of epistles, or, more rarely, select passages from some or all epistles.[1] *The Comprehensive Epistle* is different, first, in that it claims to form an integral part of the *Epistles*; it is the final epistle that unveils the inner meanings of the original work. This relation between the two works is apparently confirmed by the fact that the *Epistles* also contain a number of references to the *Comprehensive Epistle* that make the same claim. Second, in its assertion that it explicates the real meaning of the original epistles, the work proposes a comprehensive interpretation of the *Epistles*. More than a summary, it is then an exegetic commentary of the original work. In medieval times, the two works were commonly attributed to the same author or authors. The *Comprehensive Epistle* thereby became very influential in mediating the influence of the *Epistles*, at least for an Ismāʿīlī audience. Thus, the Ṭayyibites of Yemen adopted both works into their corpus. However, al-Ḥāmidī (d. 1162), the first known Ismāʿīlī writer to cite the Brethren of Purity, relies primarily on the *Comprehensive Epistle*.[2]

Most modern scholars have accepted the theory of common authorship, either explicitly or implicitly.[3] For example, Netton admits "some dispute" concerning the authorship of the *Comprehensive Epistle* but concludes that "in view of the similar vocabulary, phraseology, and other resemblances it is highly unlikely that its authorship differs from that of the *Rasāʾil*."[4] The theory is

1 The latter is the case for Tübingen MS MaVI86 (dated 612 AH/1215 CE); its first part (fols. 1–41) contains a summary of the first twenty-two epistles.

2 De Smet, "La Risāla al-Ǧāmiʿa," 288; Idem, "L'auteur des *Rasāʾil*." For the Ṭayyibite adoption of the *Comprehensive Epistle*, see also Ḥusayn F. al-Ḥamdānī, "*Rasāʾil Ikhwān aṣ-Ṣafāʾ* in the Literature of the Ismāʿīlī Ṭaiyibī Daʿwat," *Der Islam* 20 (1932): 284–285; De Smet, "Henry Corbin," 114–115; Ivanow, *A Guide to Ismaili Literature*, 20.

3 See, for example, Baffioni, "Uso e rielaborazione," 6; Hamdani, "The Arrangement of the *Rasāʾil Ikhwān al-Ṣafāʾ*," 90–91; Krinis, "*Al-Risāla al-Jāmiʿa*."

4 Netton, *Muslim Neoplatonists*, 2.

DOI:10.1163/9789004690141_006

founded on two main arguments: (1) references proceeding in both directions and (2) relative stylistic and doctrinal coherence between the two works. At the same time, many scholars have noted the clearly more esoteric and Ismāʿīlī character of the *Comprehensive Epistle*.[5] As a result, some scholars have been skeptical that the two treatises were composed by the same authors.[6] Both claims can of course be true at the same time. The treatise alleges that it reveals the inner meanings of the *Epistles*. If the authors of both works were Ismāʿīlīs, it is hardly surprising that these meanings correspond to Ismāʿīlī doctrine. The question here is then whether we accept at face value the claim of the *Comprehensive Epistle*'s author that it forms a part of the original composition of the *Epistles*.

Besides clarifying the nature of the *Comprehensive Epistle*, the answer also has consequences for our understanding of the *Epistles*. If the claim of common authorship is accepted, then the original epistles should be understood through the prism of the *Comprehensive Epistle*. It provides an authoritative guide for reading the *Epistles*. If the claim is discarded, then the case is instead that the *Comprehensive Epistle* represents a later, probably an Ismāʿīlī, interpretation. In that case, the work is of value for the story of the later adoption and transformation of the *Epistles*, but its readings should not be projected onto the original work.

In this second part of the book, I will take a number of steps to answer this question. As was the case for the *Epistles* in the first part, the answer will be based entirely on internal evidence. In contrast to the previous part, however, here the problem concerns the relationship between two different works, rather than the internal evolution of a single work. Thus, only a small part of the evidence concerns manuscript variation. Instead, my aim is to demonstrate that a specific relation prevails between the two texts; based on this relation, the *Comprehensive Epistle* must postdate the *Epistles*. In Chapters 5 and 6, the argument is founded on structural and doctrinal comparisons of the two works. The current chapter addresses specific aspects of the textual relationship in three sections. The first two sections analyze the passages explicitly addressing that relation: in the first section, the passages are statements in the *Comprehensive Epistle* and the *Epistles* that explicate the former treatise's alleged status as the final epistle. In the second section, an analysis of the references from the *Epistles* to the *Comprehensive Epistle* shows that this claimed relation is improb-

5 See, for example, Baffioni, "Ibdāʿ"; de Callataÿ, "Magia," 304–306; Ebstein, *Mysticism*, 45–46; Krinis, "*Al-Risāla al-Jāmiʿa*"; Pines, "Shīʿite Terms," 171–172, 174–177, 229.

6 Carusi, "Alchimia islamica," 277–278; De Smet, "La Risāla al-Ǧāmiʿa," 295–297.

able. The third section, finally, demonstrates that the *Comprehensive Epistle* depends textually on the *Epistles* in a specific way that makes it highly likely to be a later work. The implications of the evidence provided in the three chapters will be taken together at the end of Chapter 6.

1 Final Epistle

The *Comprehensive Epistle* contains a number of passages that address its relation with the *Epistles*. All of these passages convey an overall claim: the former work constitutes an integral and essential part of the latter work. Thus, it very much pertains to the original plan of the composition of the *Epistles*. In order to assess whether this claim is true, we should first see more precisely what the author's statement of intent consists of. The overall claim can be analyzed into three interrelated theses. These present the *Epistles* and the *Comprehensive Epistle* by way of contrast: (1) the former is an introduction; the latter is its conclusion; (2) the former is persuasive; the latter is demonstrative; (3) the former is implicit; the latter is explicit.

The first thesis may be illustrated by the following example:

> For we made the previous epistles an introduction (*muqaddima*) to this epistle; they are like bodies, and this epistle is like spirit. Through the spirit, the state of the body attains its completion (*bi-l-rūḥ tamām amr al-jasad*), and in the same way, this epistle is the completion of the epistles (*tamām al-rasāʾil*).[7]

The *Epistles*, then, form an introduction to the *Comprehensive Epistle* in the sense that the former treatise is only truly completed in the latter treatise, although the latter builds on the former. The relationship between the body and the soul provides an analogy: both works are indispensable but unequal parts of a whole; the *Comprehensive Epistle* represents an end; the *Epistles* are a means toward that end. The upshot of this claim is that the value of the *Epis-*

---

7 *Al-Jāmiʿa*, ep. 39; ed. Ghālib, 472; ed. Kacimi, 617. For the two editions employed, see the introduction. In the references to the *Comprehensive Epistle*, the epistle number indicates the ordinal number of the epistle (in the original *Epistles*) of which the section in the *Comprehensive Epistle* is a commentary. In the editions and most manuscripts of the *Comprehensive Epistle*, the forty-fifth epistle is transferred to the position of the final epistle; thus, after that epistle, the ordinal numbers of epistle commentaries in the *Comprehensive Epistle* do not correspond to those of the *Epistles*.

tles is in effect downgraded to the instrumental one of preparing students to understand the ultimate truth expressed in the *Comprehensive Epistle*. We have seen that the *Epistles* follow a pedagogical progression toward gradually higher forms of knowledge; the *Comprehensive Epistle* is the culmination of that progression.

The two further claims specify the precise sense in which the *Epistles* provide an introduction to the *Comprehensive Epistle*. The following passage provides an example for the second thesis:

> Our *Comprehensive Epistle* is the ultimate goal (*al-gharaḍ al-aqṣā*), for in it we show by clear demonstrations (*bi-l-barāhīn al-shāfiya*) all that of which we explained a part (*sharaḥnā baʿḍahu*), by way of persuasion (*bi-ṭarīq al-iqnāʿ*), in the *Epistles*.[8]

It is consequently the case that the *Epistles* are introductory in the sense that they only employ persuasive arguments for their doctrinal claims; the *Comprehensive Epistle*, in contrast, demonstrates the truth of these claims. In the *Epistles*, the pedagogical progression proceeds along the curriculum of sciences. The transition from the *Epistles* to the *Comprehensive Epistle*, in contrast, is not merely from one class of objects of knowledge to another. Rather, it is an epistemic one from a lower to a higher degree of certainty. This is the most surprising of the three theses. It is true that the *Epistles* often give at best persuasive arguments, such as the many allegorical stories, in support of their doctrinal claims. But the *Comprehensive Epistle* contains no demonstrative proofs, either.

The following example finally presents all three theses taken together:

> For in this subject (*faṣl*), we did not express ourselves clearly (*lam nakun nufṣiḥ al-qawl bihi*), nor did we speak of it freely (*lā aṭlaqnā al-kalām fīhi*) or refer to it openly and unequivocally (*al-dalāla ʿalayhi bi-l-taṣrīḥ al-shāfī*), but only with signs (*talwīḥ*), symbols (*ramz*), and allusions (*ishāra*), in our previous epistles. There we stated that those epistles are introductions (*muqaddimāt*) to what we want to state and explain by demonstration (*burhān*) in this epistle, which we made the locus of goals (*makān al-aghrāḍ*) and the perfection of speech (*kamāl al-qawl*). For what precedes was in potentiality, and in this locus, it is as if the potential becomes actual, and thereby completion (*tamām*) is attained. For that reason, we

8 Ibid, ep. 1; ed. Ghālib, 65; ed. Kacimi, 61.

> said that this epistle is the "epistle of demonstration" (*risālat al-burhān*), and it is the completion and conclusion (*khātima*).[9]

The passage reiterates the idea that the *Epistles* constitute an introduction to the *Comprehensive Epistle*, because only the latter is demonstrative in nature; thus, it is accorded the title "epistle of demonstration" (*risālat al-burhān*). Beyond this, the passage states that the introductory nature of the *Epistles* also consists of the fact that they refer to their subject matter by means of symbols and veiled allusions. The *Comprehensive Epistle*, in contrast, explains every subject in explicit terms. Again, the difference is not that the *Comprehensive Epistle* expands on the knowledge contained in the *Epistles*. It is rather the case that the same objects of knowledge are conveyed implicitly in the former work and explicitly in the latter work. Accordingly, as the analogy employed in the passage indicates, the progression from the *Epistles* to the *Comprehensive Epistle* is one from potentiality to actuality.

In these passages, the idea is that the two works constitute a whole: the *Comprehensive Epistle* completes the *Epistles*, but one should first study the latter before proceeding to the former. This supposed unity of the two works is not supported by the manuscript tradition. There seem to be no manuscripts that convey the two treatises as a single whole; instead, the two works constitute separate manuscript traditions.[10] This could be explained by the *Comprehensive Epistle*'s claim to esoteric truth, which should not be conveyed to a novice. But the work also provides a different, and somewhat contradictory, rationale for its existence:

> This epistle [34], O brother, may God support you and us with His spirit, contains wisdom that we may not refrain from mentioning and repeating in this *Comprehensive Epistle*. For it contains a comprehensive discourse (*qawl jāmiʿ*), which points to knowledge by things insofar as they are those things (*yadull ʿalā al-maʿrifa bi-l-ashyāʾ bi-mā hiya bihi al-ashyāʾ*), in order for this [comprehensive] epistle to be a substitute for what is needed (*qāʾima bi-mā yuḥtāj ilayhi*) should all of the original epistles be absent. Thus, the *Comprehensive Epistle* takes the place of all these epistles and adds to them.[11]

---

9 Ibid, ep. 1; ed. Ghālib, 71; ed. Kacimi, 68–69.

10 A possible exception is Tübingen MS MaVI86 (dated 612/1215): the manuscript's first part (fols. 1–41) contains a summary of the first 22 epistles; the second part consists of the latter half of the *Comprehensive Epistle*. Even in this case, the placement of the two works within a single manuscript is probably accidental.

11 *Al-Jāmiʿa*, ep. 34; ed. Ghālib, 352–353; ed. Kacimi, 468–469.

In this passage, the *Epistles* seem superfluous; a student may gain all required knowledge from the *Comprehensive Epistle* if the original epistles are lost.

If only the *Comprehensive Epistle* made these claims, with the *Epistles* oblivious of the existence of a further treatise explaining their ultimate meaning, the claims would not be as credible. In the end, the most important argument for the theory of common authorship is that the *Epistles* make precisely the same claim. In the *Epistles*, the most elaborate statements concerning the *Comprehensive Epistle* occur in the introductory part of the *Fihrist*:

> For the *Comprehensive Epistle* is restricted to demonstrations (*maqṣūra ʿalā al-barāhīn*) of what we have pointed to (*asharnā ilayhi*) in our epistles, which are like an introduction (*madkhal*) and a token (*ʿanwān*) for it.[12]

> The epistles are followed by an epistle that comprises what is contained in all of these introductory epistles (*al-risāla al-jāmiʿa li-mā fī hādhihi al-rasāʾil al-mutaqaddima kullihā*) and embraces all of their truths (*al-mushtamila ʿalā ḥaqāʾiqihā bi-asrihā*). Its goal is to clarify the truths to which we hinted and pointed at in these epistles in the clearest and most evident way possible. It presents in detail what is contained in the epistles (*yaʾtī ʿalā mā fīhā*), and their essential truths and meanings are clarified there in a complete, correct, and thorough manner (*yatabayyan ḥaqāʾiquhā wa-maʿānīhā mulakhkhaṣa mustawfāt muhadhdhaba mustaqṣāt*) by certain geometrical demonstrations (*barāhīn handasiyya yaqīniyya*), true philosophical proofs (*dalāʾil falsafiyya ḥaqīqiyya*), scientific explanations (*bayyināt ʿilmiyya*), intellectual arguments (*ḥujaj ʿaqliyya*), logical propositions (*qaḍāyā manṭiqiyya*), syllogistic testimonies (*shawāhid qiyāsiyya*), and persuasive methods (*ṭuruq iqnāʿiyya*). Their essence and true meaning cannot be grasped and attained to any extent whatsoever except by someone who has been trained in what we have presented before (*irtāḍa bi-mā qaddamnā*), so that he has gained proficiency, knowledge, training and expertise in them or what resembles them (*bi-mā yushākiluhu*). Because all of these epistles are like introductions (*muqaddimāt*) and preambles to it, and the abstruse and concealed secrets of its proofs (*adilla*) and tokens (*namūdhaj*) are only opened and revealed to those who have been educated (*tahadhdhaba*) by these fifty-two epistles or similar books. For the *Comprehensive Epistle* is the ultimate

12 *Rasāʾil*, vol. 1, 39.

> goal (*muntahā al-gharaḍ*) and utmost objective (*ghāyat al-murād*) for what we have provided before it.[13]
>
> After studying (*naẓar*) the epistles and grasping all their meanings, one is entitled (*istaḥaqqa*) to study the *Comprehensive Epistle*, which comprises the final objective (*nihāyat al-murād*), pure goal (*nazihat al-murtād*), and triumph in this life and the afterlife (*al-fawz fī al-maʿāsh wa-l-maʿād*). This is because the *Comprehensive Epistle* may be reached by means of the *Epistles* (*bi-hinna al-tawaṣṣul ilayhā*), and through understanding the *Epistles*, that epistle may be comprehended (*bi-fahmihinna al-wuqūf ʿalayhā*).[14]

These statements reiterate the three theses conveyed in the *Comprehensive Epistle*; the middle passage makes the claim of the supposedly demonstrative nature of that work even more emphatically. All three passages emphasize the idea that the *Epistles* and the *Comprehensive Epistle* together constitute a program of study, within which the latter treatise is the final objective. It is thus both the case that (1) one should not study the *Comprehensive Epistle* unless one has first mastered the *Epistles*, and that (2) one can only come to understand the true meaning of the *Epistles* after studying the *Comprehensive Epistle*. These statements assume a particular significance due to their placement in the introductory part of the *Epistles*. If the *Fihrist* presents the authors' general plan for the *Epistles*, clearly the *Comprehensive Epistle* forms an integral part of that plan.

Based on the first part of this book, we now know that the text of the *Epistles* evolved in time. More specifically, the textual relation of the *Fihrist* with the *Epistles* has been addressed above.[15] There it was shown that the manuscripts convey several, very different, variants of the *Fihrist*. As it happens, the three cited passages only appear in full in the latest variant contained in the Beirut edition. Besides this testimony, four of the studied manuscripts contain a reference to the *Comprehensive Epistle*; two manuscripts carry the second cited passage; two further manuscripts contain a more concise reference indicating

---

13 Ibid, 42–43.

14 Ibid, 46. For an analysis, and full citations and translations, of the references from the *Epistles* to the *Comprehensive Epistle*, see Mattila, "Authorship." Besides the three cited references, the *Fihrist* includes a fourth, implicit, reference (*Rasāʾil*, vol. 1, 43): "This is the *Fihrist* of the *Epistles* of the Brethren of Purity, Friends of Loyalty, People of Justice, and Sons of Praise, which consist of fifty-two epistles and an epistle on the purification of souls and rectification of character traits."

15 See pp. 71–77.

the existence of a further epistle that should only be studied after the *Epistles*.[16] Twelve manuscripts carrying the *Fihrist* contain no reference to the *Comprehensive Epistle* at all. The *Fihrist* in general is likely to postdate the *Epistles*, and this is almost certainly the case for its later variants. Clearly, the references to the *Comprehensive Epistle* were only gradually incorporated into its text. In the text as it appears in most manuscripts, there is no indication whatsoever that the authors composed a further epistle which is indispensable for understanding the meaning of the original epistles.

As a final note, if the alleged relationship between the two works were true, presumably it should be reflected in the *Epistles*, beyond the scant explicit references to the *Comprehensive Epistle*. As regards the three theses, however, while they appear in the original epistles, they make no reference to an additional epistle. Regarding the first thesis, the *Epistles* often characterize an epistle as an introduction, but this is in the sense that each epistle provides an introduction to its subject matter, such as astronomy, as opposed to its being aimed at a specialist in that subject.[17] No mention is made that the reader will later be provided with conclusions in a separate epistle. As for the second thesis, the idea that the *Epistles* employ a persuasive, as opposed to a demonstrative, method appears to be entirely absent in the work. One exception is the *incipit* of Epistle 44 (*Beliefs of the Brethren of Purity*), which states that the epistle in question shows the immortality of the soul "by a persuasive method but not by way of demonstration" (*bi-ṭarīq muqniʿ lā bi-ṭarīq al-burhān*). But the *incipit* is only present in the Beirut edition variant, while it is absent in most manuscripts.[18] The fact that this statement is an apparent interpolation supports the case that the idea that the original epistles correspond to a persuasive method was introduced later. Regarding the third thesis, the *Epistles* frequently incite the reader to gain an understanding of hidden mysteries and subtle allusions, whether made by prophets, ancient philosophers, or the authors them-

16 Manuscripts ر and E contain the second cited passage at the end of the *Fihrist*; manuscripts J and K instead provide the following reference after the summaries of the fifty-one epistles: "Know, O brother, that we summarized the essence of these fifty-one epistles in an epistle that we called comprehensive (*jamaʿnā zubad hādhihi al-rasāʾil al-aḥad wa-khamsīna fī risāla sammaynāhā jāmiʿa*). But it is of no benefit for you to read it before you have studied the epistles presented here. For these epistles are an introduction (*madkhal*) to the knowledge contained in the *Comprehensive Epistle*. The [proper] beginning is for you to first read these epistles in [the assigned] order."

17 See pp. 8–9.

18 *Rasāʾil*, vol. 4, 14; *On Companionship and Belief*, 30. The phrase is perhaps taken from the *Fihrist* entry of the corresponding epistle, which employs the exact same expression in its later variants.

selves.[19] But apart from the few references to be discussed below, there is no mention that these allusions will be explained in a further epistle. In general, apart from the explicit references, nothing in the *Epistles* anticipates the *Comprehensive Epistle*. The original epistles seem entirely unaware that a further epistle is needed to explain their meaning.

2 References from the *Epistles* to the *Comprehensive Epistle*

In the end, the existence of references from the *Epistles* to the *Comprehensive Epistle* seems to be the strongest argument for the theory of common authorship. In the case of the *Fihrist*, we have seen that the great majority of the examined manuscripts contain no references to the *Comprehensive Epistle*; thus, clearly the references present in some manuscripts were inserted later. The *Fihrist* is a case apart, however, since its textual transmission appears to be separate from that of the main body of the work. The epistles themselves also contain references to the *Comprehensive Epistle*. If these references pertain to the original text of the *Epistles*, then the two works are likely to have been composed by the same authors. The other possibility is that the references are interpolations that were introduced into the manuscripts after the *Comprehensive Epistle* was composed.[20] An examination of these references shows that it is indeed the case that most of these references are interpolations; while some are not, they provide little support for the theory of common authorship.[21]

The text of the Beirut edition contains fourteen references to the *Comprehensive Epistle* (see Table 17). If the treatise were an integral part of the original plan of composition, this is not a high number. Moreover, the references are unevenly divided over the epistles: the *Fihrist* contains four, epistles in the first section two, and epistles in the fourth section eight references, respectively; the second and third parts have no references at all. When these references are compared with the corresponding passages in the most ancient manuscript (ع), the result is that only six of them are present in any form. Also, the imbalance of distribution is now more striking: all six references occur in the fourth section of the *Epistles*; moreover, they are distributed between three of the final four epistles (49, 50, 52b). These statistics alone make it clear that the cross-

19 See, for example, *On Companionship and Belief*, 42.

20 Although the author does not examine the references, this possibility was suggested already in Carusi, "Alchimia islamica," 288.

21 For an in-depth analysis of the references, see Mattila, "Authorship."

references provide much less support for the theory of common authorship than it initially seemed.

A more thorough examination of the references further weakens their testimonial value. Based on their presence in MS ع, the references may be divided into three groups. The first group consists of eight references that are obvious interpolations; that is, they do not occur in MS ع. A second group consists of the three ambiguous cases where a reference to the *Comprehensive Epistle* is in some sense present in MS ع, but the passages in question show highly revealing differences in manuscripts. For questions regarding the *Comprehensive Epistle*'s genesis, these might be the most interesting cases. A third group consists of the four references where the passages of the Beirut edition and MS ع have an unambiguous correspondence; all four occur in Epistle 52b.

The first group is clear-cut: the Beirut edition and MS ع provide more or less identical passages, apart from the latter's exclusion of a reference to the *Comprehensive Epistle*. Besides the *Fihrist* (where the textual relation between the testimonies is more complex), such references occur in Epistles 7 (*Theoretical Sciences*), 9 (*Ethics*), and 42 (*Opinions and Doctrines*). The reference in Epistle 7 provides an example:

> We have produced an epistle for each branch (*faṣl*) of the abovementioned sciences and mentioned in them some of those meanings (*ṭarafan min dhālika al-maʿnā*), **and we completed them in the *Comprehensive* (*Epistle*)** (***wa-atmamnāhā bi-l-jāmiʿa***), to awaken the negligent (*tanbīhan li-l-ghāfilīn*) and guide beginners (*irshādan li-l-murīdīn*), excite the interests of students (*targhīban li-l-ṭālibīn*) and serve as a path for those who would learn (*maslakan li-l-mutaʿallimīna*).[22]

Both testimonies provide an identical passage, apart from the section in bold which includes the reference to the *Comprehensive Epistle*. The reference is present neither in MS ع nor in any variant of the critical edition. The bolded section is clearly an interpolation, which is simply inserted into the middle of the original passage. It is probably the context of the *Epistles*' overall goal of producing an epistle for each science that inspires the editor to complement this with the claim that their meanings are explained more fully in the *Comprehensive Epistle*. The cases of Epistles 9 and 42 are similar in that they also include a concise reference to the *Comprehensive Epistle* in the Beirut edition variant, which is not included in the otherwise identical passage of MS ع. In all

22 *Rasāʾil*, vol. 1, 274; *On Composition*, 95; tr. de Callataÿ, 119.

three cases, it is safe to assume that the references are interpolations. However, it is not entirely clear why the references were placed in these precise locations, or why they were not scattered more widely over the epistles.

The second group is the textually most interesting; in particular, because these passages may provide clues for how the idea of a comprehensive epistle first arose. The group consists of three references in Epistles 49 (*Spiritual Beings*) and 50 (*Governance*), as well as a fourth passage of interest in Epistle 50. In each passage, the textual relation between the different testimonies is considerably more complex than in the previous group. I will therefore cite each passage in full. Epistle 49 contains a single reference to the *Comprehensive Epistle*. This is the first case in the *Epistles* where the reference to the *Comprehensive Epistle* in the Beirut edition appears to be confirmed by MS ع, at least to the extent that all variants now refer to something "comprehensive":

> In our comprehensive epistles (*rasāʾilinā al-jāmiʿa*) [**the *Comprehensive Epistle* (*al-risāla al-jāmiʿa*)**; our comprehensive books (*kutubinā al-jāmiʿa*)], we have mentioned the knowledge of these degrees (*rutab*) and grades (*manāzil*), I mean the praiseworthy and blameworthy among them, and we mentioned their different substrates and individuations (*mawāḍiʿahā wa-ashkhāṣahā*), such as the earth, minerals, plants, animals, and the human being, and that the last of the minerals is connected with the first of the plants, the last of the plants with the first of the animals, and the last of the animals with the first degree of human beings, and the last degree of human beings with the angels, that is, when human beings are pure (*idhā ṣafā*).[23]

The passage occurs in variant 49b of the critical edition and is concerned with human psychical actions and the degrees of the Universal Soul as instantiated in sublunar and supralunar existents. The Beirut edition variant refers explicitly to the *Comprehensive Epistle*, which appears to be supported by MS A's reference to "our Comprehensive Epistle" (*risālatinā al-jāmiʿa*). Even though all variants carry the term "comprehensive," in other variants, the reference is far from unambiguous. The rest of the manuscripts employed in the edition refer to "our comprehensive epistles" (*rasāʾilinā al-jāmiʿa*) in the plural; a single manuscript (غ) even contains a reference to "our comprehensive books" (*kutubinā al-jāmiʿa*). The plural form does not initially seem to make sense. Yet, this is the reading in the majority of the manuscripts, including MS ع. Moreover, as we

23 *Rasāʾil*, vol. 4, 237–238; *On God and the World*, 212.

will see shortly, this precise reading is reiterated in a later passage where the context does not allow one to read it in the singular. For now, it is sufficient to conclude that the Beirut edition reference to the *Comprehensive Epistle* is only unambiguously confirmed by a single testimony; most manuscripts contradict this reading.

Epistle 50 contains two references to the *Comprehensive Epistle*, as well as a third passage, which is of utmost relevance. The first reference might be a key passage for our understanding of the *Comprehensive Epistle*. It occurs at the very beginning of Epistle 50:

> Know, O brother, may God support you and us with His spirit, that [this chapter is the intended purpose of these epistles of ours and their aim on which one should rely (*hādhā al-faṣl huwa al-gharaḍ al-maqṣūd ilayhi min rasāʾilinā hādhihi wa-huwa ghāyatuhā wa-ʿalayhi al-iʿtimād*). That is because][24] we have placed in each of our epistles a chapter (*faṣl*) which we intended to be its core and pure essence (*jaʿalnāhu lubbahā wa-khalāṣatahā*), so that one who successfully understands it and acts in accordance with it (*wuffiqa lahu min fahmihi wa-ʿamila bihi*) will gain happiness in this world and the next. **And we have summarized (*lakhkhaṣnā*) what we have presented in our fifty-one epistles in a single epistle separate from the epistles (*mufrada ʿan al-rasāʾil*) which we called comprehensive (*al-jāmiʿa*), and which does not belong to the totality of the epistles (*khārija min jumlat al-rasāʾil*), where we provided an explanation (*bayān*) of what we have related in the other epistles in the most particular way possible (*bi-akhaṣṣ mā amkannā minhu*). For all our epistles can hardly be grasped by one man (*laysa takād tajtamiʿ rasāʾilunā kulluhā ʿinda rajul wāḥid*) unless he is guided by God in it (*illā man sahhala Allāh taʿālā lahu dhālika*), and we therefore made that epistle to represent its sisters (*li-tanūb ʿan akhawātihā*). But the most correct and best thing in our opinion (*al-aṣwab wa-l-ajwab ʿindanā*) is for you to read the *Comprehensive Epistle* only after you have read our fifty-one epistles. For when one reads it after them, its benefit will be greater and the hidden mysteries of our epistles will be revealed (*infataḥa ʿalayhi mā inghalaqa min rasāʾilinā*), even though one who discovers this epistle without all or some of the epistles will still not be devoid of its benefits. As for this epistle, we entitled it governance and leadership (*wa-samnāhu bi-l-siyāsa wa-l-riʾāsa*) so that it would induce**

24 The bracketed passage is omitted in MSS ح, ل, ن, and the Beirut edition.

> **your soul to follow its requirements** (*li-taḥmil nafsaka ʿalā mūjibihā*) **and that you might read it under the guidance of one of our noble brethren** (*taqraʾahā ʿalā man yakhuṣṣuka min ikhwāninā al-kirām*) — **may God have mercy on them—and to confer with them during your shared times of activity** (*fī awqāt nashāṭika wa-nashāṭihim*), **for then you will not remain devoid of its benefits.**[25] And we order you, O blissful brother, may God support you and us with His spirit, after learning what is in [this chapter of] this epistle to follow what we have ordered you to do in it,[26] **and then you will gain supreme happiness in religious and worldly matters if God so wishes.**[27] [For we called it the comprehensive chapter (*al-faṣl al-jāmiʿ*), because it combines the most eminent of the blisses and benefits (*ajall al-saʿādāt wa-l-manāfiʿ*), **if God so wishes.**][28]

The bolded sections represent additions in the Beirut edition, which are supported by three manuscripts with some variation; the bracketed sections indicate passages that are present in most manuscripts but absent in the Beirut edition and/or some manuscripts. It is immediately clear that the passage bears the marks of heavy re-editing. First, all variants state that the authors have composed for *each* epistle a "chapter" (*faṣl*) that conveys the essential message of that epistle. In most manuscripts, however, this statement is preceded by a further claim that this particular chapter presents the goal (*al-gharaḍ al-maqṣūd*) for *all* of the epistles; this claim is omitted in the Beirut edition variant. The end of the passage confirms this by stating that this chapter is called the "comprehensive chapter" (*al-faṣl al-jāmiʿ*), a statement that is included in the Beirut edition, but omitted in at least one manuscript. Second, the lengthy section containing the reference to the *Comprehensive Epistle* only occurs in the Beirut edition and in three manuscripts, with some variation, while most manuscripts make no reference to the *Comprehensive Epistle*. Third, all variants contain an injunction to follow the advice that is presented after this introductory passage; in most manuscripts, however, these injunctions are contained in "this chapter of this epistle," as opposed to in "this epistle" in the Beirut edition variant.

---

25 The bolded section is contained in the Beirut edition, and with minor differences in ح, ل, and ن.

26 The passage "after learning … in it" is omitted in MS ق.

27 The bolded section is contained in the Beirut edition, and MSS ح, ل, and ن.

28 *Rasāʾil*, vol. 4, 250–251; *On God and the World*, 237–238. The final bracketed section is omitted in MS ل but is present in the Beirut edition. *Ajall al-saʿādāt wa-l-manāfiʿ* appears slightly differently in the manuscripts; it is *aṣl saʿādāt al-manāfiʿ* ("foundation of the beneficial blisses") in the Beirut edition.

As regards the reference to the *Comprehensive Epistle*, then, again, most manuscripts do not confirm the reference in the Beirut edition.

As we will soon see, the "comprehensive chapter" appears to refer to the entirety of Epistle 50. Perhaps this can be explained by the redaction history of the *Epistles*. In the previous part of this book, we saw that Epistle 50 might initially not have been an independent epistle.[29] If that is the case, the references to a "comprehensive" chapter could be attributed to the whole of Epistle 50; that is, within the original epistle, the part that was later split into the separate Epistle 50 constituted its "comprehensive" part. If this is so, the term "comprehensive" is an attribute of Epistle 50.

If this is the case, then the most striking aspect of the cited passage is that Epistle 50 is both characterized as "comprehensive" and accorded the role of providing an ultimate explication of the *Epistles*. Clearly, these attributes should pertain to the *Comprehensive Epistle*. Yet, in most manuscripts, any reference to the *Comprehensive Epistle* is absent. The Beirut edition variant instead inserts a lengthy addition and some changes that seem to serve the end of according the *Comprehensive Epistle* the role that in the original passage pertains to Epistle 50.

The impression that Epistle 50 is indeed the original *Comprehensive Epistle* is confirmed by a further passage. This occurs at the end of Epistle 50, which includes the sole further reference to a "comprehensive chapter" in all of the *Epistles*:

> The sacrifice (*qurbān*) of the Brethren of Purity combines all of these characteristics in their totality, religious (*sharʿī*) and philosophical (*falsafī*) [and follows their traditions (*sunan*) and passes them on to those to whom they pertain and gives them to those who deserve them].[30] Since this action (*fiʿl*) [**chapter** (***faṣl***)][31] comprises the psychical virtues (*kāna jāmiʿan li-l-faḍāʾil al-nafsāniyya*), and we learned that when you follow **its** directions (*imtathalta* ***fīhi***[32] *al-waṣiyya*), your angelic form will be perfected (*kamulat laka al-ṣūra al-malakiyya*), and it will be prepared for you to reach and descend upon it in your afterlife (*kānat laka fī maʿādika*

29 See pp. 81, 94.

30 The bracketed section is omitted in the Beirut edition, and in MSS ح, ز, ک, and ل; the Beirut edition, ر and ن instead contain a lengthy addition that relates this sacrifice to Abraham.

31 In the Beirut edition: "this chapter" (*hādhā al-faṣl*). The critical edition of this epistle gives no manuscript variants that would support the Beirut edition reading.

32 *Fīhi* added in the Beirut edition, ح, ز, ل, and ن.

> *muhayya'a li-wuṣūlika ilayhā wa-nuzūlika bihā/ʿalayhā*), **we concluded the epistle with this chapter** (***khatamnā al-risāla bi-hādhā al-faṣl***) and called this chapter in this epistle [**it** (***sammaynāhu***)[33]] the chapter comprising useful benefits (*al-faṣl al-jāmiʿ li-l-fawāʾid al-nāfiʿa*) [among our comprehensive epistle(s) with diverse benefits (*min rasāʾilinā al-jāmiʿa dhāt al-funūn al-nāfiʿa*)],[34] as it is to the epistles [*minhā*] as heart and head are to the body, and it is the ultimate goal after learning what is contained in them (*huwa nihāyat al-gharaḍ baʿda al-wuqūf ʿalā mā fīhā*), following all that we have stipulated (*al-irtisām bi-jamīʿ mā rasamnā*) and relying on what we have prescribed (*al-iʿtimād ʿalā mā waṣafnā*).[35]

The beginning of the passage refers to the "instructions" (*waṣiyya*) that prepare one for virtues as well as for the afterlife, as was the case in the previous passage. However, now these instructions are, in most manuscripts, attributed to the "sacrifice of the Brethren of Purity,"[36] and only in the Beirut edition variant are they attributed to "this chapter." The most remarkable aspect of the passage, however, is that all variants call it the "chapter that comprises useful benefits" (*al-faṣl al-jāmiʿ li-l-fawāʾid al-nāfiʿa*). Most manuscripts also add "among our comprehensive epistles containing the beneficial branches of knowledge" (*min rasāʾilinā al-jāmiʿa dhāt al-funūn al-nāfiʿa*), which is omitted in the Beirut edition and some manuscripts. Only a single manuscript (MS A) gives the singular "our comprehensive epistle," which in any case makes no sense here. These attributes are particularly notable because they correspond precisely to the attribute in the full title of the *Comprehensive Epistle of Useful Benefits* (*al-Risāla al-jāmiʿa dhāt al-fawāʾid al-nāfiʿa*).[37]

In this case, no variant includes an explicit reference to the *Comprehensive Epistle*. However, the Beirut edition variant again gives the impression of correcting the situation by including a passage which explains that the reference is to the current chapter within an epistle, while it omits the reference to all epistles as "comprehensive."

The conclusion for this and the previously cited passages, then, is that the term "comprehensive" bears various kinds of signifance in the *Epistles*. First, in

33 The variant in the Beirut edition is not supported by any manuscript variant in the critical edition.

34 The bracketed section is omitted in the Beirut edition and in MSS ح, ز, ک, ل, and ن.

35 *Rasāʾil*, vol. 4, 271; *On God and the World*, 288–289.

36 For the different forms or worship and sacrifice in this epistle, see Janne Mattila, "The Philosophical Worship of the Ikhwān al-Ṣafāʾ," *Journal of Islamic Studies* 27, no. 1 (2016).

37 *Al-Jāmiʿa*, ep. 1; ed. Ghālib, 15; ed. Kacimi, 1.

two passages, all epistles are "comprehensive," perhaps in the sense that they are supposed to comprise all knowledge. Second, Epistle 50 is "comprehensive" in the sense that it comprises the benefits of, and is the ultimate goal for, *all* epistles. It seems plausible that these uses were the original impetus for composing a separate *Comprehensive Epistle*.

Epistle 50 finally contains a second explicit reference to the *Comprehensive Epistle*, which occurs in the context of the introduction of the idea of "philosophical worship," that is, a form of ritual worship addressed specifically to philosophers:[38]

> As for the second form of worship, which is philosophical-theological worship (*al-ʿibāda al-falsafiyya al-ilāhiyya*) and consists of acknowledging the unity of God, we have discussed it previously at the beginning of this epistle of ours (*fī ṣadr risālatinā hādhihi*), I mean the comprehensive (*aʿnī al-jāmiʿa*),[39] [in explanation (*sharḥ*) of the epistle on arithmetic],[40] [and the beginning of the epistle on arithmetic also mentions some of it].[41] But the prayers and sacrifices [that are accepted and answered][42] [we saved for this chapter].[43]

Once again, the variants manifest considerable variance. Most manuscripts refer the discussion of philosophical worship to the beginning of "this epistle of ours," which is identified as "comprehensive." A single manuscript contains a variant in which the reference is to "these epistles of ours" in the plural, which are also identified as comprehensive. Only the Beirut edition variant, supported by two manuscripts, contains an explicit reference to the *Comprehensive Epistle*. In most manuscripts, the reference is qualified by a reference to the section or epistle on arithmetic; some manuscripts, however, omit this mention. As it happens, philosophical worship in this sense is discussed neither at the beginning of Epistle 50 nor in the section of the *Comprehensive Epistle* devoted to the epistle of arithmetic, nor even in the epistle of arithmetic itself. That is, unless the reference is to something very general, such as the philosophical

38 See Mattila, "The Philosophical Worship."

39 In MS ح: *ṣadruhā al-risāla al-jāmiʿa*; ق: *ṣadr rasāʾilinā hādhihi aʿnī al-jāmiʿa*; the Beirut edition, ل, ن: *sadr al-risāla al-jāmiʿa*.

40 The bracketed section is omitted in MS أ.

41 The bracketed section is omitted in MS ز.

42 The bracketed section is omitted in MS ع.

43 *Rasāʾil*, vol. 4, 262; *On God and the World*, 266. The bracketed section is omitted in the Beirut edition and MSS ح, ز, ل, and ن.

understanding of the unity of God (*tawḥīd*) discussed in both Epistle 1 and the *Comprehensive Epistle*.[44] As a result, since all the other passages in the fiftieth epistle support the view that the term "comprehensive" is an internal reference within the *Epistles*, it seems likely that this reference also refers to either Epistle 50, or else to all epistles, which are given the attribute "comprehensive," as is suggested by a single manuscript variant.

These passages in Epistle 50 appear to be of extreme importance for understanding the genesis of the *Comprehensive Epistle*. First, every passage related to the *Comprehensive Epistle* contains considerable manuscript variation. The impression is that these passages were edited in some manuscripts once the *Comprehensive Epistle* was introduced as a separate epistle. Second, it seems likely that the term "comprehensive" was initially not an attribute of a separate epistle but of either Epistle 50 or even the *Epistles* as a whole. Third, Epistle 50 consequently must be of special significance for the genesis of a separate *Comprehensive Epistle*.[45] The cited passages accord both the role as the final end of the study of the epistles and the specific attributes pertaining to the *Comprehensive Epistle* to this particular epistle. The subject matter of the epistle is "governance" (*siyāsa*) in a broad sense which includes both character formation and religious worship. Insofar as these serve the end of purification of the soul, it makes some sense to postulate philosophical and religious self-improvement as the ultimate message of the *Epistles*. If this is the case, Epistle 50 might be the original "comprehensive epistle." Later on, this idea was extrapolated into a separate epistle that not only presents an ethical-eschatological goal but also summarizes the contents of each epistle.

Finally, a third group consists of four references to the *Comprehensive Epistle* in Epistle 52b (*Magic*).[46] In each case, the passages in the Beirut edition and MS ج variants are largely identical. Thus, the references are unambiguous in the sense that both variants clearly refer to the *Comprehensive Epistle* as separate from the *Epistles*, and that no variant suggests that the references are later additions. The third reference provides an example:

> Know, O brother, that God's proof among the created beings and His trustee on the Earth over the world of animals is the human form, His

44 See *On Arithmetic & Geometry*, 19–24; *al-Jāmiʿa*, ep. 1; ed. Ghālib, 19–20, 27–29; ed. Kacimi, 8–9, 17–19.

45 This link between the *Comprehensive Epistle* and Epistle 50 is also noted in Baffioni's introduction to the critical edition. See *On God and the World*, 128–131.

46 *Rasāʾil*, vol. 4, 340, 356, 378, 425; MS Atıf Efendi 1681, fols. 545^{b}, 550^{a}, 557^{a}, 570^{b}.

> vice-regent on the Earth over plants and animals. Similarly for the minerals, as we stated in the epistle on the actions of the spiritual beings that the activities and effects of the wide circle forever emerge and become apparent in what is below it. Know that there are virtuous and noble substances in the mineral sphere, and likewise among plants and trees and what emerge and come to be from them. Similarly, among animals there are kings, [nobilities,] and leaders, as we have stated in the *Comprehensive Epistle of Useful Benefits* [our *Comprehensive Epistle*].[47]

Apart from minor variation, the passages are identical, and all variants now clearly refer to the *Comprehensive Epistle*. The *Comprehensive Epistle*'s section on Epistle 22 (*Animals*) is one of the longest, and it treats the nobility of some animals over others in various passages.[48] The reference to the *Comprehensive Epistle* seems superfluous, however, since the same ideas are also expressed in Epistle 22.[49] This is the case for all four references in Epistle 52b; none of them refers to a doctrine conveyed *only* in the *Comprehensive Epistle*, but in each case the reference could equally have been to the *Epistles*. In any case, the contrast to the references in the two previous groups is striking. The result of the examination of these references is that Epistle 52b is the *only* epistle that refers unambiguously to the *Comprehensive Epistle*. It is also the case that neither of the other two variants for Epistle 52 carries a reference to the *Comprehensive Epistle*. The fact that all unambiguous references to the *Comprehensive Epistle* occur in a specific variant of Epistle 52 provides no evidence for the theory of common authorship. This is the case in particular because there are strong indications that Epistle 52b is itself spurious in the sense that it postdates all or most other epistles. Thus, the fact that all unambiguous references to the *Comprehensive Epistle* occur in a probably spurious epistle supports the view that the *Comprehensive Epistle* also postdates the *Epistles*, apart from Epistle 52b. Based on the examination of references, it seems, then, likely that at the time of the composition of the *Epistles*, a separate *Comprehensive Epistle* did not exist. Once this treatise came to form a part of the Ikhwānian corpus, several manuscripts were edited to accommodate its new role as the final part of the *Epistles*.

---

47 *Rasāʾil*, vol. 4, 378; MS Atıf Efendi 1681, fol. 557[a].

48 See, for example, *al-Jāmiʿa*, ep. 22; ed. Ghālib, 170–172, 178–183; ed. Kacimi, 212–214, 223–229.

49 *The Case of the Animals*, 66–86.

3 The *Comprehensive Epistle* and the *Fihrist*

Of the two main arguments for the theory of common authorship, the evidence on interpolations refutes the first argument based on cross-references. The second argument of doctrinal and stylistic unity will be addressed in the next two chapters. Before that, a further piece of evidence concerns the purely textual, as opposed to doctrinal or stylistic, relationship between the two works. It turns out that large sections of the *Comprehensive Epistle* depend on a specific variant of the *Epistles*' introductory section, the *Fihrist*. As shown in Chapter 1, the manuscripts carry at least three variants of the *Fihrist*; the *Comprehensive Epistle* depends on the latest variant (*Fihrist c*). This is also precisely the variant that includes the four references to the *Comprehensive Epistle* that we discussed above. It seems, then, that there is a textual relationship between the *Comprehensive Epistle* and *Fihrist c*. This has consequences for the former text's dating since the *Fihrist* is likely to postdate the *Epistles*, and this is almost certainly the case for *Fihrist c*.

A connection between the *Fihrist* and *Comprehensive Epistle* seems initially surprising; there is no particular reason why the two texts should be related. This relation may be first approached by means of the *Comprehensive Epistle*'s two explicit references to the *Fihrist*.[50] The first reference occurs at the beginning of the section corresponding to Epistle 1 (*Arithmetic*), where the epistle's "intended goal" (*al-gharaḍ al-murād*) is introduced by the phrase: "We stated in the *Fihrist* of the *Epistles*" (*qulnā fī Fihrist al-Rasāʾil*).[51] The summary's division into title and goal sections already suggests that the subsequent text is indebted to the *Fihrist*. The second *Fihrist* reference occurs toward the end of the *Comprehensive Epistle* in its commentary on Epistle 45 (*Companionship of the Brethren of Purity*): "The position (*mawḍiʿ*) of this epistle's commentary (*sharḥ*) within the *Comprehensive Epistle* comes from the *Fihrist* of the *Epistles of the Brethren of Purity* that we composed and delivered to you so that you would know each of its epistles by its characteristic/name (*sima*/*ism*), appellation (*laqab*), and statements (*dhikr*)."[52] In the *Comprehensive Epistle*, Epistle 45 is transferred to the position of the final epistle. Apparently, the author justifies this change by the authority of the *Fihrist*. The claim is puzzling as Epistle 45 does not in fact occupy the final position in any consulted manuscript of either the *Epistles* as a whole or the *Fihrist* in particular. Of course, it is possi-

50 For references from the *Epistles* to the *Fihrist*, see pp. 72–73.

51 *Al-Jāmiʿa*, ep. 1; ed. Ghālib, 18; ed. Kacimi, 6.

52 Ibid, ep. 45; ed. Ghālib, 537; ed. Kacimi, 709.

ble that the author employed a variant of the *Fihrist*, where Epistle 45 was the final epistle. For the present purposes, the main import of both references is that the *Fihrist* was of some importance for the *Comprehensive Epistle*'s author. Kacimi's critical edition reports no manuscript variants in either refence to the *Fihrist*. As a result, it is safe to assume that the *Comprehensive Epistle*'s author was familiar with some variant of the *Fihrist*.

The textual dependence goes much deeper than these two references indicate, however. A textual comparison between the two texts shows that each of the fifty-two epistle commentaries in the *Comprehensive Epistle* makes use of the *Fihrist*. Moreover, each epistle entry depends specifically on *Fihrist c*, and their relation is very similar to that prevailing between the *Fihrist* variants.[53] This relation may be illustrated by the example of Epistle 3 (*Astronomy*), which was employed for the *Fihrist* variants in Chapter 1:

> **Then it is followed by** the third **epistle from the mathematical epistles** [which] **is** an epistle on the stars, and it is like an introduction **and preamble. In it we stated about astronomy that which we made into a token** (*dalāla*) **of that which we pointed** (*asharnā*) **to and intended in the discourse on written signs and appointed tokens of horizons and souls, and heaven and the Earth, among** the composition of the spheres, attributes of the zodiac, and trajectory of the planets, and on the knowledge of their influences on this world, and the manner in which the elements and generated beings are affected by them in origination and decay, in generation and corruption.
>
> The **intended** goal was to arouse desire in the **untainted**, pure, **clean, and luminous** souls (*tashwīq al-nufūs* ***al-zakiyya*** *al-ṣāfiya* ***al-ṭāhira al-nayyira***) **illuminated by the light of wisdom and tokens of knowledge** (***dalāʾil al-maʿrifa***) to ascend to the world of the spheres and the layers of the heavens, **and arrival at the degree of perfection which is the degree of the angels that are separated from physical place, corporeal pleasures, and worldly desires. Then they despise death and desire it, for it is blissful for them and they do not fear it, nor are they saddened to leave their accustomed world and its delights and possessions. For they come to know and become certain that the place to which they will arrive and come is greater in sublimity and higher in degree, and that they will witness what no eye has seen, no ear has heard, and has**

53 For an analysis of this textual relation for two passages, and the full Arabic text, see Mattila, "*Fihrist*."

> **not occurred to human heart. They will rise to the** degrees of spiritual beings and **accompany** the closest angels, highest abode, and supernal substances; **they will join the prophets, messengers, believers, knowers, witnesses, and the virtuous who preceded them, and** arrive at the **sphere** of sanctity and trusted spirit (*al-wuṣūl ilā* ***sāḥat*** *al-qudus wa-l-rūḥ al-amīn*), **and the proximity of the Lord of the worlds in spirit, bliss, paradise, and felicity.**[54]

In this passage, the plain text is shared by *Fihrist c* and the *Comprehensive Epistle*; the bolded words indicate the latter text's additions. In the original Arabic, the *Comprehensive Epistle* retains the text of *Fihrist c* in its precise wordings with only minor modifications. The shared text is expanded by words and phrases inserted between the shared text. This is precisely the same method by which *Fihrist c* expanded *Fihrist b*, and *Fihrist b* expanded *Fihrist a*. As a result, we have a chain of four texts, each of which expands an earlier variant while retaining the wordings of the original passage.

The additions in the *Comprehensive Epistle* are much more substantive than those between the *Fihrist* variants; as a result, they radically change the tone of the passage. In *Fihrist a*, both the title and goal sections were extremely concise, so that the former presented Epistle 3 as an introduction to astronomy dealing with the spheres, the zodiac, and the planets; the latter presented its goal as arousing in the human soul a desire to ascend toward the world of the spheres. In *Fihrist b* and *c*, in the title section, the content of the epistle is expanded to include the celestial influences on our world; in the second section, the world of the spheres is qualified as being the abode of spiritual beings and angels. In each variant of the *Fihrist*, then, the content of Epistle 3 is identified as being essentially scientific in nature, even if the astronomical content is related to the inner meaning of inciting a desire for the soul's ascent. At the beginning of the cited passage, the additions in the *Comprehensive Epistle* make the astronomical content reveal signs of "horizons and souls" (*al-āfāq wa-l-anfus*), an expression frequently employed in the work, as we will see in the next chapter. The additions of the latter part expand on the goal related to the soul's ascent with its eschatological aspects so that it now constitutes most of the passage, whereas in the formulation of *Fihrist a* it was a brief sentence. As a result of the successive edits, the original description of Epistle 3 as an essentially technical treatise is completely transformed into an essentially esoteric treatise. Since

54 *Al-Jāmiʿa*, ep. 3; ed. Ghālib, 95–96; ed. Kacimi, 102–103. The corresponding *Fihrist* passage is at *Rasāʾil*, vol. 1, 22.

the cited passage constitutes the entire section devoted to Epistle 3 in the *Comprehensive Epistle*, almost all of the technical content of the epistle is gone from the work's representation of that epistle.

Epistle 17 (*Generation and Corruption*) provides another example.[55] In *Fihrist a*, the description of its contents corresponds more or less to Aristotle's *On Generation and Corruption*, with the added Neoplatonic element of the agency of the Universal Soul. *Fihrist c* relates the Universal Soul to angels, and thus makes the passage slightly more religious in character. The extensive additions in the *Comprehensive Epistle* transform the original context of Aristotelian physics into a detail of its larger esoteric meaning. This involves various new inner meanings ascribed to the epistle that are also omnipresent in the *Comprehensive Epistle*, such as the release of particular souls from the material world; the "great error" of Adam that resulted in the particular souls' descent; the souls' imprisonment in nature; and the prophetic cycles of manifestation and occultation. As in the case of Epistle 3, the entry for Epistle 17 only contains a section based on the *Fihrist*. Thus, the concise entry in the *Comprehensive Epistle* provides no further explanation of exactly how all of these themes relate to the original epistle.

The *Comprehensive Epistle*'s author applies precisely the same method for each of its fifty-two epistle entries. Each commentary starts with a section that depends on the corresponding epistle entry in *Fihrist c*. None of the epistle commentaries simply reproduces the original text unchanged, but each expands on it with additions of varying length. For some epistles, the *Comprehensive Epistle* only inserts occasional words or phrases without substantially changing the *Fihrist c* entry's content. In most cases, the additions are more substantial, sometimes amounting to several edited pages. In these cases, the end result is completely different in content from the original.

As a result, extensive sections of the *Comprehensive Epistle*, including the entire entry for some epistles, such as Epistles 3 and 17, depend textually on *Fihrist c*. Why should there be such a relationship between the *Comprehensive Epistle* and *Fihrist c*? Both texts could have been authored by the same person. This does not seem likely, however, since the *Comprehensive Epistle* gives very different meanings to the epistles from those of *Fihrist c*. A more likely answer is that the *Comprehensive Epistle*'s author employed *Fihrist c* as the base text for composing the treatise. This makes sense: the *Fihrist* is itself a summary of the

---

55 *Rasāʾil*, vol. 1, 27; *al-Jāmiʿa*, ep. 17; ed. Ghālib, 147–148; ed. Kacimi, 171–172. The Arabic text and English translation of the corresponding passages in the *Fihrist* and the *Comprehensive Epistle* are cited in Mattila, "*Fihrist*."

Epistles; thus, it provides the perfect starting point for writing a more extensive summary. For each epistle, the *Fihrist* text was expanded by adding elements that conveyed the author's esoteric understanding of its subject matter. The method of editing still seems curious. Why should the author retain the precise wordings of the original, even in cases where, due to extensive additions, the end result is unrecognizable? Perhaps each expansion is the result of marginal notes in a manuscript, and these were later incorporated into the main text. A more far-fetched explanation is that the author accorded special value to the original wordings, perhaps because the original text was attributed, for example, to a Shīʿīte imām. In the manuscript tradition, Ikhwānian texts are often attributed to imāms.

This textual relation provides a new tool for comparing the *Comprehensive Epistle* with the *Epistles*. First, each epistle entry in the *Comprehensive Epistle* can now be analyzed into two separate sections: (1) a *Fihrist* passage and (2) the author's independent contribution. For many epistles, such as Epistle 3, the entry consists exclusively of the first section. It seems that, in these cases, the author deemed an expanded *Fihrist* entry to provide a sufficient commentary for that epistle. In the case of other epistles, the author followed the *Fihrist* section with an independent discussion of the subject. Clearly, the author viewed these epistles to be of greater importance.

As a result, the epistle entries can be divided into two categories based on whether they include a second, independent, section. In general, as shown in Table 18, the epistle commentaries are of very different lengths. The two longest ones are devoted to Epistles 1 (*Arithmetic*) and 22 (*Animals*), and span seventy-five and seventy pages, respectively, in the Ghālib edition. Most epistle entries, in contrast, are very concise, often consisting of a single edited page. This discrepancy now receives a technical explanation: the shortest epistle commentaries consist solely of the first section based on the *Fihrist*. Among the fifty-two entries in the *Comprehensive Epistle*, eighteen only contain the first section; the remaining thirty-four epistle entries complement the *Fihrist* section with an independent section. Most of the scientific epistles, for example, pertain to the former group. It seems safe to assume, then, that the author of the *Comprehensive Epistle* did not view them to be of central importance.

Second, the title and goal sections of the *Comprehensive Epistle* can now be compared with those provided in the *Fihrist*. In the case of Epistles 3 and 17, we have seen that the additions transformed a mainly technical description of their contents in the *Fihrist* to a mainly esoteric one in the *Comprehensive Epistle*.

Tables 6 and 7 summarize the themes of the title and goal sections, respectively, in the *Fihrist* variants versus the *Comprehensive Epistle*. Overall, the latter work's additions may be divided into two larger categories with regard to the extent of their expansions. The first category consists of neutral additions, which qualify and expand the themes mentioned in *Fihrist c* without attributing essentially novel subjects to the epistle. Thus, for example, the title section of Epistle 11 (*Categories*), in both *Fihrist* variants, attributes to the epistle the subject matter of ten Aristotelian categories; the *Comprehensive Epistle* expands this by providing a list of the ten categories. Many additions are more complex in that they go beyond the subject matter provided in *Fihrist c* but are still restricted to a more complete description of that epistle. For example, the *Fihrist a* entry for Epistle 31 (*Differences of Languages*) states that its subject matter consists of differences of languages; *Fihrist c* complements this with scripts, religions, and doctrines; and the *Comprehensive Epistle* adds the celestial causes for the transformation of doctrines. All these subjects are indeed discussed in this epistle.

The epistle entries in the second category, in contrast, introduce a clearly novel interpretation to the one provided in *Fihrist c*. Many of these additions have an esoteric flavor in the sense that they expand the original entry with its alleged inner meaning. Four themes appear particularly prominent in these additions. First, theological additions relate an epistle to God, divine providence, cosmological hierarchy, and the nature of the spiritual world. A second group relates to the eschatological themes of liberation from matter, the soul's ascent, paradise and hell, and resurrection and "second creation." A third group consists of additions related to prophets, imāms, and the cycles of occultation and manifestation. A fourth group consists of additions that encourage the contemplation of God's signs in the world. Such additions appear particularly conspicuous when they are attributed to Aristotelian or technical treatises, as was the case with Epistles 3 and 17.

As regards the title entries (Table 6), in the *Fihrist*, these mostly present a neutral description of an epistle's subject matter. The additions in the *Comprehensive Epistle* often esoterize the description. Thus, all *Fihrist* variants attribute to Epistle 5 (*Music*) the subject matter of music, its effect on the soul, and "the music of the spheres," while the *Comprehensive Epistle* inserts an elaborate passage that discusses the spiritual world and the soul's ascent. As for the goal sections (Table 7), *Fihrist c*, in particular, in many cases already contains a reference to an epistle's inner meaning; but the *Comprehensive Epistle*'s additions both expand these and introduce novel esoteric readings where none were present in the *Fihrist*. Thus, for Epistle 14 (*Posterior Analytics*), in *Fihrist a*, the goal is knowledge about syllogisms and demonstration; in *Fihrist c*, it is also the "balance of justice" (*al-wazn bi-l-qisṭ*). The *Comprehensive Epistle* expands these by adding items that appear surprising in this logical context: the mysteries of creation, the afterlife, and the necessity of following a guide (*murshid/hādī*). It is true that many of these additions fit well doctrinally with the original *Epistles*. But the additions in the *Comprehensive Epistle* clearly greatly expand the presence of Ismāʿīlī terms and ideas in comparison with the epistle summaries of the *Fihrist*. Moreover, the additions shift the perspective by reading the epistles as primarily esoteric in content.

A comparison between the *Comprehensive Epistle* and the *Fihrist* thus gives the surprising result of strong textual dependence: the beginning of all fifty-two, and the entirety of eighteen, epistle entries depend on *Fihrist c*. In consequence, a considerable part of the *Comprehensive Epistle* employs the *Fihrist* as its source text. Thus, if Epistle 50 might have provided an impetus for composing a separate comprehensive epistle, the *Fihrist* provided the author with a textual source. This is of obvious relevance for the dating of the *Comprehensive Epistle*. Clearly, the *Comprehensive Epistle* must postdate *Fihrist c*. Since *Fihrist c* further postdates *Fihrist a*, and even *Fihrist a* is likely to postdate the *Epistles*, this further supports the view that the *Comprehensive Epistle* is a later addition to the corpus. This clearly contradicts the theory of common authorship. As a final note, the *Comprehensive Epistle* employs specifically the *Fihrist* variant for which the only testimony so far is the Beirut edition. The manuscript source of this edition is probably of Ismāʿīlī provenance.[56] Thus, this textual dependence provides further evidence for the claim that the *Comprehensive Epistle* was composed in an Ismāʿīlī context. In the next chapters, we will see whether doctrinal grounds confirm this conclusion.

56 See p. 33.

CHAPTER 5

Structure of the *Comprehensive Epistle*

The second argument for the theory of common authorship is that the two works are of sufficient doctrinal and stylistic unity to have been composed by the same authors. The evidence presented in the previous chapter concerned the textual relation between the *Epistles* and the *Comprehensive Epistle* regardless of their doctrinal content. The doctrinal aspect was only touched on in the *Comprehensive Epistle*'s additions to the *Fihrist*. Most of the ideas expanding on the *Fihrist* are also present in the *Epistles*. Thus, they do not necessarily prove that the *Comprehensive Epistle* is doctrinally different from the *Epistles*. After all, the *Comprehensive Epistle* claims to provide the inner meanings contained implicitly in the *Epistles*; it should therefore add novel content to its explanation of the epistles. A proper disproof of the second argument is to show that the *Comprehensive Epistle* diverges from the doctrinal positions of the *Epistles*. This task is pursued in Chapters 5 and 6: the next chapter will focus on specific doctrinal positions; the current chapter addresses the structure and themes of the *Comprehensive Epistle*.

Besides the theory of common authorship, both chapters relate to this book's narrative of the *Epistles*' gradual evolution from a philosophical encyclopedia to a theological treatise. The first stage of that evolution took place within the redaction of the *Epistles*; the *Comprehensive Epistle* represents a second stage. The questions of this chapter are the following: how does the *Comprehensive Epistle*'s structure compare with the arrangement of the *Epistles*; what does this reveal about the relationship between the two works? That is, the inquiry concerns the precise ways in which the *Comprehensive Epistle* employs and elaborates on the *Epistles*, and whether these support or contradict the theory of common authorship. I will address this problem in two sections which discuss (1) the epistle commentaries and their relation with the original epistles and (2) the major themes of the *Comprehensive Epistle*.

1 Epistle Commentaries

Starting with the general structure of the *Comprehensive Epistle*, as already mentioned, the epistle commentaries are of very unequal scope (see Table 18). The epistles may be classified into four groups based on their length in the Ghālib edition: (1) more than seventy pages: Epistles 1 (*Arithmetic*) and 22 (*Ani-*

mals); (2) more than twenty pages: Epistles 25–26, 36, and 38; (3) ten to twenty pages: Epistles 9, 14, 30–32, 34–35, 37, 39–40; and (4) less than ten pages: the remaining thirty-six epistles. It is, then, clear that the *Comprehensive Epistle* is not a balanced commentary of all the epistles but a highly selective one. As shown in the previous chapter, the discrepancy is in part due to the fact that many epistle entries consist solely of an expanded *Fihrist* entry with no genuine commentary. This raises the question of the grounds of the author's choice to focus on certain epistles and ignore others.

An inspection of these groups immediately reveals that the author's primary interest does not lie in the more technical epistles. Apart from three exceptions (Epistles 1, 14, and 22), the sections related to mathematics, logic, and physics, that is, the first half of the work roughly corresponding to the Aristotelian curriculum, pertain to the very shortest in the *Comprehensive Epistle*. Moreover, even in the three cases, their technical aspects are almost completely faded out in the commentary. A preliminary conclusion is, then, that the *Comprehensive Epistle*'s author was not particularly interested in the scientific stratum of the *Epistles*. This is not a surprising conclusion, however, nor does it necessarily contradict the theory of common authorship. After all, the *Comprehensive Epistle* is supposed to reveal the inner meanings of the *Epistles*; supposedly, these meanings have little to do with the technicalities of number theory or Aristotelian logic, for example. It therefore makes perfect sense that the technical aspects of the *Epistles* are given the least emphasis.

Still, the author's lack of interest in scientific subjects appears to be contradicted by the fact that by far the longest sections in the *Comprehensive Epistle* are devoted to Epistles 1 (*Arithmetic*) and 22 (*Animals*). In both cases, however, there are factors that explain why this is the case. As regards the first epistle, the length is explained, first, simply by the fact that the commentary on this epistle initiates the *Comprehensive Epistle*. Thus, the section is an introduction to the work, and therefore addresses the central themes of the *Comprehensive Epistle* in general, whether or not they are relevant for the subject of arithmetic. This is true, in particular, of the esoteric interpretation of the story of Adam, which occupies much of the section's latter part.[1] Second, the weight given to this epistle is also due to the subject of arithmetic in the Pythagorean sense of the ontological and epistemic priority of numbers, which is prominent in both works.[2] The idea is that natural numbers correspond to ontological levels of reality, and

1 *Al-Jāmiʿa*, ep. 1; ed. Ghālib, 65–90; ed. Kacimi, 61–92.

2 See *On Arithmetic & Geometry*, 8–9, 11–12, 22–24; *al-Jāmiʿa*, ep. 1; ed. Ghālib, 18–21, 24–31; ed. Kacimi, 7–10, 14–21.

arithmetic thereby represents the proper starting point for knowledge. Insofar as all numbers are derived from number one, which corresponds to God, arithmetic furthermore guides the student toward an affirmation of God's oneness and unity (*tawḥīd*).[3] For the *Comprehensive Epistle*'s author, this presumably constitutes the esoteric meaning of the epistle on arithmetic. Even here, the lack of interest in the scientific content is affirmed; besides the Pythagorean idea, the *Comprehensive Epistle* completely ignores the specifics of number theory present in the *Epistles*.

In the case of Epistle 22, the *Comprehensive Epistle* shows some interest in the specifically zoological subject matter as regards various animal shapes and kinds or the position of animals within the chain of being.[4] But the author's main focus is again on esoteric themes that have at best a tangential relation with the subject of animals. Thus, the commentary mainly addresses human beings as regards their position between animals and angels; the presence of animal traits in human beings; human beings' microcosmic nature; their status as God's vice-regents on Earth; as well as celestial causation of earthly events and the descent and reascent of being. Beyond this, the author is clearly fascinated by the animal fable, for which a metaphorical interpretation is offered.[5] The author also elaborates on the section of the story that concerns enmity between humans and *jinn*s.[6]

These epistle entries thus suggest that the *Comprehensive Epistle* has a rather loose relation with the epistles on which it comments. Examples of two other entries further illustrate this relation. Epistle 14 (*Posterior Analytics*) is the third epistle with technical subject matter that is given any weight in the *Comprehensive Epistle*. This seems a curious choice since the Aristotelian theory of demonstration does not seem to fit with the author's esoteric inclinations. When the epistle and its commentary are compared, it turns out that the contents of the *Comprehensive Epistle* are even less related to the original epistle than was the case with Epistles 1 and 22. Epistle 14 is by no means faithful to the Aristotelian work on which it is based.[7] But its subject matter is still more or less contained within the confines of the theory of demonstration. The beginning of the treatise addresses four methods for acquiring knowledge: (1) division (*taqsīm*), (2) analysis (*taḥlīl*), (3) definition (*taḥdīd*), and (4) demonstration

3 *Al-Jāmiʿa*, ep. 1; ed. Ghālib, 29–31; ed. Kacimi, 19–21.

4 See Ibid, ep. 22; ed. Ghālib, 155–156, 215–221; ed. Kacimi, 190–191, 274–283.

5 Ibid, ep. 22; ed. Ghālib 191–215; ed. Kacimi, 241–273.

6 Ibid, ep. 22; ed. Ghālib, 193–198, 204–214; ed. Kacimi, 244–251, 258–272.

7 See Baffioni's introduction in *On Logic*, 16–24.

(*burhān*).[8] From there, the treatise proceeds to discuss causes for error in syllogistic and analogical reasoning, the relation of sense knowledge to intelligible knowledge, sound and unsound syllogisms, the first intelligible principles, and other aspects related to the main subject.[9] The treatise drifts away from logic only in its final chapters, first, by discussing the doctrines of void versus plenum and pre-eternity versus the createdness of the world.[10] Even here, these presumably represent case studies of why philosophers have sometimes arrived at faulty conclusions in their reasoning.[11] Only at the very end does the epistle introduce clearly religious subjects: the great resurrection (*al-qiyāma al-kubrā*) occurring at the end of time, the necessity for the soul's purification, the human being's perfection as resemblance with angels, and the soul's ascent to the spiritual world.[12] These are stock themes that permeate the *Epistles*, yet their introduction is motivated by the treatise's actual subject. The great resurrection is related to the temporal finitude of the world; demonstrative knowledge, along with purification of the soul, is what enables human beings' resemblance with angels and thus ascent to the spiritual world. While Epistle 14, then, contains a religious or esoteric layer, as do all epistles,[13] these are limited to concise sections within an otherwise technical discussion. Moreover, the esoteric themes are to some extent at least connected with the treatise's subject.

In contrast, in the section devoted to Epistle 14 in the *Comprehensive Epistle*, the subject of demonstration is almost entirely absent. Only at the beginning of the epistle commentary, are the terms 'syllogism' or 'demonstration' even mentioned.[14] This is the section that expands on *Fihrist* c; thus, the inclusion of these terms is due to their presence in the *Fihrist*. Even in the *Fihrist* section, however, demonstration becomes a minor detail once the *Comprehensive Epistle* expands the epistle's goal by introducing the additional aspects of (1) the examination of "horizons and souls," (2) the mysteries of creation, (3) the resurrection and second creation, and (4) the necessity of a divinely inspired guide.

8 *On Logic*, 129–139.

9 Ibid, 139–184.

10 Ibid, 184–190.

11 In Epistle 42 (*Opinions and Doctrines*), faulty inferences are given as one reason for fallacious beliefs, such as the belief in the world's pre-eternity. For this epistle, see Mattila, "The Ikhwān al-Ṣafāʾ on Religious Diversity."

12 Ibid, 189–192.

13 See pp. 9, 11–12 above.

14 *Al-Jāmiʿa*, ep. 14; ed. Ghālib, 131–132; ed. Kacimi, 147–148.

In the textually independent part of the epistle commentary, the focus is on the second of these inner meanings; it consists of two chapters on spiritual and corporeal creation (*al-khalq al-ruḥānī/jismānī*), respectively.[15] The chapters are introduced by a bridge from the epistle's subject to the discussion that follows: "For we set this epistle before others (*qaddamnā*), and made them [the epistles?] introductions (*muqaddimāt*) to what comes after, in imitation of God's wisdom … For He created all things at once in potentiality in His first and most perfect creation (*ibdāʿihi al-awwal wa-khalqihi al-akmal*); then He brought them, one by one, from potentiality to actuality."[16] The idea is, then, perhaps that just as corporeal creation is contained potentially in God's first creation of the Intellect, so Epistle 14 contains in potentiality the doctrine of spiritual and corporeal creation that the *Comprehensive Epistle*'s author has decided to address in this context. The actual discussion presents the emanationist theory of creation from God downwards. This is done, in particular, in terms of the two primary hypostases of Intellect and Soul, which are identified with the Ismāʿīlī terms "Preceder" (*sābiq*) and "Follower" (*tālī*). The second chapter presents the order of creation in the sublunar world from minerals upwards as the inverse of that of the spiritual world; addresses religion (*sharīʿa*) in terms of its outer (*ẓāhir*) and inner (*bāṭin*) aspects, which are related to the spiritual and corporeal spheres, respectively; as well as giving the order of scientific knowledge proceeding from mathematics onwards. The common theme that presumably ties all of this together is the progression from potentiality to actuality: the Intellect contains all creation in potentiality; similarly, the basic creed (*shahāda*) of divine unity and prophecy contains in potentiality all religion, and mathematics embraces every science.

Thus, the discussion in the *Comprehensive Epistle* bears almost no relation to the contents of Epistle 14 that the author is supposed to comment on. It is not only the case that the author omits its technical content concerning the theory of demonstration; it is also the case that the religious subjects introduced at the end of Epistle 14 are similarly ignored. That is, even the inner meanings ascribed to this epistle are different in the *Epistles* versus the *Comprehensive Epistle*. In the former, they relate to the soul's eschatological end; in the latter, to the order of creation. As a result, the section devoted to Epistle 14 in the *Comprehensive Epistle* seems entirely independent from the treatise on which it is supposedly based.

Moreover, apart from the tenuous bridge sentence joining the *Fihrist* section with the independent discussion, the two sections of the epistle commentary

15 Ibid, ep. 14; ed. Ghālib, 132–142; ed. Kacimi, 148–161.

16 Ibid, ep. 14; ed. Ghālib, 132; ed. Kacimi, 148.

are also independent from each other. The first section is textually based on *Fihrist c* but adds four inner meanings to its explanation of Epistle 14. Of these, only the "mysteries of creation" are addressed in the commentary; the three other subjects' relation with Epistle 14 is never explained.

Epistle 25 (*Fall of the Sperm*) may be provided as another case study.[17] The epistle's main subject is the human embryo's development in the womb within an astrological context.[18] The epistle is initiated by a general discussion of the planets and their spiritual powers, as well as their effects on sublunar beings as intermediaries of the Creator and the Universal Soul.[19] Most of the treatise is concerned with a technical discussion of the specific effects of each planet on an embryo's development during each month of pregnancy, as well as on a human being at birth and during different stages of human life.[20] Beyond this, the treatise presents an analogy between an embryo in the womb and a human soul in the material world: just as the embryo stays in the womb in order to become sufficiently developed to enter this world, so the human soul stays in our world to prepare itself for the supernal world.[21] Presumably, this is the treatise's inner meaning, which it occasionally extends by addressing eschatology, the contrariety between the lower and supernal worlds, and the necessity of prophecy for the soul's purification.

In the *Comprehensive Epistle*, the *Fihrist* section is considerably expanded to emphasize the treatise's esoteric aspects.[22] *Fihrist c* states that Epistle 25's subject matter consists of the fall of sperm; the soul's attachment to the body; the body's transformation in the womb; and the spheres' effects on bodily composition. The epistle's goal is to explain a particular soul's state before its attachment to a body, and to show that the reason why bodies stay in the womb is to perfect their forms. The *Comprehensive Epistle* expands the title section by introducing the subject of the soul's descent and reascent, and thus its eschatological fate. Moreover, the author inserts a lengthy statement claiming that the reason for which the epistle was composed is to show the human being's transient nature in this corporeal world, and that one's fate in the afterlife is determined by the state attained in this life. In the goal section, the *Compre-*

17 *Rasāʾil*, vol. 2, 417–455.

18 For the astrological theory of embryology in the *Epistles* and other texts, see Charles Burnett, "The Planets and the Development of the Embryo," in *The Human Embryo: Aristotle and the Arabic and European Traditions*, ed. G.R. Dunstan (Exeter: University of Exeter Press, 1990), 95–112.

19 *Rasāʾil*, vol. 2, 417–420.

20 Ibid, 421–455.

21 Ibid, 453–454.

22 *Al-Jāmiʿa*, ep. 25; ed. Ghālib, 232–236; ed. Kacimi, 298–304.

hensive Epistle's expansion mainly concerns the Sun's pivotal role during the fourth month of pregnancy with regard to the spiritual powers flowing into a body. None of this really distorts the original epistle's intentions, even if the emphasis is now firmly in the big picture rather than the specifics of astrological embryology.

The independent section in the *Comprehensive Epistle* drifts far from the original subject. First, the author presents a prolonged injunction on the importance of secrecy.[23] The author is concerned that the treatise might end up in the hands of those who reject it due to a misunderstanding of its contents. For this reason, the author was obliged to conceal its secrets from the undeserving. Thus, secret letters, along with a conversion key, are discussed, although it is not clear where these are used. In the subsequent two chapters, the underlying theme is the idea of two "faces" or "aspects" (*wajh*) related to the supernal and sublunar worlds.[24] The first of these chapters addresses the cosmological hierarchy, and then revelation in the form of a succinct account of prophets from Adam to Muḥammad. The latter chapter is concerned with the two "sides" (*ṭaraf*) that the soul faces, again set in the cosmological context of the spiritual and material aspects of creation.[25] The commentary is concluded by three short chapters more closely connected with the treatise's original subject matter.[26] The first deals with the astrological context of the embryo's development, in particular the Sun's role; the second focuses on the analogy between an embryo in the womb and a particular soul in the material world; and the third on the Sun and Moon as "spiritual parents" for the soul's "spiritual birth" (*al-wilāda al-rūḥāniyya*).

In sum, the *Comprehensive Epistle* again omits the epistle's technical subject matter, that is, the astrologically determined phases in an embryo's development, apart from the final section emphasizing the Sun's role in this process. As regards the epistle's inner meaning, the original treatise and its commentary agree that it concerns the analogy between an embryo in the womb and a particular soul in this world. However, the *Comprehensive Epistle* accords this theme surprisingly little emphasis; instead, it focuses on the emanationist account of creation and the position of cosmic and particular souls within a cosmological context. In this case, there is less of a rift between the commen-

23 Ibid, ep. 25; ed. Ghālib, 236–239; ed. Kacimi, 304–308.

24 Ibid, ep. 25; ed. Ghālib, 239–251; ed. Kacimi, 309–325.

25 For the idea of soul's "two faces" in Avicenna and Ismāʿīlī thought, see Daniel De Smet, "La doctrine avicennienne des deux faces de l'âme et ses racines ismaéliennes," *Studia Islamica* 93 (2001).

26 *Al-Jāmiʿa*, ep. 25; ed. Ghālib, 251–254; ed. Kacimi, 326–332.

tary's two sections, as the theme was included in the *Comprehensive Epistle*'s expanded *Fihrist* section. This is not the case with the sections on secret letters and the Sun and Moon as spiritual parents, however, which again seem to bear no relation with either the original epistle or its alleged inner meanings as presented in the expanded *Fihrist*.

These examples show that the relationship between the original epistles and their commentaries in the *Comprehensive Epistle* is at best loose. Since each commentary starts with a section based on the *Fihrist*, there is a textual relation between the beginning of that commentary and the *Epistles*. The discussion in the independent section should presumably explain the inner meaning of the corresponding epistle. Sometimes this is the case: a metaphorical explanation is given to the animal story in Epistle 22; numbers are related to levels of reality; an embryo's stay in the womb is presented as an analogy of a soul in the world. Apart from the first, however, these ideas were already present in the *Epistles*. For the most part, the discussions seem to address themes that have no particular relation with the epistle in question even in the sense of explaining their inner meaning. Nor is it often related to the epistle's alleged meaning in the *Epistles*. This is also the impression one gets from the lack of continuity between the two commentary sections: the first section attributes several inner meanings to an epistle; the second section mostly discusses entirely unrelated subjects. None of this supports the theory of common authorship; the epistles and their commentaries do not seem to reflect a single author or group who first composed the epistles and then explained their meaning in a separate epistle.

2 Central Themes

Apart from individual epistles and their commentaries, the relation between the two works may be approached at the level of the *Comprehensive Epistle* as a whole. As was shown above, the work's division into epistle commentaries does not really prevent the author from discussing subjects with no clear relation to the epistle in question. As a result, a number of themes assume a prominent role throughout the *Comprehensive Epistle*. If we accept the theory of common authorship, these represent the ultimate meaning of the *Epistles*; if not, they instead represent the *Comprehensive Epistle*'s core message, which might be different from that of the *Epistles*.

Most of these central themes are also present in the *Epistles*. In the introduction, many such subjects were attributed to the esoteric layer of the original epistles: the Pythagorean correspondence between levels of reality and numbers; descent of being from God; the soul's descent into matter and reascent into the spiritual world; prophecy; cycles of occultation and manifestation; and Judgment Day. Since the *Comprehensive Epistle* mostly dispenses with the *Epistles'* scientific content, its focus is on these esoteric subjects. A closer examination of the *Comprehensive Epistle*'s discussion of these themes seems to contradict its function as the final epistle in two ways. First, some themes are prominent in both works, but the *Comprehensive Epistle* in fact offers nothing new to the explanations provided in the *Epistles*. Second, other themes are more prominent in the *Comprehensive Epistle*, but its explanations do not align with those in the *Epistles*.

The doctrine of resurrection (*baʿth*/*qiyāma*) is an example of the first category. For many epistles, the *Comprehensive Epistle*'s author states explicitly that its subject constitutes a core doctrine. This is also the case for resurrection: knowledge about its reality is the "ultimate goal" (*al-gharaḍ al-aqṣā*/*al-ghāya al-quṣwā*) for all epistles.[27] But since this is "hidden knowledge" (*ʿilm ghāmiḍ*) and a "subtle mystery" (*sirr laṭīf*), the previous epistles could only explain a "part of it" (*ṭarafan minhu*) and point to it by "symbols" (*rumūz*); only in the *Comprehensive Epistle*, the truth is expressed "openly" (*bi-l-taṣrīḥ*).[28] The claim that the *Epistles* only explain a part of the doctrine of resurrection is also present in Epistle 38 (*Resurrection*).[29] In this epistle, the explanation of resurrection consists of knowledge of the soul's spiritual and heavenly substance;[30] the bodily and spiritual afterlives of good and bad souls, respectively;[31] and resurrection of bodies and souls.[32] The point of the soul's spiritual afterlife is further highlighted by a number of allegorical stories, which is perhaps what is meant by symbolic allusions. As for complete knowledge about this "subtle mystery," readers are guided to seek the savants of this branch of knowledge, "our noble and virtuous brethren" (*ikhwānunā al-kirām al-fuḍalāʾ*).[33]

The *Comprehensive Epistle* makes the same point concerning resurrection as the soul's post-bodily life, and of punishment for bad and reward for good souls.[34] But this is now firmly attached to the notion of Judgment Day, a description of which covers most of the commentary.[35] It is, then, true that the *Compre-*

27 *Al-Jāmiʿa*, ep. 38; ed. Ghālib, 428–429; ed. Kacimi, 563.
28 Ibid, ed. Ghālib, 429, 435–436; ed. Kacimi, 563, 570–571.
29 *Rasāʾil*, vol. 3, 288.
30 Ibid, 288–290.
31 Ibid, 290–299.
32 Ibid, 301–302.
33 Ibid, 302–303.
34 *Al-Jāmiʿa*, ep. 38; ed. Ghālib, 436–437; ed. Kacimi, 572–573.
35 Ibid, ep. 38; ed. Ghālib, 437–453; ed. Kacimi, 573–592.

hensive Epistle complements the *Epistles* by its focus on Judgment Day. However, both works agree on the meaning of resurrection: a small resurrection as the soul's spiritual ascent or return to the body; and a great resurrection of the Universal Soul's return to the Intellect and the destruction of the material world.[36] Thus, the *Comprehensive Epistle* does not seem to reveal a secret that was left implicit in the *Epistles*. In any case, there is no apparent reason to conceal this particular notion of Judgment Day.

Another example is the concept of microcosm (*al-ʿālam al-ṣaghīr*), that is, the idea that the human soul or body is in some sense parallel to the macrocosm of the universe.[37] In the *Comprehensive Epistle*'s commentary on Epistle 9 (*Ethics*), the author states that the explanation of this idea was "one of our greatest goals" (*akthar al-gharaḍ minnā*) for the composition of the treatise.[38] This is a surprising claim since the *Epistles* devote two whole treatises to the subject. In the commentary on the main microcosmic treatise (Epistle 26), the claim is instead that the aim was merely to complement the *Epistles*' discussion by "jots of wisdom (*nukat min al-ḥikma*) and benefits of grace (*fawāʾid min al-niʿma*)," as well as by providing "additional clarity" (*ziyādat al-sharḥ wa-l-bayān*).[39] In the commentary on the second microcosmic epistle (Epistle 34), the author states that the fact that the human being is a "summary" (*mukhtaṣar*) of the cosmos was shown in Epistle 26, but that it should be explained here that the constitution (*bunya*) of the world resembles the human being.[40] This explanation was of course already contained in the *Epistles*. However, the *Comprehensive Epistle* adds an additional element: God is a Writer, and the world is a "preserved tablet" (*al-lawḥ al-maḥfūẓ*) that contains the forms of things for those who can read His signs. The microcosm then becomes a special case of divine signs in the cosmos. However, the *Comprehensive Epistle* adds nothing substantive to the doctrine of microcosm in the *Epistles*; the minor additions merely relate the doctrine to a religious term prevalent in Ismāʿīlī discourse. As in the case of resurrection, the only difference is, then, that the *Comprehensive Epistle* shifts the discussion to a more explicitly religious context.

As for themes that are clearly more prominent in the *Comprehensive Epistle*, the notion of divine signs present in the cosmos is a prominent instance.

---

36 For the notion of resurrection in the *Epistles*, see de Callataÿ, *Ikhwan al-Safa'*, 28–33; Marquet, *La philosophie des Iḫwān al-Ṣafāʾ*, 383–406.

37 For microcosmism in the *Epistles*, see Nokso-Koivisto, "Microcosm-Macrocosm Analogy."

38 *Al-Jāmiʿa*, ep. 9; ed. Ghālib, 119; ed. Kacimi, 132.

39 Ibid, ep. 26; ed. Ghālib, 256; ed. Kacimi, 335.

40 Ibid, 34; ed. Ghālib, 347; ed. Kacimi, 462.

An exhortation to reflect on such signs appears often enough in the *Comprehensive Epistle* to constitute its doctrinal feature. The general idea is based on the Qurʾān: "We will show them our signs in the horizons and their souls so that the truth will become clear to them."[41] This precise expression of "signs" (*āyāt*) present in "horizons" (*āfāq*) and "souls" (*anfus*) appears throughout the *Comprehensive Epistle*. Moreover, the author presents the explanation of divine signs as a central inner meaning that the treatise aims to convey:

> Since we named this epistle the *Comprehensive* [*Epistle*], we must gather to it meanings (*maʿānin*) by concise discourse and succinct description. We explain in it what we left unclear (*aʿjamnāhu*) in our previous epistles that are at your disposal … In this epistle, we wish to explain by clear tokens (*bi-wāḍiḥ al-dalāla*) the intended goals (*al-aghrāḍ al-maṭlūba*) that we alluded to (*asharnā ilayhi*), indicated (*dallalnā ʿalayhi*), and signaled (*lawwaḥnā bihi*) in all the epistles. Then this epistle will be balanced (*muwāzin*) with what precedes it and the epistles that came before it and add to them truthful proofs (*al-barāhīn al-ṣādiqa*) and just propositions (*al-qaḍāyā al-ʿādila*) that attest to their veracity and support their argument (*taqūm bi-ḥujjatihā*). Its method (*maḥajja*) will clarify the signs written in horizons and souls (*al-āyāt al-maktūba fī al-āfāq wa-l-anfus*), which the noble Pen traced on the apparent Tablet in writing (*khaṭṭahu al-qalam al-karīm fī al-lawḥ al-mubīn musaṭṭaran*), and established in the world in knowledge (*athbatahu fī al-ʿālam mukhbiran*), as signs for those who reflect on the kingdoms of Earth and heaven (*āyāt li-l-nāẓirīna fī malakūt al-arḍ wa-l-samāʾ*). God's book is what He wrote by His hand and in which He gathered all things as a summary of the higher world (*mukhtaṣara min al-ʿālam al-aʿlā*).[42]

The beginning of the passage reiterates some of the *Comprehensive Epistle*'s self-proclaimed goals: it explicates and proves what was left implicit and unproven in the *Epistles*. The passage also adds the new attribute of succinctness to the *Comprehensive Epistle*. More importantly for the current subject, it expands on the *Epistles* by clarifying the "signs in horizons and souls."

Clearly, the reference is to natural revelation: God has revealed Himself by signs in His creation. The concept of divine sign is vague. It could mean anything from an exhortation to study the natural sciences, and thus better under-

41 Qurʾān, 41:53.

42 *Al-Jāmiʿa*, ep. 1; ed. Ghālib, 91; ed. Kacimi, 92–93.

stand God's creation, to grasping the hidden meanings of the natural world. Both ideas are present in the *Epistles*. But the Qurʾānic phrase "signs in horizons and souls" is only mentioned in the *Fihrist* and four epistles.[43] In each case, it is a passing reference rather than a central doctrine.

In contrast, in the *Comprehensive Epistle*, this precise phrase provides the ultimate meaning for two epistles with technical subject matter. The commentaries on Epistles 3 (*Astronomy*) and 5 (*Geography*) state that their ultimate goal is to provide "pointers" (*dalālāt*) to divine signs.[44] In the former, the reference is an addition to the *Fihrist*; in the latter, the phrase is also present in the *Fihrist*. In the case of five further epistles (Epistles 13, 14, 43, 46, and 48), the *Comprehensive Epistle* adds this phrase to the corresponding section of the *Fihrist*. Clearly, the idea is then that the technical subjects of many epistles present signs of God. This does not necessarily contradict the *Comprehensive Epistle*'s position as the final epistle. Indeed, from the perspective of providing inner meanings, the idea that astronomy and geography refer to the theological idea of natural revelation makes sense. However, both epistle commentaries are very concise. Neither epistle commentary thus provides any substance for the notion that their subjects are divine signs, apart, perhaps, from the statement that Earth is a living being and thus analogous to an animal.[45]

Another example of a theme that is more prominent in the *Comprehensive Epistle* is the story of Adam's fall from grace and his seduction by Iblīs. Again, the *Comprehensive Epistle*'s author is explicit that this is one of the primary subjects of the treatise:

> When we stipulated (*sharaṭnā*) that we will present in this epistle what is necessary for the one who wants to know the mysteries of hidden knowledge (*asrār al-ʿulūm al-khafiyya*) and religious symbols (*al-marmūzāt al-sharʿiyya*), and what is indispensable for those with understanding (*dhawī al-albāb*), among the foremost things we meant and indicated was to answer the one who asks: "What is the state of Iblīs like in his continuous and constant existence from Adam's time, seducing the nations one generation after another, and one tribe after another, along with the passing of ages and the dominion of times, and persisting in that way until the day of the appointed time (*waqt maʿlūm*)? Does he persist in that body in which he seduced Adam or in another body, what is its composition

---

43 *Rasāʾil*, vol. 1, 23; vol. 4, 408; *On the Natural Sciences*, 244; *On God and the World*, 104, 290.

44 *Al-Jāmiʿa*, eps. 3, 5; ed. Ghālib, 95–96, 103; ed. Kacimi, 102, 111.

45 Ibid, ep. 5; ed. Ghālib, 101; ed. Kacimi, 109.

> (*tarkīb*) like, and is it similar (*mushākil*) to the human form or of a different form?"[46]

The passage goes on to emphasize the centrality of this mystery for those who seek knowledge and wisdom, and aspire to reach the "foundations of religion" (*uṣūl al-dīn*) and the "reality of knowledge" (*ḥaqāʾiq al-maʿrifa*), upon which the savants and sages founded their wisdom. It is indeed the case that the *Comprehensive Epistle* accords much more space to Adam's story than the *Epistles*.

Both Adam and Iblīs are still very much present in the *Epistles*. However, the referent of the name Adam is somewhat ambiguous.[47] First, there is the archetypal human form, of which Marquet employs the term the "celestial Adam."[48] The *Epistles* do not in fact employ this term. Epistle 9 (*Ethics*) instead speaks of the "universal absolute human" (*al-insān al-muṭlaq al-kullī*), identified with the "universal human soul" (*al-nafs al-kulliyya al-insāniyya*).[49] This is the complete potentiality of the human form which is only partially instantiated in individual human beings. It is this absolute human being that is God's vice-regent (*khalīfa*) on Earth, and it is the absolute humanity that was attached to Adam's body upon his creation.[50] It seems, then, to be the case that Adam, unlike ordinary humans, embodied absolute humanity, as did later prophets,[51] even if the *Epistles* do not appear to state this explicitly.

In the *Epistles*, explicit references to Adam are either to the first human being in general, or to the Qurʾānic Adam, in particular. As regards the former, it is told that Adam and Eve were created after non-human animals in the equatorial zone due to its suitability for ideal temperamental constitution.[52] God created Adam in the "best possible form" (*aḥsan al-ṣuwar*),[53] or even "in His form and image" (*ʿalā ṣūratihi wa-mithālihi*).[54] A few epistles also tell the Qurʾānic story of Adam and Eve's betrayal by Iblīs and their expulsion from paradise in some detail.[55] More often, the story is referred to in passing,[56] sometimes in disap-

---

46 Ibid, ep. 9, ed. Ghālib, 126; ed. Kacimi, 140–141.

47 See de Callataÿ, *Ikhwan al-Safa'*, 26–28; Marquet, *La philosophie des Iḫwān al-Ṣafāʾ*, 208–218.

48 Marquet, *La philosophie des Iḫwān al-Ṣafāʾ*, 208–213.

49 *On Ethics*, 28.

50 *Rasāʾil*, vol. 3, 475; *The Case of the Animals*, 73; *On Ethics*, 28.

51 See Marquet, *La philosophie des Iḫwān al-Ṣafāʾ*, 209–210. Marquet locates the concept in the context of Ismāʿīlī thought.

52 *The Case of the Animals*, 7–8, 148; *On Life, Death, and Languages*, 73–75, 97–98.

53 *Rasāʾil*, vol. 3, 414, 427; *On Life, Death, and Languages*, 97. See also *On Ethics*, 14–16.

54 *On God and the World*, 88.

55 *Rasāʾil*, vol. 3, 18–19; vol. 4, 376–377; *The Case of the Animals*, 72–76; *On Life, Death, and Languages*, 73–75.

56 See, for example, *Rasāʾil*, vol. 3, 458–459; vol. 4, 101; *The Case of the Animals*, 197, 228; *On*

proval of the "pseudo-philosophers" (*mutafalsifa*) who deny its reality.[57] These passing references are often related to the motif of the sin (*jināya*) committed by "our father Adam," which is the cause for our current material condition.[58] In Epistle 48 (*The Call to God*), the nature of Adam's sin is said to be a "hidden mystery" (*sirr khafī*) whose nature one is not allowed to reveal.[59] Yet, in another epistle, Adam's sin is attributed to embodied souls,[60] an idea that is not explained further. This suggests the possibility that Adam's sin in fact refers to a lapse within the emanationist account of creation. Adam's vice, which is identified with greed (*ḥirṣ*),[61] could then concern the Universal Soul whose greed causes its fall into matter.[62] If this is true, the *Epistles* do not say so explicitly.

Regarding the subject of Iblīs and demons (*shayāṭīn*), the Brethren of Purity are entirely overt with their allegorical reading. The real nature (*ḥaqīqa*) of the devil again constitutes "subtle" esoteric knowledge, which the saints (*awliyāʾ*), in particular, possess.[63] But the authors are explicit in rejecting the idea that Iblīs is an evil entity who exists independently of God's volition and continually brings contrition to the world. The devil is instead identified with the internal psychical aspects of bad character traits, corrupt beliefs, and evil actions; it thereby relates to the dominance of appetitive and irascible souls, in opposition to the virtue and knowledge attributed to the rational soul.[64] Nor is this a particularly outrageous reading of the Qur'ān; the authors state that many religious scholars hold this view. Besides the internal enemies of human nature, the devil has a referent in the external world: the unbelievers and opponents of true religion.[65] The world of generation and corruption, in contrast to the angelic spiritual world, is characterized by the proximity of demons and "sol-

---

Astronomia: An Arabic Critical Edition and English Translation of Epistle 3, ed. and trans. Jamil F. Ragep and Taro Mimura (Oxford: Oxford University Press and the Institute of Ismaili Studies, 2015), 109; *On Companionship and Belief*, 16; *The Call to God*, 86, 95, 231.

57 *On Ethics*, 176–178.

58 *On Arithmetic & Geometry*, 118; *On the Natural Sciences*, 57; *On Companionship and Belief*, 81.

59 *The Call to God*, 231–234.

60 *On Life, Death, and Languages*, 30.

61 See *On Companionship and Belief*, 102; *On Ethics*, 146.

62 See Marquet, *La philosophie des Iḫwān al-Ṣafāʾ*, 211–212. Marquet identifies the "celestial Adam's" sin with his desire to know God directly, as opposed to by the intermediacy of the Intellect. Marquet's reading clearly arises from his Ismāʿīlī hypothesis of authorship, while nothing of the sort is stated explicitly in the text.

63 *On Ethics*, 176–177.

64 Ibid, 181–183.

65 Ibid, 183–186.

diers of Iblīs" (*junūd Iblīs*);[66] the "demonic brothers" (*ikhwān al-shayāṭīn*) or followers of Iblīs and their character traits are contrasted with the virtuous individuals of the brethren of purity.[67]

The theme of Adam and Iblīs is, then, very much present in the *Epistles*, but it is so by means of passing references rather than an extensive discussion. The *Comprehensive Epistle*, in contrast, devotes long sections to the subject, both in the first part that precedes the commentary on Epistle 1 but also later. Starting with the idea of Adam as the archetype of humanity, the author speaks of the beginning of "corporeal creation" (*al-khalq al-jismānī*) as the first Adam, which is further identified with the "Adamite nature" (*al-fiṭra al-ādamiyya*).[68] The *Epistles* employed the terms "absolute human" (*al-insān al-muṭlaq*) and "universal human soul" (*al-nafs al-insāniyya al-kulliyya*) in this context; neither occurs in the *Comprehensive Epistle*. Instead, there is an ambiguous passage on the "universal human" (*al-insān al-kullī*), or "complete universal human" (*al-insān al-tāmm al-kullī*), which is contrasted with the "particular human" (*al-insān al-juzʾī*).[69] The former is depicted in terms of spiritual perfection; she is the "pure holy spirit" (*al-rūḥ al-quds al-khāliṣa*), the "degree of justice subsisting in truth" (*manzilat mīzān al-qisṭ qāʾim bi-l-ḥaqq*), and is entirely virtuous (*fāḍil*) and obedient to her Creator. It is through this person that knowledge of the true nature of things, and of the first creation (*ibdāʿ al-khalq al-awwal*) and divine unity (*tawḥīd*), is attained. As stated above, in the *Epistles*, it is possible to interpret the human archetype as being instantiated in prophets and imāms. Here, however, the reference is clearly to the Ismāʿīlī notion of a *qāʾim* appearing at the beginning of each prophetic cycle. Thus, a few pages later, the *Comprehensive Epistle* explicates the meaning of "complete human" by introducing the notion of seven "virtuous individuals" (*ashkhāṣ fāḍila*) through whom complete knowledge and the holy spirit are conveyed to our world.[70]

As noted above, it is possible that the *Epistles* hint at an allegorical reading that relates Adam's fall to the cosmological scheme of creation. The *Comprehensive Epistle* presents a cosmological explanation explicitly; the author also maintains that a "part" (*ṭaraf*) of this explanation was revealed in the previous epistles.[71] Thus, at the end point of the descent of existence, particular souls fell in error (*khaṭīʾa*/*zalla*) and sin (*jināya*); as a result, they fell from the spir-

---

66 *On Arithmetic & Geometry*, 117.
67 *On Ethics*, 145–147, 149–150.
68 *Al-Jāmiʿa*, ep. 1; ed. Ghālib, 16; ed. Kacimi, 4.
69 Ibid, ep. 26; ed. Ghālib, 276–279; ed. Kacimi, 366–371.
70 Ibid, ep. 26; ed. Ghālib, 285; ed. Kacimi, 380.
71 Ibid, eps. 14, 22; ed. Ghālib, 141–142, 166–168; ed. Kacimi, 160–161, 204–208.

itual to the corporeal world and were united with sublunar bodies. The souls' fall is followed by their division into the categories of (1) repentant (*tāʾib*) and acceptant (*rājiʿ*) and (2) unrepentant (*ʿāṣin/muṣirr*) and arrogant (*mustakbir*) souls, both of which are divided into two further classes based on whether they are aware of the error they committed. It is the class of knowingly unrepentant souls, who purposefully spread corruption in the world, that are now identified with Iblīs and demons. Thus, all ensouled existents of the sublunar world, apparently from minerals to humans, are divided into the categories of "praiseworthy" (*maḥmūd*) and "blameworthy" (*madhmūm*). A state of perpetual enmity (*ʿadāwa*) prevails between these two groups.

With regard to the meaning of Iblīs, here also the *Comprehensive Epistle* is in partial agreement with the *Epistles*. At each time in history, demons (*shayāṭīn*) and devils (*abālisa*) are the people who oppose the message provided by the prophets, imāms, and caliphs of that time.[72] The spiritual Iblīs (*Iblīs al-rūḥānī*), in turn, refers to the appetitive and irascible souls, a doctrine that the author attributes specifically to Epistle 9 (*Ethics*).[73] However, the *Comprehensive Epistle* adds a novel level of metaphor by introducing the cycles of occultation (*dawr al-satr*) and manifestation (*dawr al-kashf*). Thus, Iblīs's refusal to bow before Adam is explained in terms of his ignorance of the cycle of occultation (*dawr al-satr*); he wanted to be the first *qāʾim*, that is, an initiator of a prophetic cycle, instead of Adam.[74] For that end, he went on to lure Adam concerning the cycle of manifestation and its "intellectual benefits" (*al-fawāʾid al-ʿaqliyya*). Iblīs is then an individual (*shakhṣ*) who opposes the acceptance of truth and the obedience of God's commandments. It is only when the cycle of occultation ceases that Iblīs and demons will disappear altogether.[75] These cycles are also related to the specifics of the story of Adam and Eve, in particular the forbidden fruit, which is hardly mentioned in the *Epistles*.[76] According to the *Comprehensive Epistle*, then, God forbade approaching the tree until a set time when tasting its fruit became permitted.[77] The tree thus represents the

72 Ibid, eps. 1, 22, 27, 31, 38, 40, 49; ed. Ghālib, 75–81, 201–203, 209–210, 220, 224, 293, 330–331, 445, 451, 476, 533–534; ed. Kacimi, 73–80, 255–256, 264, 281, 287, 390, 439, 582, 589–590, 621, 703.

73 Ibid, ep. 1; ed. Ghālib, 75–81; ed. Kacimi, 73–80.

74 Ibid, ep. 1; ed. Ghālib, 74–75; ed. Kacimi, 72–73.

75 Ibid, ep. 1; ed. Ghālib, 74–75; ed. Kacimi, 73.

76 For passing references, with no suggestion of an allegorical meaning, see *Rasāʾil*, vol. 4, 377; *The Case of the Animals*, 75; *On Companionship and Belief*, 16; *On Life, Death, and Languages*, 74, 112–113.

77 *Al-Jāmiʿa*, ep. 1; ed. Ghālib, 66–71; ed. Kacimi, 63–68.

"remainder of the first cycle of unveiling" (*baqiyyat dawr al-kashf al-awwal*). Adam was the first "deputy" (*mustakhlif*) of the cycle of concealment (*dawr al-satr*); the cycle of manifestation (*dawr al-kashf*) will begin at the appearance of the messianic figure of "pure soul" (*al-nafs al-zakiyya*) at the day of judgment. In consequence, God forbade Adam to eat the fruit of the tree. The devil seduced him with a trick (*ḥīla*), which consisted of the claim that the secret knowledge revealed by the tree would concern resurrection (*al-nash'a al-ukhrā*/*qiyāma*).

Thus, as regards the central themes of the *Comprehensive Epistle*, in many cases, its contribution seems to be restricted to giving more space to ideas that were already present in the *Epistles*. In the case of the allegorical interpretations of Adam and Iblīs, however, the *Comprehensive Epistle* fulfills the promise of a final epistle that reveals hidden meanings of the original epistles. The identification of Iblīs and demons with the opponents of prophetic truth, or the irrational parts of the soul, is already present in the *Epistles*. The equation of Adam's sin with that of cosmic or particular souls is possibly hinted at in the *Epistles*, and is only explicated in the *Comprehensive Epistle*, along with the elaboration of specific classes of good and bad souls present at every level of the sublunar hierarchy of beings. The entirely novel aspect of interpretation consists in relating the story of Adam's fall to the cycles of manifestation and occultation. However, nothing in the *Epistles* suggests that this is the meaning that its authors assigned to the story of Adam and Eve.

In this case, the ideas that the *Comprehensive Epistle* adds to the *Epistles* are of a clearly Ismāʿīlī nature. Many scholars have noted that the allegorical interpretation of the story of Adam in the *Comprehensive Epistle* is reminiscent of Ismāʿīlī thought.[78] This is true of relating Adam's sin with a lapse in the order of creation; the Ismāʿīlī context is even more obvious in the identification of the forbidden fruit with the cycle of manifestation. If it is agreed that these Ismāʿīlī ideas are already implicit in the *Epistles*, then the *Comprehensive Epistle* fulfills its self-acclaimed aim of explicating what was hinted at in the *Epistles*. If not, the *Comprehensive Epistle* interpolates Ismāʿīlī concepts into the *Epistles*. The problem, then, in effect reverts to the question of the Ismāʿīlī nature of the *Epistles*. In the following chapter, this question will be addressed with regard to the doctrinal contents specifically related to Ismāʿīlī thought.

---

78 Corbin, *Histoire*, 126–127; De Smet, "La Risāla al-Ǧāmiʿa," 276–283; Idem, *La philosophie ismaélienne: un ésotérisme chiite entre néoplatonisme et gnose* (Paris: Les Éditions du Cerf, 2012), 108–111.

CHAPTER 6

Doctrine of the *Comprehensive Epistle*

If the theory of common authorship were true, then the *Epistles* and the *Comprehensive Epistle* should reflect the ideological commitment of a single author or a group. As discussed in the introduction, much of the debate on the Brethren of Purity has revolved around their alleged Ismāʿīlī affiliation. Some aspects of the *Epistles* suggest an Ismāʿīlī influence, and the hypothesis of their Ismāʿīlī identity has been dominant in scholarship. The preceding chapters have shown that many of the inner meanings that the *Comprehensive Epistle* attributes to the *Epistles* are more overtly Ismāʿīlī in character than those present in the *Epistles*. If we agree that the *Epistles* were composed by Ismāʿīlī authors, then this is to be expected. Perhaps, then, the Brethren of Purity wrote their epistles in an environment that was hostile to the Ismāʿīlī mission, such as in Būyid Iraq,[1] and thus shrouded their more overtly Ismāʿīlī ideas in symbolism. The *Comprehensive Epistle*, in turn, was meant for a restricted audience, and was able to express these ideas freely. If this were the case, the two treatises should still not contradict each other in key doctrines. In contrast, if the claim of common authorship is false, the *Epistles* were at some point adopted by an Ismāʿīlī author who read into them ideas that were not initially there. In that case, we would expect there to be a doctrinal evolution from the *Epistles* to the *Comprehensive Epistle*.

Some Ismāʿīlī perspectives present in the *Comprehensive Epistle* have been mentioned in previous chapters. This chapter, however, addresses the Ismāʿīlī doctrine in a more systematic way. Doctrinal aspects of the two works are compared by means of three case studies: (1) God and creation, (2) cyclical history, and (3) the identity of the Brethren of Purity. The first two sections discuss the primary doctrinal elements that distinguish Ismāʿīlī thought from most schools of Islamic theology; the third section addresses the question of the two works' ideological commitment by analyzing the referent of the term "Brethren of Purity." The previous chapters have provided strong evidence against the theory of common authorship; it is highly likely that the *Comprehensive Epistle* postdates the *Epistles*. This chapter supports that claim by the further argument of doctrinal evolution: whether or not the original Brethren of Purity were affiliated with the Ismāʿīlīs, the *Comprehensive Epistle* presents an Ismāʿīlī doctrine that is not entirely in line with the *Epistles*.

---

1 For the intellectual context of Būyid Iraq, see Kraemer, *Humanism*.

1 God and Creation

Perhaps the most distinctive doctrine of Ismāʿīlī theology is the understanding of God's uniqueness and unity (*tawḥīd*) in terms of absolute transcendence.[2] Among the major Islamic schools of theology (*kalām*), the Muʿtazila also followed the way of negative theology. But the Ismāʿīlīs went beyond that to claim that nothing at all, not even being, can be attributed to God. As for the philosophers, the Arabic Plotinus was a central influence on Ismāʿīlī theologians in this regard. Only the Ismāʿīlīs, however, adopted Plotinus's position of the first principle's absolute transcendence; the philosophers instead identified God with pure being. Thus, the Ismāʿīlī theologian al-Kirmānī accuses even the Muʿtazila of assocating God with His creatures (*tashbīh*),[3] since negation of the attributes of God is also a form of associationism. Al-Kirmānī's major theological treatise, the *Rest of the Intellect* (*Rāḥat al-ʿaql*), then, subcribes to double negation: God is neither an existent (*ays*) nor a non-existent (*lays*).[4] Ismāʿīlī theology is highly philosophical, and in many respects resembles the theology of Islamic philosophers during this period.[5] The major difference is that God is not an Aristotelian first cause nor a self-intellecting intellect. Instead, as with Plotinus, God is beyond both being and rationality. The real Creator is instead the first hypostasis, the Intellect, which is the first cause of existence; it is to the Intellect that the Qurʾānic attributes refer rather than to God.

The Ismāʿīlī position on creation is closer to that of the philosophers. Since al-Nasafī (d. 943) at least, the creative process was formulated in Neoplatonic terms as a higher being's overflowing or emanation of its being to a lower entity.[6] Here also, there are some crucial differences. First, *contra* both Plotinus and Muslim philosophers, Ismāʿīlī theologians preserve God's creative act (*ibdāʿ*) or command (*amr*) in the sense that the Intellect does not follow necessarily from the perfection of God's essence but from a voluntary act of creation.[7] Second, Ismāʿīlī authors tend to employ their own technical terms: the first two created beings are identified with Qurʾānic terms, such as the Pen (*qalam*) and

---

2 See De Smet, "Ismāʿīlī Theology," 313–317; Idem, *La philosophie ismaélienne*, 55–64; Walker, *Early Philosophical Shiism*, 72–80.

3 De Smet, "Ismāʿīlī Theology," 314.

4 Al-Kirmānī, *Rāḥat al-ʿaql*, 129–134.

5 For the theology of the philosophers, see Peter Adamson, "Philosophical Theology," in *The Oxford Handbook of Islamic Theology*, ed. Sabine Schmidtke (Oxford: Oxford University Press, 2018), 297–312.

6 Daftary, *The Ismāʿīlīs*, 223–234.

7 See De Smet, "Ismāʿīlī Theology," 317–318; Walker, *Early Philosophical Shiism*, 81–86; De Smet, *La quiétude de l'intellect*, 101–146.

Tablet (*lawḥ*), the Preceder (*sābiq*) and Follower (*tālī*); the creative act is identical with the imperative 'Be!' (*kun*). Third, as regards the problem of the origin of evil, Ismāʿīlī authors account for it by means of a lapse in the creative process. Al-Sijistānī identifies the defect with the Universal Soul and al-Kirmānī with the third Intellect.[8] It is this defect in creation that leads to the particular soul's material condition. This contradicts the mainstream philosophical view, such as that of al-Fārābī and Avicenna, where divine providence proceeds in an uninterrupted fashion down to the sublunar world by the intermediacy of secondary causes. Thus, there is no specific place within the cosmological hierarchy with which the origin of evil could be identified.

Starting with the question of divine essence, the *Epistles* contain no treatise devoted to God; nor is there a systematic discussion of divine unity (*tawḥīd*), essence, or attributes within any epistle. This is surprising given the *Epistles*' theological ambitions. As we saw in Chapter 1, this also contradicts Epistle 7's philosophical curriculum, where the theological part of philosophy should be initiated by a discussion about God. Moreover, this diverges from the standard structure of theological treatises, which typically start with the subject of God and divine unity.[9] Both the *Epistles* and the *Comprehensive Epistle* still address divine unity. If the theory of common authorship is true, they should have a similar position concerning the question of God's transcendence.

The absence of a systematic discussion makes the *Epistles*' theological doctrine difficult to assess.[10] The subject may be approached by means of a set of key terms: divine unity (*tawḥīd*), transcendence (*tanzīh*), and attribute (*ṣifa*). These are brought up to different extents in a great number of epistles. As regards the first two, the *Epistles* are unsurprisingly emphatic on God's unity and uniqueness,[11] as well as on the transcendence of His essence.[12] Yet, it is only on rare occasions that the terms are given any specific meaning in relation to God's essence and attributes. With regard to transcendence, the attributes of created beings, both corporeal and spiritual, are negated of God.[13] In Epistle 31

8 De Smet, *La quiétude de l'intellect*, 243–260; Walker, *Early Philosophical Shiism*, 96–101.

9 See, for example, Ulrich Rudolph, "Reflections." The article argues for the influence of *kalām* on al-Fārābī's *Virtuous City* by comparing its structure with Ashʿarite theological works. For Ismāʿīlī theology, al-Kirmānī's *Rest of the Intellect* provides an example.

10 For an attempt at such an assessment, see Marquet, *La philosophie des Iḫwān al-Ṣafāʾ*, 49–63.

11 See, for example, *Rasāʾil*, vol. 4, 387, 411; *On God and the World*, 266; *On Life, Death, and Languages*, 107, 123–124.

12 See *Rasāʾil*, vol. 4, 387, 396, 411; *On God and the World*, 56, 59, 63–64, 84, 93–95, 97–100, 185.

13 *Rasāʾil*, vol. 3, 402–403; *Sciences of the Soul and Intellect, Part 1*, 48–49, 113; *On God and the World*, 86–94, 98–105; *On Life, Death, and Languages*, 100.

(*Differences of Languages*), however, the authors state that most savants agree on the principles (*uṣūl*) of divine unity and the attribution of "appropriate characteristics" (*mā yalīq bihi min ṣifātihi*) to God,[14] even if they may disagree on the particulars (*furūʿ*). Many passages, moreover, give a non-exhaustive list of God's attributes, such as Hearing (*samīʿ*), Knowing (*ʿalīm*), and Compassionate (*raʾūf*).[15] One passage states that knowledge of God's attributes increases by acquisition of knowledge about the attributes of His creation;[16] another states that most savants (*ʿulamāʾ*) have erred in this question, not due to the obscurity (*khafāʾ*) or subtlety (*diqqa*) of the subject, but because they pursued this knowledge in the same manner as they would seek knowledge about sensible things.[17] Thus, Epistle 43 (*The Way to God*) states that one should not discuss God's essence and attributes by way of "conjecture" (*ḥazr/takhmīn*) before one's soul has been purified; this only leads to "perplexity" (*ḥayra*) and "perdition" (*ḍalāla*).[18] Finally, in contrast to the Ismāʿīlī reticence of describing God as a cause, in the *Epistles*, God is repeatedly identified with the cause (*ʿilla*) of all existents.[19] In sum, on the one hand, the authors of the *Epistles* are emphatic about divine transcendence; it is clear that God is not the Aristotelian intellect of the philosophers. On the other hand, the authors allow the predication of positive attributes of God, including describing God as the First Cause, which goes against the double negation of divine attributes in Ismāʿīlī theology.

The *Epistles* also seem to allow an epistemic access to God. In Epistle 42 (*Opinions and Doctrines*), the authors state that God has placed knowledge of His essence (*huwiyya*) in the soul's substance (*jibillat al-nufūs*), which is neither acquired nor learned (*min ghayr taʿallum wa-lā iktisāb*).[20] In an epistemological context, the authors sustain that while knowledge about God cannot be attained by the same methods as other kinds of knowledge, human beings have access to God by means of (1) instinctive natural knowledge (*al-maʿrifa al-gharīziyya*) and by (2) attribution (*waṣf*), abstraction (*tajrīd*), transcendence

14 *On Life, Death, and Languages*, 107.

15 *Rasāʾil*, vol. 3, 402–403, 506; vol. 4, 130, 408; *On Composition*, 88–89; *On Life, Death, and Languages*, 37.

16 *Rasāʾil*, vol. 3, 506.

17 Ibid, 513.

18 *On Companionship and Belief*, 15.

19 *Rasāʾil*, vol. 4, 130; *On Logic*, 35; *On Arithmetic & Geometry*, 9; *Sciences of the Soul and Intellect, Part I*, 113; *Sciences of the Soul and Intellect, Part III*, 74–75; *On Composition*, 88–89; *On God and the World*, 87, 98–101.

20 *Rasāʾil*, vol. 3, 516.

(*tanzīh*), and unification (*tawḥīd*), as well as by (3) demonstrative methods (*bi-ṭuruq al-burhān*).[21] Epistle 28 (*Human Capability for Knowledge*), on the other hand, seems more pessimistic; the human intellect is only able to conceive (*taṣawwur*) objects that are neither too sublime (*jalāla*) nor too obscure (*khafāʾ*), and hence fails to perceive (*idrāk*) God in His "essential sublimity and intense manifestation" (*bi-māhiyyat dhāt jalālatihi wa-shiddat ẓuhūrihi wa-wuḍūḥ bayānihi*).[22] In Epistle 39 (*Movements*), God is said to be "veiled" (*muḥtajab*) from the view of most people, who can only reach Him by the intermediacy of His creation; the saints (*awliyāʾ Allāh*), however, may attain a vision (*mushāhada*) of God and see (*yarawna*) Him by the "eye of their insight" (*ʿayn al-baṣīra*), in apparent reference to some kind of mystical perception.[23] The passages do not necessarily contradict each other. It may be that complete perception of God is beyond the human intellect, and thus most human beings; yet, humans have some intellectual access to God, and, in the case of elect individuals, may even attain a supra-rational vision of God's essence. Thus, the result seems to be that human beings can have knowledge of God's essence; to some extent, such knowledge is indeed innate to the human soul. In the latter case, the reference might be to an awareness of God's existence rather than actual knowledge about His essence. A more profound kind of knowledge about the divine essence could then be accessed either by demonstrative means or by a form of mystical intuition.

In the *Comprehensive Epistle*, the terms *tawḥīd* and *tanzīh* are more frequent, despite the work's shorter length. An immediately apparent difference is that these terms are often contrasted with another pair of terms, *tashbīh* and *taʿṭīl*,[24] which denote the association of divine essence with created beings versus denuding it of all content. In the *Epistles*, by contrast, the latter pair of terms is hardly mentioned. The author also forwards the claim that *tawḥīd* is the goal of either theoretical sciences in general[25] or of the *Comprehensive Epistle*, in particular;[26] the *Epistles* allege no such thing. Clearly, then, the *Comprehensive Epistle* lays more emphasis on the subject of divine unity.

21 *Sciences of the Soul and Intellect, Part I*, 108.

22 *Rasāʾil*, vol. 3, 22.

23 *Sciences of the Soul and Intellect, Part III*, 41–45. For this notion in the *Epistles*, see Baffioni, "From Sense Perception to the Vision of God."

24 *Al-Jāmiʿa*, eps. 1, 14, 26, 27, 39; ed. Ghālib, 27–28, 59, 114, 140, 260, 279, 296, 458–459; ed. Kacimi, 17–18, 55, 125–126, 158–159, 342, 371, 393, 598–599.

25 *Al-Jāmiʿa*, ep. 7; ed. Ghālib, 114; ed. Kacimi, 125–126.

26 Ibid, ep. 25; ed. Ghālib, 236–237; ed. Kacimi, 304.

As for the actual content of the concept, like the *Epistles*, the *Comprehensive Epistle* rejects the predication of the attributes of created things of God.[27] But the treatise also denies God's attributes in general.[28] Moreover, the divine essence is said to be beyond definition (*ḥadd*) and attribution (*waṣf*),[29] or beyond attributes and relations (*iḍāfa*), apart from those employed by Himself;[30] the essence is not encompassed by any qualities, and can thus only be described by approximation (*taqrīb*).[31] It also seems to be the case that the *Comprehensive Epistle* makes no positive claims about God's attributes; nor does it provide a list of divine names, apart from the mention that the property "Knowing" (*ʿālim*), but not "Intellecting" (*ʿāqil*), may be attributed to God.[32] There are, however, at least two instances where the Creator (*al-bārī*) is described as a cause,[33] even if this occurs much less frequently than in the *Epistles*. Yet, even in these cases, it is not entirely clear that the first cause of the world is God; at least the latter passage seems to allow that the world is actually derived from the Intellect.[34] This idea is formulated in clearer terms in the following passage:

> The purely one thing (*al-shayʾ al-wāḥid al-maḥḍ*) is indivisible, for what is divided consists of parts (*aqsām*). And when a thing is divided into parts, it comes to have many names and different attributes (*ṣifāt*). The Creator, may He be exalted, is beyond that; He is above the attributes they ascribe to Him. As for the one that is described (*mawṣūf*) by sublimity (*jalāla*) and magnificence (*ʿiẓma*), and to whom existence is attributed, it is the beginning of every existent that accepts the emanation of goodness, and degrees of existence (*ḥudūd*) revert to it. This is the first Intellect, and its

27 Ibid, eps. 1, 2, 26, 27, 36, 39; ed. Ghālib, 19–20, 94–95, 268, 284, 296, 406, 462–463; ed. Kacimi, 8, 99, 355, 379, 393, 534, 604.

28 Ibid, eps. 22, 25, 35; ed. Ghālib, 165–166, 248, 369–370; ed. Kacimi, 203–204, 321, 491–492.

29 Ibid, ep. 39; ed. Ghālib, 458–459; ed. Kacimi, 598–599.

30 Ibid, ep. 26; ed. Ghālib, 286; ed. Kacimi, 382.

31 Ibid, ep. 26; ed. Ghālib, 279; ed. Kacimi, 371.

32 Ibid, ep. 26; ed. Ghālib, 280; ed. Kacimi, 372.

33 Ibid, eps. 2, 39; ed. Ghālib, 94, 458–459; ed. Kacimi, 99, 599.

34 The passage (*al-Jāmiʿa*, ep. 39; ed. Ghālib, 458–459; ed. Kacimi, 598–599) states that some sages (*ḥukamaʾ*) claim that the "origin of existents" (*aṣl al-mawjūdāt*) is the Creator (*al-bārī*); others state that it is the Intellect because they want to preserve God's transcendence (*tanzīh*). Both are said to pertain to the "way of truth" (*ṭarīq al-ḥaqq*). Later, it is stated that the Creator is the "originator and cause of the principle of the world" (*mubdiʿ aṣl al-ʿālam wa-ʿillatuhu*). This formulation seems to preserve God from being a direct cause of all existents in the Aristotelian sense.

> Creator is beyond the attributes of those who describe Him and properties (*naʿt*) of those who ascribe properties to Him. For it is said by belief and submission that it is He and there is no other God but He. This statement is the affirmation of divine unity (*ithbāt al-tawḥīd*), and it hence became the basis (*aṣl*) on which every religion (*sharīʿa*) and creed (*dīn*) is founded. This is because the Intellect denied divinity (*ulūhiyya*) of its essence and affirmed it (*athbatahā*) for its Creator. And it said: "There is no God but He," and thus found its creator. This is the meaning of the affirmation of pure unity (*ithbāt al-waḥda al-maḥḍa*), and it is due to its connection of divine support (*ittiṣāl al-taʾyīd*), which is continuous and never ceases or expires but continues in eternity.[35]

The passage states explicitly that the attributes traditionally predicated of God, such as sublimity and magnificence, in fact pertain to the Intellect. It is, then, the Intellect of which the Qurʾān speaks when it attributes divine names to God. God, in contrast, is beyond attribution. This is not the doctrine of the *Epistles* which, despite its emphasis on the unity and transcendence of God's essence, assigns several attributes to God. In other words, unlike the *Epistles*, the *Comprehensive Epistle* appears to adopt the Ismāʿīlī position of God's absolute transcendence.

With regard to the possibility of human knowledge of God, the *Comprehensive Epistle* does reiterate the *Epistles*' claim that such knowledge may be either innate or acquired through "description by abstraction and transcendence" (*al-waṣf bi-l-tajrīd wa-l-tanzīh*), or even by the "demonstrative method" (*bi-ṭarīq al-burhān*).[36] The claim contains a major difference, however: the *Comprehensive Epistle* does not say that the divine essence can be described by attributes but only by the negation of all attributes. In other passages, the author seems to limit knowledge about God to the indirect evidence gained from creation,[37] or else the Intellect as the first created being.[38] There is also the interesting case of Epistle 28 (*Human Capability for Knowledge*), devoted to epistemology. In *Fihrist a*, the epistle's goal is stated to be "instigation to knowledge about the Creator" (*al-tanbīh ʿalā maʿrifat al-bāriʾ*).[39] In the *Comprehensive Epistle*, by contrast, the goal consists in "knowledge of the limits of human knowledge"

35 *Al-Jāmiʿa*, ep. 1; ed. Ghālib, 28; ed. Kacimi, 18–19.
36 Ibid, ep. 35; ed. Ghālib, 370; ed. Kacimi, 492–493.
37 Ibid, eps. 14, 26, 36; ed. Ghālib, 140, 275–276, 400; ed. Kacimi, 158–159, 366–367, 528.
38 Ibid, eps. 26, 34; ed. Ghālib, 286, 350; ed. Kacimi, 382, 466–467.
39 MS Atıf Efendi 1681, 4ᵃ.

(*mā yantahī ilayhi al-insān fī al-ʿulūm*) with no mention of knowledge about God.[40] Thus, even if both works are somewhat ambiguous with regard to the subject of divine essence and attributes, there is a clear shift in the *Comprehensive Epistle* toward a doctrine that restrains the human epistemic access to God.

With regard to creation, both the *Epistles* and the *Comprehensive Epistle* attest to the Pythagorean notion of an ontological correspondence between levels of reality and numbers. Thus, number one corresponds to God and numbers two to nine to distinct levels of reality.[41] In both works, the main idea is that just as natural numbers are derived from number one, so distinct levels of being are derived from God. A closer inspection reveals some differences, however. First, in the *Epistles*, the manner of this derivation is often expressed in terms of God being the first cause of all existents;[42] the *Comprehensive Epistle* does not employ this formulation. Instead, the relation is expressed as God being the creator (*mubdiʿ*),[43] an expression also used in the *Epistles*, or as God preceding existents (*qabla al-mawjūdāt*) the way number one precedes other numbers.[44] Moreover, there is the larger discrepancy that some passages in the *Comprehensive Epistle* refrain from equating God with number one. At the beginning of the *Comprehensive Epistle*, it is stated that the correspondence with number one indicates the Creator's "pure creation" (*al-ibdāʿ al-maḥḍ*), for which there is "no similarity (*shibh*) or resemblance (*mathal*)".[45] The impression is, then, that it is the act of creation rather than God that corresponds to number one. This is confirmed by a subsequent passage:

> Its [God's] existence is established (*yataqarrar*) in the intellects of corporeal things by its being one (*wāḥid*) for which there is no second. For it is the one that is one in number (*wāḥid al-ʿadad*), which indicates (*yadull*) its pure creation (*ibdāʿuhu al-maḥḍ*) for which there is no second, as it is in similarity (*shibh*) and likeness (*mithāl*). This is because for [number] one it is not possible for anyone to imagine that there is some number that precedes it. It is the first existent from which existence emanates, and the existent whose existence does not precede it is the thing of the

40 *Al-Jāmiʿa*, ep. 28; ed. Ghālib, 305; ed. Kacimi, 406.

41 The main discussions are contained in Epistles 1, 32, and 33 in the *Epistles*, and within the corresponding sections of the *Comprehensive Epistle*.

42 *On Arithmetic & Geometry*, 9, 23; *Sciences of the Soul and Intellect, Part 1*, 6, 8–9, 10–11, 13.

43 *Al-Jāmiʿa*, eps. 1, 32; ed. Ghālib, 19–20, 24–25, 334; ed. Kacimi, 8–9, 14, 446.

44 Ibid, eps. 1, 32; ed. Ghālib, 19–20, 341; ed. Kacimi, 8–9, 455.

45 Ibid, ep. 1; ed. Ghālib, 20; ed. Kacimi, 8.

> one (*shayʾ al-wāḥid*), and from it the second one emanates. That which is at the degree (*manzila*) of number one (*wāḥid al-ʿadad*) is the first creation (*al-ibdāʿ al-awwal*), and the singular that establishes [the letter] alif (*al-aḥad bi-ithbāt al-alif*) is the Creator (*al-mubdiʿ*). For alif is the first in letters, and alif has been established (*thabuta*) in the singular (*al-aḥad*) twice through two different forms (*bi-ṣūratayni mukhtalifatayni*). The first is single in its essence (*mufrada bi-dhātihā*), and the second opposes the form of the first (*mukhālifa li-ṣūrat al-ūlā*) because it is alif that is bound to itself (*munʿaṭif baʿḍuhā ʿalā baʿḍ*). The first alif that comes from the singular is singularity (*fa-l-alif al-ūlā min al-aḥad hiya al-aḥadiyya*), and the second alif is oneness (*wāḥidiyya*), and its oneness lies in its singularity (*fa-wāḥidiyyatuhu fī aḥadiyyatihi*). There is no definition (*ḥadd*) for it by which it could be described or its essence referred to (*wa-lā yushār ilayhi anniyya*), for it is the singular (*aḥad*), and the first creation (*awwal mubdaʿ*) that it created is the one (*wāḥid*) that proceeds from singularities (*al-munbaʿitha min al-āḥād*), the beginning of the spring of springs of even and odd numbers (*mabdaʾ aṣl yanbūʿ yanābīʿ al-azwāj wa-l-afrād*).[46]

In this confusing passage, the *Comprehensive Epistle*'s author introduces a distinction between two concepts of unity, which I render as "one" (*wāḥid*) and "singular" (*aḥad*). The former, that is the number one, is now identified with the first creation, which is the Intellect. Above it, there lies a higher degree of singularity, which supposedly corresponds to the divine essence. Such a distinction is entirely absent in the *Epistles*, which unambiguously identifies God with number one. It appears that the *Comprehensive Epistle*'s author is uncomfortable with this identification in terms of its ability to convey God's absolute transcendence. It is, then, the Intellect that corresponds to number one in the sense that it is the first existent from which all other existents are derived. In a later passage, the author accordingly states that the science of numbers (*ʿilm al-ʿadad*) precedes all other sciences in the same sense as the Intellect precedes all things; just as all things exist potentially in the Intellect, so all sciences exist potentially in the science of arithmetic.[47] In the *Comprehensive Epistle*, the absolutely transcendent God would therefore lie outside the scheme of Pythagorean correspondences.

The shift from the *Epistles* to the *Comprehensive Epistle* is even more evident in the terminology employed. As stated above, Ismāʿīlī authors often identify

46 Ibid, ep. 1; ed. Ghālib, 20; ed. Kacimi, 8–9.

47 Ibid, ep. 1; ed. Ghālib, 29; ed. Kacimi, 19.

the hypostases and creative progression with terms of Qurʾānic origin. These do not automatically imply a doctrinal shift but the frequency of their use attests to the author's Ismāʿīlī commitment. Thus, for the couple of Intellect and Soul, Ismāʿīlī authors commonly use a number of terms. The first of these are Preceder (*sābiq*) and Follower (*tālī*/*lāḥiq*). These are present in both the *Epistles* and the *Comprehensive Epistle*. The former employs them six times as synonyms for the two cosmological hypostases;[48] the latter also uses them six times.[49] Their frequency would, then, appear to be balanced in the two works. However, besides the fact that the *Epistles* are a much longer work, the distribution of these terms in the *Epistles* is revealing. Two of the occurrences are in Epistle 49b (*Spiritual Beings*); for the first occurrence, the critical edition shows that there is considerable manuscript variation in its formulation. Of the remaining four instances, two occur in the *Fihrist* and two in Epistle 52b (*Magic*). The former two are present only in *Fihrist c*; the first pertains to the additions that *Fihrist c* makes to the epistle entry of Epistle 52 in *Fihrist a*; the latter is placed within *Fihrist c*'s added cosmological section.[50] The manuscripts also manifest variation with regard to these precise terms in the case of Epistle 31 (*Differences of Languages*): the terms are present in the Beirut edition but in only one of the manuscripts employed in the critical edition.[51] All of this suggests that there was a tendency to insert these terms into the manuscripts at a later stage. Since Epistle 52b is likely to postdate the *Epistles*, the fact that it contains two of the occurrences only reinforces this conclusion. Moreover, *Fihrist c*, Epistle 52b, and the *Comprehensive Epistle* all contain an interesting elaboration of the Ismāʿīlī couplet where nature, matter, and non-existence (*ʿadam*) are given the rhyming attributes of "driving" (*sāʾiq*), "desiring" (*shāʾiq*), and "annihilating" (*māḥiq*), respectively.[52] Since this formulation occurs nowhere else in the *Epistles*, the impression is, again, that these texts are connected, perhaps because all three were composed after the *Epistles* were adopted to an Ismāʿīlī context.

The second pair of Ismāʿīlī terms corresponding to the Intellect and the Soul, the Pen (*qalam*) and the Tablet (*lawḥ*), goes back to the Qurʾān; the second

48 *Rasāʾil*, vol. 1, 41, 42; vol. 4, 409, 410; *On God and the World*, 61, 97–98.

49 *Al-Jāmiʿa*, eps. 14, 18, 25, 32, 52; ed. Ghālib, 133, 134, 248, 335, 536–537; ed. Kacimi, 149, 150, 178, 320, 447, 708. The third reference only occurs in the longer variant of the commentary on Epistle 18 (ed. Kacimi, 172–181); thus, it is not present in the very short variant conveyed by the Ghālib edition (ed. Ghālib, 149).

50 See p. 76.

51 *Rasāʾil*, vol. 3, 144; *On Life, Death, and Languages*, 100.

52 *Rasāʾil*, vol. 1, 41; vol. 4, 410; *al-Jāmiʿa*, ep. 32; ed. Ghālib, 335; ed. Kacimi, 447.

term occurs in verse 85:21–22: "Verily, this is a glorious Qur'ān in a preserved tablet (*qur'ān majīd fī lawḥ maḥfūẓ*)." In the *Epistles*, the term Pen is employed of the Intellect on only one occasion; even then, it occurs as a part of a long list of terms by which different authors or groups have designated the two primary principles below God.[53] The Ismāʿīlī terms are accorded no particular significance; they occur in the middle of the list, and, like all other terms, are attributed to an anonymous group: "And some of them say (*wa-minhum man qāla*): Tablet and Pen." In addition to this passage, the term Tablet is used three times but always in the precise form "preserved tablet" (*al-lawḥ al-maḥfūẓ*), and in the specific microcosmic context where the human being constitutes an "abridgement" (*mukhtaṣar*) of the "preserved tablet."[54] The term is, then, not given the specific cosmological sense where it constitutes the counterpart of Pen. In the section corresponding to the passage in the *Epistles*, the *Comprehensive Epistle* reiterates the list of terms that are equivalent to the Intellect and the Soul,[55] as well as the statement that the human being constitutes an abridgement of the "preserved tablet."[56] But both terms are much more prominent in the *Comprehensive Epistle* and are given explicitly Ismāʿīlī connotations. One passage elaborates on the meaning of the "preserved tablet": it is divided into seven "lines" (*suṭūr*), among which the first is the beginning and the seventh the end.[57] This is a clear reference to the Ismāʿīlī doctrine of seven prophets or *nāṭiq*s. Another passage states that "what is written on the noble tablet" (*mimmā huwa maktūb fī al-lawḥ al-karīm*) is a "subtle mystery and deep sea" (*sirr daqīq wa-baḥr ʿamīq*).[58] On at least three occasions, the Pen-Tablet couple is employed in the straightforward cosmological sense that is not present in the *Epistles*.[59]

The case of a third couplet, Throne (*ʿirsh*) and Footstool (*kursiyy*), is even more revealing. The first term is based on two Qur'ānic contexts: 25:59 (cf. 7:54; 10:3; 13:2; 20:5; 32:4; 57:4): "He who created the heavens and the Earth and what is between them in six days, and then the Compassionate was established on His throne (*istawā ʿalā al-ʿirsh*)"; 69:17 (cf. 39:75; 40:7): "And the angels will appear at its ends, and above them eight will on that day carry the throne of your Lord (*ʿirsh rabbika*)." Neither passage explains what the throne is. The first

53 *Sciences of the Soul and Intellect, Part I*, 18.
54 *Rasā'il*, vol. 1, 29; vol. 2, 460; vol. 4, 12.
55 *Al-Jāmiʿa*, ep. 32; ed. Ghālib, 336; ed. Kacimi, 449.
56 Ibid, eps. 23, 26; ed. Ghālib, 225, 259, 261; ed. Kacimi, 289, 340, 343.
57 Ibid, ep. 41; ed. Ghālib, 498; ed. Kacimi, 653–654.
58 Ibid, ep. 22; ed. Ghālib, 159; ed. Kacimi, 195.
59 Ibid, ep. 1; ed. Ghālib, 16, 25, 91; ed. Kacimi, 4, 15, 93.

passage relates it to the creation of the world, and thus enables a cosmological interpretation of its meaning. The second passage relates it to angels and the Judgment Day. The use of the latter term is based on verse 2:255: "And His footstool embraces (*wasiʿa kursiyyuhu*) the heavens and the Earth." This again fits a cosmological interpretation.

Both the *Epistles* and the *Comprehensive Epistle* resort to these two terms, and the related Qurʾānic verses, repeatedly, mostly without explicating their meaning. But when the two works do equip the terms with a meaning, they contradict each other. In the *Epistles*, angels are related to the celestial spheres. Since the Qurʾān identifies angels with carriers of God's throne, it makes sense to equate the throne with the spheres. Thus, the *Epistles* provide a specific cosmological sense of the Throne and the Footstool. All variants of the *Fihrist* entry for Epistle 16 (*Heaven and World*) promise, among other things, to provide an explanation for the terms "great throne" (*al-ʿirsh al-ʿaẓīm*) and "wide footstool" (*al-kursiyy al-wāsiʿ*).[60] In that epistle, the Footstool is indeed identified with the eighth sphere of the fixed stars, which is "wide" (*wāsiʿ*) because it encompasses the seven planetary spheres, and thus "encompasses the heavens and the Earth" (*wasiʿa al-samawāt wa-l-arḍ*).[61] The Throne, in turn, is the outermost ninth sphere, which "embraces" (*muḥīṭ*) the other eight spheres, and which the "eight will carry on that day." This specific cosmological identity for the two Qurʾānic terms is repeated in other epistles.[62] Remarkably, the Throne and Footstool are never identified with the Intellect and Soul.

The beginning of the *Comprehensive Epistle*'s commentary on Epistle 16, which is based textually on the *Fihrist*, states that the "cream" (*zubda*) of the epistle includes, among other things, knowledge of the "wide footstool" and the "encompassing throne" (*al-ʿirsh al-muḥīṭ*).[63] The concise epistle entry gives no explanation whatsoever for the meaning of these two terms, however. Nor are the terms identified with the eighth and ninth spheres elsewhere in the *Comprehensive Epistle*, either. Instead, the *Comprehensive Epistle*'s author equates

60 *Rasāʾil*, vol. 1, 27.

61 *On the Natural Sciences*, 75. The edition employs the attribute "intermediary" (*wāsiṭ*), as opposed to "wide" (*wāsiʿ*), for the sphere of the fixed stars. As shown in the critical apparatus, however, the majority of the manuscripts gives the latter term, which is clearly the correct one given the Qurʾānic verse. The edition also states that the eighth sphere of the fixed stars encompasses the eight spheres, which makes no sense. Moreover, the Throne is identified with the eighth sphere, while the reference to the Footstool is omitted. Again, the critical apparatus shows that most manuscripts include the reference to the Footstool and identify the eighth sphere with the Footstool and the ninth sphere with the Throne.

62 *On the Natural Sciences*, 235, 388, 403; *On God and the World*, 159.

63 *Al-Jāmiʿa*, ep. 16; ed. Ghālib, 145; ed. Kacimi, 168.

the Throne with the Intellect and the Footstool with the Soul, either explicitly or implicitly, on several occasions.[64] For these two terms, then, it is not only the case that the *Comprehensive Epistle* expands their employment in comparison with the *Epistles*, but it is also the case that there is a contradiction in the meaning that the two works assign to them.

There are many further cosmological terms with Ismāʿīlī connotations that are prominent in the *Comprehensive Epistle*, but either they do not occur at all in the *Epistles* or they possess a different meaning. First, two more couplets referring to the two primary hypostases are employed. The *Comprehensive Epistle* refers several times to the Intellect and the Soul as the first and second limit (*al-ḥadd al-awwal/thānī*), respectively.[65] In the *Epistles*, these terms only occur in Epistle 52b;[66] even there, the context is astrological, and the reference is not to the two cosmological hypostases. The terms first and second face (*al-wajh al-awwal/thānī*) are similarly used repeatedly for the Intellect and the Soul in the *Comprehensive Epistle*, but never in the *Epistles*.[67] There is also the interesting fact that the *Comprehensive Epistle* at times speaks of the "first Intellect" (*al-ʿaql al-awwal*).[68] This formulation also occurs twice in the *Epistles*, but there the reference is clearly to the human intellect.[69] In the *Comprehensive Epistle*, in contrast, the term is employed in reference to the hypostasis of the Intellect. This formulation seems strange for both works because, in the Ikhwānian cosmology, there is only one cosmic intellect. Of course, in Fārābīan cosmology, adopted among Ismāʿīlī authors by al-Kirmānī, the first Intellect initiates a vertical series of cosmic intellects. In the *Comprehensive Epistle*, the formulation may be a vestige of an Ismāʿīlī doctrine with several intellects; or more probably, it aims to emphasize the Intellect's ontological priority with respect to the human intellect.

The situation is similar with respect to Ismāʿīlī technical terms that designate the creative process. The *Comprehensive Epistle* often employs the term "word" (*kalima*) in the specific sense of God's creative action that produces the

64 Ibid, eps. 1, 25, 26, 34, 38; ed. Ghālib, 25, 243, 261, 351, 453; ed. Kacimi, 15, 315, 343, 466–467, 592.

65 Ibid, eps. 25, 31; ed. Ghālib, 240, 241, 242, 243, 248, 250, 253, 323; ed. Kacimi, 310, 311, 313, 315, 320, 324, 330, 427.

66 *Rasāʾil*, vol. 4, 349, 429.

67 *Al-Jāmiʿa*, eps. 25, 31; ed. Ghālib, 240–241, 241, 242, 243, 249, 325; ed. Kacimi, 310–311, 312, 313, 314, 314–315, 322, 431.

68 Ibid, eps. 1, 14, 25, 26, 32, 36, 40; ed. Ghālib, 28, 134, 335, 404, 483; ed. Kacimi, 19, 150, 311, 384, 447, 448, 532, 631. There is some manuscript variation in the use of this formulation; the third and fourth occurrences (ed. Kacimi, 311, 384) have just *ʿaql* in the Ghālib edition.

69 *Rasāʾil*, vol. 4, 396; *On Composition*, 105.

Intellect;[70] the *Epistles* never use the term in this sense.[71] In a similar vein, the term denoting God's creative command (*amr*) is present in a more vague form in some passages of the *Epistles*,[72] but has a more explicitly cosmological connotation in the *Comprehensive Epistle*.[73] Many more terms related to creation and cosmological hierarchy could be explored in the same way.

The *Comprehensive Epistle*, then, adheres to an Ismāʿīlī framework in both doctrine and terminology: it srongly emphasizes God's absolute transcendence, and consistently employs Ismāʿīlī technical terms to denote the hierarchy of creation. This not the case for the *Epistles*. When set against the *Comprehensive Epistle*, the conclusion is rather that neither its doctrinal nor its terminological contents are specifically Ismāʿīlī in nature. In particular, the *Epistles* do not emphasize God's transcendence in the sense of denying the predication of all attributes of His essence. Nor do the authors consistently employ Ismāʿīlī terms, and when they do, their meanings sometimes contradict those assigned to them in the *Comprehensive Epistle*. It is these occasional contradictions that appear to be most decisive for the theory of common authorship. The higher prevalence of Ismāʿīlī terms in the *Comprehensive Epistle* could be explained by its aim to provide inner meanings to the *Epistles*. It makes no sense, however, that the inner meaning of a term in the *Comprehensive Epistle* is entirely different from its explicit meaning in the epistles that are supposed to constitute its introduction.

2 Prophecy and Imāmate

Another set of Ismāʿīlī doctrines concerns prophecy and imāmate. The general Shīʿite and Ismāʿīlī tendencies of the *Epistles* relevant to this question have been discussed in the introduction. The specifically Ismāʿīlī doctrine relates, in

70 *Al-Jāmiʿa*, eps. 22, 25, 26; ed. Ghālib, 215, 240, 242, 243, 247, 248, 255, 280, 286; ed. Kacimi, 272, 309, 310, 311, 313, 314, 315, 320, 333, 372, 382. The third occurrence (ed. Kacimi, 311) is missing in the Ghālib edition.

71 See, however, *Rasāʾil*, vol. 3, 458; *On Ethics*, 169.

72 See, for example, *Rasāʾil*, vol. 4, 340, 397; *On God and the World*, 58, 67, 80, 87, 98, 111, 221.

73 *Al-Jāmiʿa*, eps. 1, 22, 25, 26, 27, 30, 31, 32, 34, 35, 36, 39, 40, 41; ed. Ghālib, 25, 71, 164, 168, 176, 241, 242, 250, 251, 253, 255, 279, 280, 286, 301, 302, 304, 321, 324, 325, 335, 338, 348, 350, 351–352, 369, 375, 404, 405, 406, 459, 460, 461, 462, 465–466, 484, 493, 496, 501; ed. Kacimi, 15, 69, 173, 202, 206, 219, 311, 313, 314, 320–321, 324, 327, 333, 372, 373, 382, 399, 401, 403, 425, 430, 431, 447, 451, 463, 465–467, 491–492, 499, 532, 533, 534, 599, 601, 602, 604, 607–608, 633, 647, 651, 657–658. One of the occurrences (ed. Kacimi, 173) is not present in the Ghālib edition because it carries a shorter variant of the section on Epistle 18.

particular, to the notion of prophetic cycles.[74] Again, the general idea is conveyed by a number of technical terms. According to the Ismāʿīlī interpretation of the prophetic history conveyed in the Qurʾān, time is divided into seven eras or cycles (*dawr*); each cycle is initiated by a speaker-prophet (*nāṭiq*) that brings a new exoteric revelation (*sharīʿa*). The prophet of each era is followed by a "spiritual legatee" (*waṣiyy/asās/ṣāmit*) charged with the revelation's esoteric interpretation, and a series of further imāms (*mutimm/atimmāʾ*) upholding both its exoteric and esoteric aspects. The final seventh imām of each cycle abrogates that cycle's religious law and initiates the next era as the new speaker-prophet. The seventh speaker-prophet (*qāʾim/mahdī*) introduces no further law but brings the esoteric truth into open, and thus initiates the eschatological era that culminates in Judgment Day. Among later Ismāʿīlīs, as the *qāʾim*'s emergence was pushed into the future, further heptads of imāms were introduced for the Islamic era. These rotated between cycles of occultation (*satr*) and manifestation (*kashf*) as regards the divulgation of esoteric truth.

Beginning with the cycles of manifestation and occultation, the idea is much less pronounced in the *Epistles* than in the *Comprehensive Epistle*. The first two parts of the *Epistles* do not refer to such cycles. In the third part, there is a single reference to a "new cycle" (*dawr jadīd*), which is equated with the "second creation" (*al-nashʾa al-thāniya*).[75] The latter term usually refers to the Judgment Day, but here the reference is to the creation of Adam and Eve. The period starting from Adam is contrasted with the preceding era of a world devoid of human inhabitants. Adam is not introduced as the first speaker-prophet, and no mention is made of whether the new era is one of manifestation or occultation. Thus, the first three parts of the *Epistles* contain no references at all to the Ismāʿīlī notion of cycles of manifestation and occultation.

It is only the fourth part that introduces the explicit idea of cycles. Epistle 48 (*The Call to God*) is initiated by a description of the Brethren of Purity as a large secret society scattered throughout different lands and represented by members of all social classes.[76] The addressee is then given instructions for recruiting new members.[77] Later, the epistle calls for building a "spiritual virtuous city" (*madīna fāḍila rūḥāniyya*) whose virtuous and wise members seek the soul's

---

74 See, in particular, De Smet, *La quiétude de l'intellect*, 355–377; Idem, *La philosophie ismaélienne*, 125–149; Walker, *Early Philosophical Shiism*, 114–142; Daftary, *The Ismāʿīlīs*, 96–98, 131–133, 164–166, 207–209, 226–228, 271–273.

75 *On Life, Death, and Languages*, 73.

76 *The Call to God*, 32–34.

77 Ibid, 34–38.

salvation through co-operation.[78] This spiritual utopia has been interpreted as a reference to Ismāʿīlī missionary activity (*daʿwa*).[79] But the epistle itself does not make any explicit Ismāʿīlī claims. Besides a generic justification for concealing their "mysteries" (*asrār*) from the unworthy,[80] neither the doctrines nor the terms make it clear that the epistle describes an Ismāʿīlī community, as opposed to a utopian philosophical-religious society. A key chapter[81] starts by stating that among the "special properties of our virtuous brothers" (*min khawāṣṣ ikhwāninā al-fuḍalāʾ*) is their knowledge of "things of religions" (*umūr al-diyānāt*), "mysteries of prophecies" (*asrār al-nubuwwāt*), and "preparatory sciences" (*riyāḍiyyāt*). In some manuscripts, the latter are explicitly defined as "philosophical" (*falsafiyya*).[82] This is followed by an exhortation:

> If you meet one of them and know them to be of proper conduct (*rushd*), announce them that which will delight them and remind them of the recommencement and awakening (*intibāh*) of the cycle of manifestation (*dawr al-kashf*), and the passing away of the cycle of slumber (*ruqūd*), and of the beginning of the dominion of good people (*dawlat ahl al-khayr*) in the lands and the lifting of sorrows from worshippers once the conjunctions are transferred from the zodiacs of the thirds of fire (*burūj muthallathāt al-nīrān*) to the thirds of plants and animals in the tenth cycle that corresponds to the house of power (*bayt al-sulṭān*) and the emergence of luminaries (*ẓuhūr al-aʿlām*).[83]

A minimal reading of this passage relates it to the messianism that is clearly present in some epistles of the fourth section. A maximal reading takes it as one of the most explicit references to the Ismāʿīlī idea of an eschatological era of justice and truth. The latter reading is reinforced by the immediately following passage, where the brethren are equated with the "people of our faction" (*ahl shīʿatinā*), and the coming era is connected with the "appearance of

78 Ibid, 58–62.

79 See, for example, Baffioni, "Epistle 48."

80 *The Call to God*, 38–39.

81 This epistle shows considerable variation in the manuscripts, including in the order of chapters, as well as in the omission of many chapters in various manuscripts (see Table 1 and Hamdani's technical introduction). Thus, the chapter in question is second in the Beirut edition and eighteenth in the Oxford edition (following MS ج and many other manuscripts).

82 *Rasāʾil*, vol. 4, 146; *The Call to God*, 155.

83 *The Call to God*, 155–156.

our imām" (*khurūj imāminā*) and the "victory of our cause" (*nuṣrat amrinā*).[84] The manuscripts, however, provide far from unanimous support for this reading. In the critical edition, only three manuscripts refer to an expected imām; most manuscripts (including MS ع, which the editor otherwise uses as the base manuscript) talk of the emergence of our "signs" (*āyātinā*) or "days" (*ayyāminā*).[85] This suggests a tendency in some manuscripts to interpret the text in a specifically Shīʿite sense.[86] The reference to the cycle of manifestation is nevertheless almost universally shared by the manuscripts.[87] But even this only shows that the author of Epistle 48, in the manuscripts we have, employs an Ismāʿīlī concept to express messianic beliefs. It does not show that the Brethren of Purity were Ismāʿīlīs. By contrast, in the *Comprehensive Epistle*'s commentary of this epistle, the "spiritual virtuous city" is equated explicitly with Ismāʿīlī ideas: its "foundation" (*bināʾ*) is the "kingdom of the greatest law-giver" (*mamlakat ṣāḥib al-nāmūs al-akbar*), and it pertains to the "command of the seventh (leader)" (*amr al-sābiʿ*).[88]

Besides this reference, cycles are mentioned in three further epistles of the fourth section. Epistle 49b (*Spiritual Beings*) refers to an eschatological era in the context of the "sphere of the human being" (*dāʾirat al-insān*): its first representative is Adam and the last the "master of the new cycle in the recommenced conjunction" (*ṣāḥib al-dawr al-jadīd fī al-qirān al-mustaʾnif*).[89] In Epistle 50 (*Governance*), the "recommencement of the final cycle" (*istiʾnāf al-dawr al-ākhir*) is connected with astrological events that determine the times of philosophical feasts.[90] The first reference connects the eschatological era with the

84 Ibid, 157.

85 Ibid.

86 That the manuscripts do not entirely support an Ismāʿīlī reading is also implicitly admitted by the editor and translator of the epistle. Hamdani has argued in various studies that the *Epistles* are an Ismāʿīlī work that predates the establishment of the Fāṭimid caliphate in 909. In the Arabic edition, he provides many untitled chapters with an explicitly Shīʿite title. In the technical introduction, he states that he uses MS ع as the base manuscript due to the wishes of the editorial committee. At the same time, he regrets the lack of a more clearly Shīʿite or Ismāʿīlī manuscript, which he hopes will be found in the possession of some Ismāʿīlī community. For Hamdani, many manuscripts instead show a later Ṣūfī appropriation of the treatise.

87 In both references to a 'cycle', MS ع has *dūn* for *dawr*, which does not make sense. Since this is the oldest and usually the most reliable manuscript, it still shows that the Ismāʿīlī reading of the passage was not self-evident for the copyists.

88 *Al-Jāmiʿa*, ep. 48; ed. Ghālib, 527; ed. Kacimi, 693.

89 *On God and the World*, 178. The relevant terms contain no significant variation in the manuscripts.

90 Ibid, 278. For the philosophical feasts, see Mattila, "The Philosophical Worship."

appearance of a savior figure; the second does not. Neither reference relates the final era to the manifestation of esoteric truth. Both seem too generic for a clearly Ismāʿīlī identification, especially since the *mahdī*'s appearance at the end of history is not an exclusively Ismāʿīlī idea. What these passages show is that some epistles in the fourth section convey messianic expectations; they do not show that the authors adhered to the Ismāʿīlī notion of alternating cycles of manifestation and occultation.

The most explicit references to the cycles of manifestation and occultation occur in two passages of Epistle 52b (*Magic*). The first reference states that prophets, imāms, and their followers are "apparent to sight" (*ẓāhirūna bi-l-ʿiyān*) and "existent in a place" (*mawjūdīna fī al-makān*) during the "cycle of manifestation" (*dawr al-kashf*); during the "cycle of occultation" (*dawr al-satr*), in contrast, they remain hidden.[91] In the cycle of manifestation, then, "their king is manifest in bodies and spirits" (*yaẓhar malikuhum fī al-ajsād wa-l-arwāḥ*); in the period of occultation, their cause "appears in souls, intellects, rulers of the earthly kingdom (*aṣḥāb al-mamlaka al-arḍiyya*), and bodily governance (*al-khilāfa al-jismāniyya*)." The second passage states that Iblīs's influence is the strongest during the cycle of occultation because the "proof of God" (*ḥujjat Allāh*) remains in hiding.[92]

Explicit references to cycles of manifestation and occultation are thus rare in the *Epistles*. They are only present in some form in four of the last five epistles of the fourth section. In Epistle 48, the coming era of manifestation is mentioned but, based on the manuscript variation, its explicitly Shīʿīte meaning appears to have emerged gradually. Epistles 49 and 50 bring up a final messianic era without relating it to the manifestation of truth. Only Epistle 52b refers clearly to cycles of both manifestation and occultation. Once again, it is then this particular variant of the final epistle that agrees with the *Comprehensive Epistle*. This gives further evidence for the lateness of Epistle 52b; it does not support the Ismāʿīlī identity of the Brethren of Purity.

In the *Comprehensive Epistle*, in contrast, references to Ismāʿīlī cycles are omnipresent.[93] As we have seen, the *Comprehensive Epistle* accords much im-

---

91 *Rasāʾil*, vol. 4, 379.

92 Ibid, 381.

93 *Al-Jāmiʿa*, eps. 1, 9, 14, 17, 18, 22, 36, 38, 39, 41, 47; ed. Ghālib, 53, 66–67, 68, 72, 73, 74, 84–85, 86, 87, 88, 89–90, 126–127, 128, 142, 148, 200, 211, 407, 408, 409, 411, 445, 464, 500, 522; ed. Kacimi, 47, 63, 65, 70, 71, 72–73, 84–85, 86, 87, 89, 90, 141, 143, 161, 172, 179, 254, 267, 535, 536, 537, 538, 539, 541, 542, 581, 606, 655, 686. One reference (ed. Kacimi, 179) is only present in the long version of the commentary on Epistle 18; thus, it does not occur in the Ghālib edition.

portance to such cycles in its interpretation of the story of Adam and Iblīs.[94] Iblīs's sin consists, in particular, in his refusal to accept the necessity of concealing the truth; the forbidden fruit represents esoteric knowledge. Many passages, then, discuss the two cycles in this context. Adam is the initiator of the cycle of concealment; the cycle of manifestation is related to the coming appearance of a messianic figure: "For the last of the admonishers (*mudhakkirīna*) and the seal of the warners (*munadhdhirīna*) is the pure soul (*al-nafs al-zakiyya*); it is through him that the beginning of the cycle of manifestation is opened, the cycle of concealment is lifted, and the truths become manifest."[95] As in the passages of the *Epistles* mentioned above, the cycle of manifestation mainly relates to the final era of truth and justice; the current era is one of concealment. Another passage, nevertheless, states that after Adam, the "affairs of religion" (*umūr al-dīn*) rotate as day and night, "at one time by manifestation, and at another time by concealment" (*marratan bi-l-kashf wa-marratan bi-l-satr*).[96] The "seventh leader" (*al-rāʾīs al-sābiʿ*) is "intermediate between the worlds of the cycles of concealment and manifestation" (*mutawassiṭ bayna ʿālamay dawr al-satr wa-l-kashf*): he exists "potentially" (*bi-l-quwwa*) in his predecessors by means of their knowledge of his future appearance.[97]

The situation is similar with the authors' commitment to precisely seven prophetic eras. In the *Epistles*, it is difficult to find a single passage that clearly denotes seven speaker-prophets or identifies the *mahdī* as the seventh in order. In Epistle 38 (*Resurrection*), the authors provide their rendering of the Qurʾānic story of seven sleepers in a cave.[98] The seven sleepers are identified with seven astronomical cycles but not with an explicit list of Qurʾānic prophets, let alone the Ismāʿīlī speaker-prophets.[99] Epistle 44 (*Beliefs of the Brethren of Purity*) relates this story to the "society" (*jamāʿa*) of the Brethren of Purity: "We were sleeping for a period of time in the cave of our father [Adam] … until the appointed time (*waqt al-mīʿād*) arrived. We awoke once the cycle of

94 See pp. 157–162.

95 *Al-Jāmiʿa*, ep. 17; ed. Ghālib, 148; ed. Kacimi, 172.

96 Ibid, ep. 36; ed. Ghālib, 411; ed. Kacimi, 541.

97 Ibid, ep. 39; ed. Ghālib, 464; ed. Kacimi, 606.

98 *Rasāʾil*, vol. 3, 315–320. For an analysis and English translation of the Ikhwānian version of the story, see Godefroid de Callataÿ, "Astrology and Prophecy: The Ikhwān al-Ṣafāʾ and the Legend of the Seven Sleepers," in *Studies in the History of the Exact Sciences in Honour of David Pingree*, ed. Charles Burnett et al. (Leiden and Boston: Brill, 2004), 758–785.

99 Yves Marquet, "La détermination astrale de l'évolution selon les Frères de la pureté," *Bulletin d'études orientales* 44 (1992): 144–145, claims that the seven cycles correspond to Adam, Noah, Abraham, Moses, Jesus, Muḥammad, and the *qāʾim*, but provides no references to support the claim.

sleep (*dawr al-ruqād*) came to an end; then we convened at the appointed time (*ijtamaʿnā ʿalā al-mīʿād*), after dispersal (*tafarruq*) in the lands, in the kingdom of the greatest law-giver (*mamlakat ṣāḥib al-nāmūs al-akbar*) and witnessed (*shāhadnā*) our spiritual city raised in the air, which we mentioned in another epistle [Epistle 48]."[100] In what follows, this "spiritual city" is first related to "Adam and his progeny," and then to Noah, Abraham, Moses, Jesus, Yazdān, and Afrīkhūn, a temple "of which Plato told," and Muḥammad.[101] The list agrees only in part with the Ismāʿīlī speaker-prophets. Apart from the unexpected occupants of the position between Jesus and Muḥammad, if Muḥammad occupies the seventh position, there is no place for the expected *qāʾim* in this scheme.

In Epistle 26 (*The Human Being is a Macrocosm*), in the context of analogies between cosmos and the human being, the authors talk of seven "virtuous persons" (*al-ashkhāṣ al-fāḍila*) in reference to planetary souls.[102] In the *Comprehensive Epistle*, this precise term often denotes the speaker-prophets. The authors of the *Epistles*, in contrast, make no analogy between planetary souls and the seven souls governing our world. In two epistles, they, on the contrary, criticize the excessive attachment to number seven. In Epistle 5 (*On Music*), the authors praise the "virtue" (*faḍīla*) of number eight, in particular, and criticize the "seveners (*al-musabbiʿa*) who are infatuated with sevenfold things, and thus set them above others; their views (*naẓaruhum*) are partial and their pronouncements (*kalāmuhum*) not universal."[103] The Brethren of Purity's universal "doctrine" (*madhhab*) instead appreciates the virtues of all ten numbers. The criticism is reiterated in the Pythagorean Epistle 32: "The seveners (*al-sabʿiyya*) went far in revealing sevenfold existents, so that wondrous things among them emerged to them. They were infatuated with them, and they exaggerated (*aṭnabū*) in mentioning them and neglected other numbers."[104] Later, the authors refuse to give a list of sevenfold existents, as they do for other numbers, because certain "savants" (*qawm min ahl al-ʿilm*) exaggerated their importance.[105]

In contrast, the *Comprehensive Epistle* is entirely explicit that there are precisely seven prophetic cycles and that the final era is initiated by the seventh

100 *On Companionship and Belief*, 38.
101 Ibid, 39–42.
102 *Rasāʾil*, vol. 2, 477.
103 *On Music*, 111; tr. Wright, 135–136, modified.
104 *Sciences of the Soul and Intellect, Part I*, 8.
105 Ibid, 25.

prophet.[106] Thus, "knowledge subsists (*qā'im*) in seven virtuous individuals (*ashkhāṣ fāḍila*), appearing in seven ages, each of whom conveys the power of the holy spirit by which all things are informed (*ikhbār ʿan al-ashyāʾ kullihā*); when each of them appears in their time, he sets out to disclose his message (*aqāma li-iblāgh risālatihi*)."[107] Unlike the *Epistles*, the *Comprehensive Epistle* abounds with analogies between sevenfold existents of the natural world and the seven prophets. Just as there are seven climes, so the "individual of the time" (*al-shakhṣ al-zamānī*) exists in seven days; just as the outermost sphere comprises all the inner spheres, or the human body and soul comprise the organs and psychical faculties, so the "seventh leader coming at the end of time" comprises the knowledge of the six leaders that precede him.[108] The *Comprehensive Epistle* also provides the analogy that the *Epistles* failed to make: the movements of the seven "virtuous individuals" in this world correspond to the spherical movements of the seven planetary bodies.[109]

As regards the composition of each cycle, the *Epistles* do not explicitly employ a concept of a law-bringing speaker-prophet (*nāṭiq*). "Speaking" (*nāṭiq*) is employed as one of many attributes for Adam, but in the sense of the first human being created with the faculty of speech rather than the first prophet to bring revelation to the world.[110] In one passage, all prophets (*anbiyāʾ*) are accorded the attributes of "living" (*aḥyāʾ*) and "speaking" (*nāṭiqūna*), and they are said to resemble angels in their "pure souls" (*anfusuhum al-zakiyya*).[111] But again the term "speaking" does not imply a superior rank within prophets; rather, these attributes apply to all prophets and messengers (*rusul*) equally. Possibly the closest resemblance with the Ismāʿīlī use occurs in Epistle 50 (*Governance*): "Our feasts (*aʿyādunā*) ... are speaking individuals (*ashkhāṣ nāṭiqa*) and active souls (*anfus faʿʿāla*) that accomplish, with their Lord's permission, the acts and deeds that were revealed to them and with which they were inspired."[112] The reference is to the four religious feasts of the Brethren of Purity presented as analogous to the corresponding philosophical and Islamic

106 *Al-Jāmiʿa*, eps. 26, 28, 36, 38, 39, 41, 44, 47, 48, 49; ed. Ghālib, 285–286, 307–308, 404, 410, 441–442, 447, 452, 463, 464, 465, 466, 469, 498, 517, 523, 524, 527, 531; ed. Kacimi, 379–380, 381, 408, 531, 540, 577–578, 584, 585, 591, 605, 606, 607, 608, 613, 653, 654, 679–680, 688, 689, 693, 699–700.

107 Ibid, ep. 26; ed. Ghālib, 285; ed. Kacimi, 380.

108 Ibid, ep. 26; ed. Ghālib, 285; ed. Kacimi, 381.

109 Ibid, eps. 36, 39; ed. Ghālib, 403–404, 410, 466; ed. Kacimi, 531, 540, 608.

110 *On Life, Death, and Languages*, 97.

111 *Rasāʾil*, vol. 3, 481–482.

112 *On God and the World*, 284; tr. Baffioni, 229–230.

feasts.[113] The first three are equated with the "appearance" (*khurūj*) of the first three of "our *qā'ims*;"[114] the fourth is a day of mourning due to "our return to the cave of dissimulation and concealment" (*kahf taqiyya wa-l-istitār*), a state of concealment that is also stated to prevail at the time of writing.[115] The general tenor seems reconcilable with the Ismāʿīlī idea of alternating cycles of manifestation and occultation, as well as the appearance of a special figure at the beginning of each era. But there are also significant differences: instead of seven speaker-prophets and a final cycle of manifestation, the terms *qā'im* and *nāṭiq* are equated, and we have three *nāṭiq*s followed by a period of concealment. Based on the particular relation between Epistle 50 and the *Comprehensive Epistle*,[116] it seems more plausible that these passages provide an initial inspiration for an Ismāʿīlī reading of the *Epistles* than that they clearly relate the work to the Ismāʿīlīs. In any case, Epistle 50 does not present the idea of precisely seven prophetic cycles of the *Comprehensive Epistle*.

Once again, the *Comprehensive Epistle* is unambiguous in its adherence to the Ismāʿīlī understanding of the term *nāṭiq*. Thus, the "uttermost of human knowledge about religious sciences (*al-ʿulūm al-sharʿiyya*) and divine laws (*al-nawāmīs al-ilāhiyya*)" is to attain the perfection of a "speaking prophet teaching the people of his time" (*nabiyy nāṭiq muʿallim li-ahl zamānihi*).[117] The "speaking individuals" (*al-ashkhāṣ al-nāṭiqa*) convey the law by means of the "power of the Soul supported by the goods of the Intellect."[118] Thus, the "speaking messenger" (*al-rasūl al-nāṭiq*) occupies in the corporeal world the position that the Intellect has in the spiritual world.[119] As regards the analogy between physical and religious movement, then, the "foundation of religious motions (*aṣl al-ḥarakāt al-dīniyya*)" is the "messenger enunciating God's command" (*al-rasūl al-nāṭiq bi-amr Allāh*) by means of his revelation or speech (*kalāmihi*).[120] In contrast to the *Epistles*, the term *nāṭiq* is now also attributed to Adam in an Ismāʿīlī sense: he is the "first speaker to announce God's commands and revelation" (*awwal nāṭiq naṭaqa bi-amr Allāh wa-waḥyihi*).[121]

113 See Mattila, "The Philosophical Worship."
114 *On God and the World*, 284–285.
115 Ibid, 285.
116 See pp. 132–137 above.
117 *Al-Jāmiʿa*, ep. 28; ed. Ghālib, 306; ed. Kacimi, 407.
118 Ibid, ep. 41; ed. Ghālib, 494; ed. Kacimi, 648–649.
119 Ibid, ep. 39; ed. Ghālib, 461; ed. Kacimi, 602.
120 Ibid, ep. 39; ed. Ghālib, 466; ed. Kacimi, 608.
121 Ibid, ep. 47; ed. Ghālib, 520; ed. Kacimi, 684.

A closely related term is present in the *Epistles*, however. In Epistle 26 (*The Human Being is a Microcosm*), the five sensing faculties with distinct objects of sensation are equated with "five prophets of determination[122] among the messengers" (*khamsa min al-anbiyāʾ ūlī al-ʿazm min al-rusul*); their "message (*mursaluhum*) is one even if their laws are different."[123] Epistle 32a (*Intellectual Principles*), within a list of fivefold things, lists the "prophets of determination" as Noah, Abraham, Moses, Jesus, and Muḥammad;[124] in a list of five-fold religious prescriptions (*sunan al-sharīʿa*), the "law-givers" (*wāḍiʿū al-sharīʿa*) are similarly said to be five.[125] This agrees with an Ismāʿīlī view where, although there are seven speaker-prophets, neither Adam nor the *mahdī* bring an exoteric law, and thus do not count as "prophets of determination."[126] It is not entirely clear that this is the case in the *Comprehensive Epistle*. Adam's followers are said to possess "determination" (*ʿazm*) because they resisted Iblīs's temptation, unlike Adam.[127] This means that Adam is not counted among the prophets of determination. Yet, in another passage, all messengers that brought a revealed book are given the attribute of "determination" without the qualification that either Adam or the seventh speaker-prophet are excluded.[128] The *Comprehensive Epistle*'s author also states that Adam is the first to "express the commands of the law" (*naṭaqa bi-awāmir al-nāmūs*).[129] In some passages, the "holders of religions" (*aṣḥāb al-sharāʾiʿ*) are said to be six in number, five of whom are "possessors of determination" (*ūlū al-ʿazm*),[130] while the "religions" (*sharāʾiʿ*) and "laws" (*nawāmīs*) are also six.[131] The position of the *Comprehensive Epistle* does not, then, seem to be entirely fixed. But in this case, some passages of the *Epistles* fit well with the specific Ismāʿīlī doctrine that limits the number of law-giving prophets to five; the *Comprehensive Epistle*, in part, also adheres to this view. This suggests an Ismāʿīlī influence on the *Epistles*. But

122 For this term as a near synonym of *nāṭiq*, see Daftary, *The Ismāʿīlīs*, 97, 131, 226.

123 *Rasāʾil*, vol. 2, 470; the passage is also present in MS ج.

124 *Sciences of the Soul and Intellect, Part I*, 27.

125 *Sciences of the Soul and Intellect, Part III*, 165.

126 According to Daftary, *The Ismāʿīlīs*, 97, 131, 226, the Qarmaṭīs, and early Ismāʿīlīs in general, recognized seven law-giving prophets, which were called *ūlū al-ʿazm*. In the scheme of al-Nasafī and Abū Ḥātim al-Rāzī, however, Adam does not count because he either introduced no law or at least did not repel an earlier law; the seventh *nāṭiq* similarly brings no new law.

127 *Al-Jāmiʿa*, ep. 9; ed. Ghālib, 124; ed. Kacimi, 138.

128 Ibid, ep. 7; ed. Ghālib, 112; ed. Kacimi, 123.

129 Ibid, ep. 36; ed. Ghālib, 406; ed. Kacimi, 534.

130 Ibid, ep. 39; ed. Ghālib, 463; ed. Kacimi, 605.

131 Ibid, ep. 39; ed. Ghālib, 467; ed. Kacimi, 609.

the passing nature of the relevant passages suggests that they serve the aim of providing examples of five-fold things rather than present a central doctrine for the authors. As we have seen, in other epistles, the authors give different lists of central prophets where their number amounts to neither five nor seven.

As regards the doctrine of imāmate, it seems clear that the *Epistles* manifest Shīʿite tendencies. But it is not clear these tendencies take a distinctly Ismāʿīlī form. In fact, neither the *Epistles* nor the *Comprehensive Epistle* formulates an explicit system of a heptad of imāms in each prophetic cycle.[132] Nevertheless, the *Comprehensive Epistle* again presents a more explicitly Shīʿite framework, for example, in a greater prevalence of Shīʿite terms and concepts. In Ismāʿīlī thought, it should in theory be imāms and their *dāʿī*s that are charged with the correct understanding of the inner meaning of revelation. Thus, the position of the two works with regard to imāmate may be approached by means of the question of the identity of the proper inheritors of prophetic knowledge.

The *Epistles* often employ the explicit phrase "inheritors of the prophets" (*warathat al-anbiyāʾ*). The proper subject of the prophetic inheritance is determined by distinct qualifications in different epistles. In Epistle 46 (*Belief*), it is the "true believers" (*al-muʾminūna al-muḥaqqiqūna*) that are the prophets' inheritors.[133] Epistle 9 (*Ethics*) states that the inheritors of prophets are "God's friends" (*awliyāʾ Allāh*) and "devout and perspicacious servants" (*ʿibādihi al-ṣāliḥīna al-mustabṣirīna*); they are "savants of the secrets of prophecies" (*al-ʿulamāʾ bi-asrār al-nubuwwāt*) who are trained in "philosophical propaedeutics" (*al-riyāḍāt al-falsafiyya*).[134] Thus, they know the true "nature" (*ḥaqīqa*) of the afterlife, and their "discipline" (*ṣināʿatuhum*) is to call people toward God and the afterlife. This involves, for example, their renouncing the world (*al-tazhīd fī al-dunyā*) and thus presenting an example to others. A distinguishing mark (*ʿalāma*/*ṣifa*) of these friends of God is that, in their meetings (*majālis*) and seclusions (*khalawāt*), they pronounce God's name and contemplate His creation, beneficence (*iḥsān*), grace (*inʿām*), and signs.[135] All this stems from their accurate knowledge of God, as they see Him through a "true vision" (*ruʾyat al-ḥaqq*).[136] Later in the same epistle, the authors talk of forty people chosen from among 400 "ascetic, knowing, and veridic" people (*al-zāhidīna al-ʿārifīna al-muḥaqqiqīna*), which are, in turn, chosen from among

132 Thus, Ismāʿīlī technical terms, such as "legatee" (*waṣiyy*), "foundation" (*asās*), "silent one" (*ṣāmit*), "fulfiller" (*mutimm*), or "door" (*bāb*), are mostly or entirely absent in both works.

133 *Rasāʾil*, vol. 4, 62, 69, 82.

134 *On Ethics*, 207.

135 Ibid, 207–208.

136 Ibid, 208–209.

4,000 "believing, repentant, and sincere" people (*al-muʾminīna al-tāʾibīna al-mukhliṣīna*).[137] These are the "inheritors of God's prophets and successors of His messengers on Earth" (*warathat anbiyāʾ Allāh wa-khulafāʾ rusulihi fī al-arḍ*), who inherited from the prophets knowledge, belief, worship (*taʿabbud*), God's support (*taʾyīd*) and inspiration (*ilhām*); as a result, they reject this world and desire the other world.[138] All of this is more reminiscent of the qualifications of a Ṣūfī saint, combined with philosophical knowledge, than of a Shīʿīte imām.

Other epistles identify "savants" or "scholars" (*ʿulamāʾ*) as the prophet's followers. Epistle 40 (*Causes and Effects*) states that it is the scholars (*ʿulamāʾ*) and sages (*ḥukamāʾ*) that constitute the inheritors of prophets; prophets, in turn, are God's mediators (*sufarāʾ Allāh*).[139] Prophets express God's meanings and thus convey them in various languages and to every nation; once the prophets are gone, savants and sages "take their place" (*qāmū maqāmahum*).[140] The beginning of Epistle 41 (*Definitions and Descriptions*) reiterates this idea: "prophets are mediators (*sufarāʾ*) between God and His creation; savants (*ʿulamāʾ*) are prophets' inheritors, and sages (*ḥukamāʾ*) are the most eminent (*afāḍil*) among savants."[141] The authors then quote a tradition that lists seven "praiseworthy properties" (*khiṣāl maḥmūda*) of a sage: (1) praiseworthy actions, (2) perfect skills (*al-ṣanāʾiʿ al-mutqana*), (3) veracious speech, (4) good character traits, (5) correct opinions, (6) pure deeds (*aʿmāluhu zakiyya*), and (7) truthful knowledge (*ʿulūmuhu ḥaqīqiyya*).[142] In his Persian rendering of the epistle, Nāṣir-i Khosrow attributes the *ḥadīth* to "people of inspiration" (*ahl-i taʾyīd*), supposedly the imāms.[143] But the traditions presenting "scholars" or "knowers" as the prophets' inheritors were employed by both Sunnite and Shīʿīte groups to legitimize their position.[144] These seven criteria do not clearly restrict religious knowledge to the imāms; they also seem entirely in place with the members of a religious-philosophical society, such as the Brethren of Purity.

137 Ibid, 211–213.

138 Ibid, 213–214.

139 *Sciences of the Soul and Intellect, Part III*, 70–71.

140 Ibid, 71.

141 Ibid, 178.

142 Ibid, 178–179.

143 Nāṣir-i Khosrow, *Kitāb Jāmiʿ al-ḥikmatayn*, ed. Henry Corbin and Muḥammad Muʿīn (Tehran and Paris: Shirkat golshan, 1953), 85, par. 86. For Khosrow's use of Epistle 41, see Poonawala's introduction, translation, and commentary in the edition of that epistle.

144 See Liyakat N. Takim, *The Heirs of the Prophet: Charisma and Religious Authority in Shīʿite Islam* (Albany: State University of New York Press, 2006), 33–36.

Finally, the position of Epistle 47 (*Divine Law*) is close to that of al-Fārābī's *Virtuous City*. There are certain prerequisites that a prophet or a law-giver must possess; in both the *Virtuous City* and the *Epistles*, the list of twelve qualifications probably goes back to Plato's *Republic*.[145] Once the law-giver is absent, these qualities are either present in a single individual or the community (*jamāʿa*) as a whole.[146] In the former case, that person is the "master of the time (*ṣāḥib al-zamān*) and imām to the people (*imām li-l-nās*) as long as he is alive."[147] Both the general idea and the employed terms again attest to the Brethren of Purity's Shīʿite tendencies. But the latter alternative does not seem particularly Shīʿite: "If these qualities (*khiṣāl*) are not combined in a single [person], but are dispersed in the community, then that community will agree on a single opinion and their hearts will be united in mutual love."[148] This contradicts the premise of the Shīʿite theory of imāmate, where a qualified, divinely guided, person must exist at all times.

The *Comprehensive Epistle* never employs the precise phrase "inheritors of prophets." Still, the idea is present, and seems to take a different form from the *Epistles*. One passage lists imāms and caliphs, alongside prophets, as being charged with "preserving prophetic stipulations" (*mustaḥfiẓūna li-l-wadāʾiʿ al-nabawiyya*).[149] They constitute the "coffin of wisdom" (*tābūt al-ḥikma*) raised by God, and are accompanied by the "coffin of divine presence" (*tābūt al-sakīna*) carried by angels that the "descendants inherit from their forebears" (*yatawārath al-khalaf ʿan al-salaf*).[150] Another passage states that, due to his physical body, Adam has only a limited time to "execute stipulations" (*nafādh al-ḥukm*); but the "degree of prophecy and ladder of messengership" (*manzilat al-nubuwwa wa-mirqāt al-risāla*) is preserved as "inherited by the pure among his descendants" (*tatawārathuhu al-ṣafwa al-ṭāhira min waladihi*).[151] Thus, it is the "seven leaders and their companions (*man yaṣḥubuhum*)" from "among the people of their houses" (*min ahl buyūtātihim*) who have "inherited their wisdom" (*warathū ḥikamahum*).[152] In contrast to the *Epistles*, the *Comprehensive*

145 *Rasāʾil*, vol. 4, 129–130; al-Fārābī, *Al-Farabi on the Perfect State*, ch. 15, §12, 246–248. For the claim that the qualifications of a ruler in both works are derived from Galen's paraphrase of the *Republic*, or its commentary by Ḥunayn Ibn Isḥāq (d. 873), see Hamdani, "The Ikhwān al-Ṣafāʾ," 199–201.

146 *Rasāʾil*, vol. 4, 125–126.

147 Ibid, 125. The Shīʿite terms are also present in MS ع.

148 Ibid.

149 *Al-Jāmiʿa*, ep. 9; ed. Ghālib, 124–125; ed. Kacimi, 139.

150 Ibid.

151 Ibid, ep. 9; ed. Ghālib, 127; ed. Kacimi, 143.

152 Ibid, ep. 35; ed. Ghālib, 410; ed. Kacimi, 540–541.

Epistle does not relate prophetic knowledge to such generic categories as true believers, scholars, sages, or friends of God; nor are moral purity and philosophical knowledge presented as its prerequisites. It does not, then, seem to be the case that the Prophet's followers could be virtuous philosophers or Ṣūfī saints. Instead, the author of the *Comprehensive Epistle* explicitly attributes the task of preserving prophetic legacy to imāms.

Moreover, the *Comprehensive Epistle* makes it clear that a divinely guided individual must always be present in the human world:

> Know, O brother, may God support you and us with His spirit, that the world as a whole with its lofty spheres and high heavens, with its spiritual lights, moving souls, and powers that flow to corporeal elements and physical bodies, all existents and the rest of the created beings, all that the heavens and Earth contain from the highest of the highest to the lowest of the lowest, all of it is a single body. That body is prepared to receive the universal emanation (*al-fayḍ al-kullī*) from its Creator, praise to Him whose sublimity is beyond speech (*taʿālā dhikruhu*). And the word of God (*kalimat Allāh*) is connected with it and supplies it with abundance (*ifāḍa*) and beneficence (*jūd*) so that it may be completed and preserved in existence. Its [the word's] first emanation and oneness (*ittiḥāduhā*) is the first created being (*al-mubdaʿ al-awwal*), which is the Active Intellect, then by its intermediacy [it flows] to the Universal Soul, which is the Passive Intellect, then by the intermediacy of the Universal Soul until the first matter, then by the intermediacy of the first matter until the absolute body (*al-jism al-muṭlaq*), and from there it spreads to the world as a whole. And it [the word of God] favors (*mukhtaṣṣa*) among virtuous human individuals (*al-ashkhāṣ al-insāniyya al-fāḍila*) the prophets, messengers, and pious servants (*al-ʿibād al-ṣaliḥīna*). For the human form is God's deputy (*khalīfat Allāh*) on His earth, charged with the governance of His lower world (*al-qāʾim bi-tadbīr ʿālamihi al-suflī*). And at each time and upon every conjunction (*qirān*) [of the spheres] human beings have a virtuous individual (*shakhṣ fāḍil*) that delivers to them God's commands (*yulqī ʿalayhi min amrihi*) that lead to the well-being (*ṣalāḥ*) of the inhabitants of that time, and that which points to Him and to His worship (*wa-dalālātuhum ʿalayhi wa-ʿibādātuhum lahu*). For He has appointed that individual (*wa-innahu mustakhlif li-dhālika al-shakhṣ*) whether by His speech, revelation, or from beyond the veil (*min warāʾ ḥijāb*), and he embraces His presence and what is behind it (*iḥāṭatuhu li-mā bayna yadayhi wa-mā khalafahu*); he is His face, tongue, and interpreter. And among the followers of his people and successors among the leaders of

> his religion (*wa-yakhlufuhu min ruʾasāʾ sharīʿatihi*) the first is the successor (*lāḥiq*), who is the closest to him … In this human world, the head of the world's time (*raʾs zamān al-ʿālam*) and the most eminent speakers (*afḍal al-nāṭiqīna*) are those informing about wisdom, the prophet and holder of religion (*nabiyy wa-ṣāḥib sharīʿa*), and his followers among the people of his religion and followers of his call (*aṣḥāb daʿwatihi*). Then the nation (*umma*) will be a single body with the prophet as its head, his follower (*khalīfatuhu*) as its heart, their companions (*aṣḥābuhum*) as its senses and organs, and his esoteric knowledge (*ʿilmuhu al-bāṭin*) as its spirit and soul.[153]

The eloquent passage highlights the cosmic necessity of imāmate. In very Ismāʿīlī terms, the author states that God's goodness is transmitted, by the intermediacy of His creative word, to the world as a whole; the world thus constitutes a single body united by the divine word. It follows that the divine word must also be conveyed to the human world by means of a "virtuous individual" who is always present, even between the sporadic appearance of prophets. Then the human world will also constitute a unified body. The *Epistles* contain no similar passage that emphasizes the continuous existence of an individual transmitting God's beneficence. On the contrary, one gets the impression that a group of individuals pursuing God's proximity by knowledge and virtue is sufficient.

Finally, the notion of a final speaker-prophet that initiates an eschatological era is explicit in the *Comprehensive Epistle*, but not clearly present in the *Epistles*. As regards the term *qāʾim*, in the *Epistles*, it never refers specifically to the seventh prophet. The passages of Epistle 50 discussed above come closest to the Ismāʿīlī use of the term: the Brethren of Purity's feasts are "speaking persons" (*ashkhāṣ nāṭiqa*); the first feast commemorates the "first that arises among us" (*awwal al-qāʾimīna minnā*).[154] Clearly, the reference here is not to the final era of human history. The *Comprehensive Epistle* employs the term in an Ismāʿīlī sense. During the era of occultation, people plead for the help of the "*qāʾim* at the time when the truth becomes manifest."[155] Iblīs's refusal to submit to the necessity of occultation is related to his desire to become the "one manifesting that affair by God's command" (*al-qāʾim bi-dhālika bi-amr Allāh*).[156] That is,

153 Ibid, ep. 26; ed. Ghālib, 287–288; ed. Kacimi, 383–385.
154 *On God and the World*, 284.
155 *Al-Jāmiʿa*, ep. 1; ed. Ghālib, 69; ed. Kacimi, 66.
156 Ibid, ep. 1; ed. Ghālib, 73; ed. Kacimi, 70.

Iblīs wants to assume the appointed *qāʾim*'s position as the one who reveals the esoteric truth to the world.

In the *Epistles*, the term *mahdī* is also not really present. In the animal fable of Epistle 22, a Khurasānian spokesman states that "we wait for the appearance in our lands of the expected imām (*al-imām al-muntaẓar*)";[157] the Beirut edition, but none of the manuscripts indicated by the critical edition, employs instead the expression "imām and savior (*al-imām al-mahdī*) that is expected from Muḥammad's family."[158] In any case, the doctrine is attributed to a specific group of people rather than to the authors. The only explicit reference to "an expected *mahdī*" (*al-mahdī al-muntaẓar*) occurs in the *Fihrist c* entry for Epistle 47 (*Divine Law*); *Fihrist a* instead uses the term "expected imām" (*al-imām al-muntaẓar*).[159] Besides the likelihood that the *Fihrist* postdates the *Epistles*, this merely attests to the progressive addition of Ismāʿīlī ideas to the manuscripts. Admittedly, the term is not common in the *Comprehensive Epistle*, either. The section on Epistle 47 reiterates the mention of an expected *mahdī* in *Fihrist c*.[160] Another passage states that the authors have assigned for "each class among nations" (*ṭāʾifa min ṭawāʾif al-umam*) individuals that "call people to us" (*yadʿūnahum ilaynā*) and proclaim the "appearance of our cause (*ẓuhūr amrinā*), the coming of our *mahdī*, and the rise of our *qāʾim*."[161]

The concept of *mahdī* is, nevertheless, very much present in the *Comprehensive Epistle*, although mostly by means of alternative terms. Among the most prominent is the expression "pure soul" (*al-nafs al-zakiyya*). In the *Epistles*, this term only occurs in Epistle 2 (*Geometry*),[162] in which a purified soul's separation from the body is related to a Qurʾānic verse (35:10): "To Him alone good words ascend, and righteous deeds are raised up by Him." The Qurʾānic passage is explained as a reference to the "pure soul"; not much context is given, but arguably the reference could be to the *mahdī*. However, according to the critical edition, both the Qurʾānic citation and a reference to the "pure soul" are absent in five manuscripts, including MS ج. In the *Comprehensive Epistle*, this expression refers consistently to the *mahdī*. Thus, the "cycle of happiness" (*dawr al-saʿāda*) begins upon the emergence of the "pure soul at the time of second creation (*al-ʿaraḍ al-thānī*/*al-nashʾa al-thāniya*)."[163] Many passages sim-

157 *The Case of the Animals*, 159.
158 *Rasāʾil*, vol. 2, 290.
159 *Rasāʾil*, vol. 1, 40; MS Atıf Efendi 1681, fol. 5a.
160 *Al-Jāmiʿa*, ep. 48; ed. Ghālib, 523; ed. Kacimi, 687.
161 Ibid, ep. 49; ed. Ghālib, 526; ed. Kacimi, 692.
162 *On Arithmetic & Geometry*, 127.
163 *Al-Jāmiʿa*, ep. 1; ed. Ghālib, 67; ed. Kacimi, 63.

ilarly relate the beginning of the cycle of manifestation to the appearance of the pure soul.[164] At the end of time, then, justice will prevail on Earth, and the "call (*daʿwa*) of the Brethren of Purity and Friends of Loyalty will emerge." God will then gather their attributes upon the "emergence (*ẓuhūr*) of the pure soul and a clean and tranquil spirit (*al-nafs al-zakiyya wa-l-rūḥ al-ṭāhira al-muṭmaʾinna*)."[165]

In sum, the two works seem to have distinct relations with the generally Shīʿīte, and specifically Ismāʿīlī, doctrine of prophecy and imāmate. Despite their Shīʿīte tendencies, the *Epistles* do not consistently insist on the necessity of imāmate; the *Comprehensive Epistle* is explicit in that there must always be a divinely guided intercessor in this world. The *Epistles*, by contrast, suggest that prophetic knowledge may be best interpreted by philosophers and saints. In the *Epistles*, the notion of prophetic cycles is at best vague; in the *Comprehensive Epistle*, it is a central and explicit doctrine. In the former treatise, number seven has no particular significance; in the latter, there are precisely seven prophetic cycles, and the eschatological savior figure is the seventh speaker-prophet. Only a few epistles refer to cycles of manifestation and occultation; the *Comprehensive Epistle* makes them a central doctrine that relates to the story of Iblīs, in particular. The contrast is also apparent at the level of terminology. The authors of the *Epistles* seem unaware of the special significance of terms such as "speaker-prophet" or "pure soul"; in the *Comprehensive Epistle*, these terms convey its central doctrines. The conclusion, then, is the same as in the question of God and creation. The *Comprehensive Epistle* gives the impression of an Ismāʿīlī treatise; the *Epistles* do not.

3 Brethren of Purity

The final section of this chapter addresses the Ismāʿīlī question indirectly. Since the *Epistles* first became known, the identity of the Brethren of Purity has formed a major mystery for their readers. It is not this book's aim to provide yet another theory about their identity. Instead, the interest here lies in the meanings that the two works assign to the term "Brethren of Purity." These are of direct relevance for the problem of the relation between the *Epistles* and the *Comprehensive Epistle*. If the two works were composed by the same

164 Ibid, eps. 1, 17; ed. Ghālib, 68, 75, 148; ed. Kacimi, pp. 65, 73, 172. See also Ibid, eps. 22, 41; ed. Ghālib, 157, 159, 493; ed. Kacimi, 192, 194, 647.

165 Ibid, ep. 22; ed. Ghālib, 203; ed. Kacimi, 257.

authors, both works should employ the term in a uniform manner. This would be the case, for example, if members of an Ismāʿīlī denomination composed both works, and the Brethren of Purity is thereby a codename for that denomination. By contrast, if the two works were authored by different people with distinct doctrinal allegiances, this should be reflected in their different ways of using the term. For instance, if the authors of the *Epistles* were a group of philosophically inclined secretaries in Baṣra, as al-Tawḥīdī claims, while the *Comprehensive Epistle*'s author was a later Ismāʿīlī scholar, clearly this should manifest itself in how the authors speak of the Brethren of Purity. The current section will, then, examine the import of the term "Brethren of Purity" for the theory of common authorship.

Given the debate on the Brethren of Purity's identity and ideological affiliation, it is surprising that there has been no systematic study of the different meanings that the *Epistles* assign to the term. It seems clear that "Brethren of Purity" has distinct referents. The expression may simply refer to (1) the authors or (2) their addressees. Often, however, the term refers to a community of people with utopian connotations. While focusing on different aspects of an ideal community, the "Brethren of Purity" may refer to (3) people with correct doctrinal beliefs or with (4) moral virtue; a group of people that (5) co-operates for the end of virtue and truth or (6) follows the prophets and/or imāms. Or else, instead of a utopian community, the term may depict the practices of (7) some concrete, factually existing, society.

The extent to which these aspects are coextensive depends on the theory of authorship. That is, if we accept al-Tawḥīdī's account of a Baṣran society with philosophical and religious aims, the first, fifth, sixth, and seventh aspects refer to the members of that society; the second aspect refers to that society's potential recruits; and the third and fourth indicate those who follow that society's specific sense of prophetic religion and ideal of virtue. If we identify a single author for the work, the first aspect represents the author's pseudonym, the third and fourth the author's definitions of prophetic religion and virtue, the fifth and sixth a utopian community, and the second a call to join the virtuous; the seventh aspect is either fictional or a reference to some group in which the author operated. If the *Epistles* represent a specific Ismāʿīlī denomination, the second aspect refers to their potential recruits and the other aspects to members of that denomination.

These seven aspects may be viewed as progressing gradually from the more generic toward the more specific. The first two levels, where the Brethren of Purity refer to the work's authors or addressees, are the most vague with no further specification of the group's nature. These represent the most common use of the term in both the *Epistles* and the *Comprehensive Epistle*. However, due to their generality, they are not particularly helpful in identifying differences between the two works. The other five aspects provide a better basis for comparison in that they furnish the term with specific content.

Starting with the third aspect, both works contain passages where the term is defined primarily by a certain kind of knowledge; that is, the Brethren of Purity are distinguished from other groups due to a doctrinal position they hold. Several kinds of such intellectualist definitions of the Brethren of Purity may be differentiated in the *Epistles*. The first kind relates the group to theoretical inquiry. Often, the group is characterized by their pursuit of theoretical knowledge in general;[166] thus, they have devoted their lives to acquiring knowledge about natural and divine things.[167] Their inquiry is marked by its "universal" (*kullī*/*ʿumūmī*) nature, as opposed to the particularistic inquiry of those who, for example, give excessive significance to a specific natural number.[168] Or their "school" (*madhhab*) is one that "examines all knowledge of existents in the world" (*al-naẓar fī jamīʿ ʿulūm al-mawjūdāt fī al-ʿālam*),[169] and investigates (*baḥth*) the causes of existents and the order in which they exist (*marātib niẓāmihā*).[170] Sometimes, the group's approach to theoretical knowledge is more qualified. They avoid logical contradictions in their discourse;[171] they do not claim knowledge of a particular science before acquiring profound knowledge of it;[172] or they focus on the investigation of natural existents (*umūr ṭabīʿiyya*).[173] Or even more specifically, they adhere to the Pythagorean claim that grades of existents correspond to natural numbers,[174] and to the claim that the soul is immortal and has an afterlife.[175] In Epistle 48 (*The Call to God*), the Brethren of Purity are, then, contrasted with people who have false opinions and are ignorant with regard to the question of the soul and its fate, in particular:

> We must, O brother, first start by revealing the layers of ignorance that enveloped us from the beginning. For it is the foundation (*aṣl*) for the

166 *Rasāʾil*, vol. 4, 386, 393, 394, 426.
167 Ibid, vol. 3, 533.
168 *On Music*, 111.
169 *On Arithmetic & Geometry*, 8–9. See also *The Call to God*, 92–93.
170 *On Geography*, 5; *The Call to God*, 83–84.
171 *On Logic*, 120–121.
172 *Rasāʾil*, vol. 4, 403.
173 *On the Natural Sciences*, 51.
174 *Sciences of the Soul and Intellect, Part I*, 16.
175 *Rasāʾil*, vol. 3, 300; vol. 4, 356; *On Arithmetic & Geometry*, 127–128.

> beginning of evils. Know that the ignorance that enveloped us prevents us from friendship and pure brotherhood. There are four kinds of ignorant people: (1) those who do not know the difference between soul and body. (2) Those who do not perceive how the soul was united with the body. (3) Those who do not know why the soul was united with the body. (4) Those who do not know how the soul is resurrected from the body. For necessarily, as long as the soul is not resurrected from the body, it will not know victory (*fawz*), salvation (*najāt*), and eternity (*khulūd*) in bliss; instead, it will stay in hell and painful torment.[176]

In other cases, the reference is to the Brethren of Purity's sources of theoretical knowledge. Thus, in their discussion of belief and actions as the two components of religion, they claim that their method is distinguished by the explanation of every aspect of each object of inquiry.[177] In general, they are characterized by their universal quest for knowledge; they "oppose no science, discard no book, and reject no doctrine (*madhhab*); for our opinion (*raʾy*) and doctrine comprises all doctrines and unites all sciences."[178] On the other hand, the group's aims are also designated as more limited for some scientific disciplines, such as mathematics; they provide the extent of technical knowledge that is sufficient.[179] The two claims are not necessarily contradictory; the Brethren of Purity's aims concern the big picture of reality for which only the general, but not the more minute, aspects of technical disciplines are relevant. All of these passages paint the Brethren of Purity as a primarily philosophical group aiming for knowledge about the nature of reality in general, and the soul and its afterlife, in particular.

Another group of passages relates the Brethren of Purity to not only theoretical but also to religious knowledge. The group's aim is, then, to unveil the true nature of reality (*ḥaqāʾiq*) in general, which requires philosophical (*ḥikmiyya*) as well as prophetic (*nabawiyya*) sciences.[180] Thus, they are distinguished by their adherence to true beliefs, as opposed to their adversaries' corrupt (*fāsida*) beliefs that harm the soul.[181] As a result, the Brethren of Purity's beliefs do not represent an illegitimate innovation (*muḥdath*/*mustaḥdath*) with respect to

176 *The Call to God*, 57–59.
177 *Rasāʾil*, vol. 3, 452.
178 *On Companionship and Belief*, 97. See also *Rasāʾil*, vol. 4, 328, 331, 368, 387.
179 *Rasāʾil*, vol. 3, 292; vol. 4, 339, 351–352, 394; *On Companionship and Belief*, 96–97.
180 *Rasāʾil*, vol. 3, 29.
181 *Rasāʾil*, vol. 3, 524; *On Companionship and Belief*, 108. For the idea of corrupt beliefs in Epistle 42, see Mattila, "The Ikhwān al-Ṣafāʾ on Religious Diversity."

the prophets but are founded on prophetic knowledge.[182] Another way of stating that the group embraces ultimate philosophical and religious knowledge is that they are in possession of the esoteric truth; that is, they understand the inner meanings of both the natural world and religious revelation. Many passages present the Brethren of Purity's distinguishing characteristic in this sense. They understand the mysteries (*asrār*) behind both the natural world and prophetic books.[183] Thus, it is their "habit" (*ʿāda*) to refer to these realities by means of images (*mithālāt*), indications (*ishārāt*), resemblances (*tashbīhāt*), and symbols (*rumūz*).[184] In consequence, the Brethren of Purity are endowed with the highest understanding of both natural and religious reality.[185]

Besides a philosophical group, the *Epistles*, then, depict the Brethren of Purity as being in possession of religious and esoteric knowledge. This is not far from al-Tawḥīdī's characterization of a group whose aim is to provide a synthesis between philosophy and religion. Still, as regards the doctrinal component attributed to the Brethren of Purity, philosophical knowledge constitutes its essential component.

The *Comprehensive Epistle* depicts a partly different picture of the group's doctrinal nature. When the work defines the Brethren of Purity in terms of knowledge, to some extent it agrees with the *Epistles*. As in the passages of the *Epistles* that emphasize esoteric knowledge, in general terms, the group is identified by their understanding of the mysteries of religious affairs; they are the "people of insight and holders of mysteries" (*ahl al-baṣāʾir min dhawī al-sarāʾir*).[186] As in the *Epistles*, in the specifics of their doctrine, they are distinguished by their Pythagorean metaphysics[187] and focus on the soul's afterlife.[188] Some of the characterizations of the Brethren of Purity's "doctrine" (*madhhab*) are different from the *Epistles*, however; for instance, their school emphasizes the harmonious nature of God's creation.[189] As regards the group's esoteric nature, they devised secret letters by which they guard the *Comprehensive Epistle* from the undeserving.[190] In the *Comprehensive Epistle*, then, the Brethren of Purity's theoretical pursuits are of an entirely religious and esoteric nature.

182 *Rasāʾil*, vol. 4, 126.
183 *Rasāʾil*, vol. 3, 302; vol. 4, 408, 411; *On Life, Death, and Languages*, 41.
184 Rasāʾil, vol. 4, 315; *On Companionship and Belief*, 42; *On Life, Death, and Languages*, 45, 123.
185 *On God and the World*, 85–86.
186 *Al-Jāmiʿa*, ep. 5; ed. Ghālib, 103; ed. Kacimi, 111.
187 Ibid, eps. 26, 33; ed. Ghālib, 260, 345; ed. Kacimi, 342, 460.
188 Ibid, ep. 5; ed. Ghālib, 103; ed. Kacimi, 111.
189 Ibid, ep. 33; ed. Ghālib, 345–346; ed. Kacimi, 460.
190 Ibid, ep. 25; ed. Ghālib, 236–239; ed. Kacimi, 304–308.

Similar characterizations are also present in the *Epistles*. But it is conspicuous that descriptions of the Brethren of Purity as a philosophical group are absent; they are not defined in terms of their pursuit of universal knowledge of all existents and their causes; nor by their eclectic qualities of drawing on every scientific and religious doctrine. Instead, they are an elect religious community with esoteric knowledge:

> I have warned you, O brother, against revealing your secrets (*kashf sirrika*) and laying open your affairs (*ẓuhūr amrika*) except for those among your believing brothers (*ikhwānika al-muʾminīna*) and accomplished peers (*aqrānika al-bālighīna*) from the Brethren of Purity and Friends of Loyalty that you trust; and [they are] the people of nobility (*ahl al-majd*) and sons of praise (*abnāʾ al-ḥamd*), purified by wisdom, blessed with grace (*niʿma*), and perfected by the form of completeness (*kamulat lahum ṣūrat al-tamām*). They came to know the religion of Islam (*ʿarafū sharīʿat al-islām*), entered the City of Peace (*dakhalū madīnat al-salām*), and were informed of the mountains of Aʿrāf; their procession (*sayr*) ended at their arrival in the abode of well-being (*manzil al-salāma*), and they settled in the house of permanence (*dār al-maqāma*).[191]

In contrast, in the fourth aspect concerning definitions of the Brethren of Purity in moral terms, the difference between the two works is not as pronounced. Countless passages of the *Epistles* depict the Brethren of Purity as the people of virtue. Often this is presented as a contrast with their vicious opponents; many passages employ the term "brethren of demons" (*ikhwān al-shayāṭīn*) as an antonym for the Brethren of Purity. The *Epistles*, then, repeatedly exhort their readers to renounce the character traits of the demonic brothers, which are antithetical to their own group's angelic character traits.[192] These two groups correspond to two fates in the afterlife: the demonic brothers will suffer from tribulations in the hell of generation and corruption; the pure brothers will enjoy an eternal bliss in the spiritual world.[193] Sometimes the group's virtue is given more specific content. One passage states that while some people deny that there is a "balance" (*mīzān*) for weighing good and bad actions, it is the "virtuous brethren" (*al-ikhwān al-fuḍalāʾ*) that possess this pre-

---

191 Ibid, ep. 1; ed. Ghālib, 79; ed. Kacimi, 78.

192 *Rasāʾil*, vol. 3, 484; *The Case of the Animals*, 142–143, 273; *On the Natural Sciences*, 413; *On Life, Death, and Languages*, 42; *On Ethics*, 101–103, 144–145.

193 *On Life, Death, and Languages*, 44–46.

cise discipline, and invoke others to adopt it.[194] In another passage, Socrates near death consoles his followers by saying that even if he is separated from his "wise and virtuous brethren" (*ikhwān ḥukamāʾ fuḍalāʾ*), he will join other "wise and virtuous brethren" that have preceded him in the afterlife.[195] In a further passage, the doctrine (*madhhab*) of the Brethren of Purity is equated with compassion (*raḥma*), sympathy (*shafaqa*), and tenderness (*taḥannun*) toward all people, even sinners and non-human animals.[196]

In the *Comprehensive Epistle*, the Brethren of Purity are also depicted as the people of righteousness. As in the *Epistles*, this is sometimes presented in the way of contrast with the vicious; on two occasions, the addressee is warned not to become one of the "brethren of demons" (*ikhwān al-shayāṭīn*).[197] The idea that the author's group will be rewarded and their adversaries punished in the afterlife is also very much present.[198] This is, of course, not very surprising given the treatise's Ismāʿīlī identity. But in the *Comprehensive Epistle*, this contrast sometimes appears in more explicitly sectarian terms; the Brethren of Purity's adversaries are, then, "soldiers (*junūd*) of Iblīs," opponents of prophets and imāms.[199] Again, the impression is, then, that the *Comprehensive Epistle* describes a religious sect; the *Epistles*, in contrast, speak of a community of pious individuals in possession of both philosophical and religious virtues.

The fifth aspect concerns a group of people working in co-operation for the ends of virtue and knowledge. When the Brethren of Purity are defined in this sense, most often it is in terms of a community that seeks the soul's salvation (*najāt*), for which knowledge and virtue constitute prerequisites. In Epistle 2 (*Geometry*), the necessity of a community is premised on the Aristotelian notion of human beings as social animals: a solitary individual can at best live a laborious (*nakid*) but not a pleasant life (*ṭīb al-ʿaysh*). Human beings are not capable of learning all the crafts they need during their short lives and therefore gather (*ijtamaʿa*) to co-operate in cities and villages.[200] As regards the aim of the soul's salvation, then, the authors state that "you are not capable of being saved (*tanjū*) by yourself from the tribulations (*miḥna*) and

194 *Sciences of the Soul and Intellect, Part I*, 85.

195 *On Companionship and Belief*, 80.

196 Ibid, 103.

197 *Al-Jāmiʿa*, ep. 1; ed. Ghālib, 66, 176; ed. Kacimi, 62, 220.

198 Ibid, eps. 21, 22, 23, 31; ed. Ghālib, 155, 176, 228–229, 331; ed. Kacimi, 190, 219–220, 293–294, 439.

199 Ibid, ep. 22; ed. Ghālib, 201; ed. Kacimi, 255.

200 *On Arithmetic & Geometry*, 115–118.

afflictions (*āfāt*) of this world caused by the sin (*jināya*) committed by our father Adam;" instead, co-operation (*muʿāwana*) of virtuous brethren, counselors (*nuṣaḥāʾ*), and friends (*aṣdiqāʾ*) is necessary to pave the way to the afterlife.[201] Many passages therefore exhort the reader to seek the companionship (*ṣuḥba*) of brothers, friends, and virtuous people (*akhyār fuḍalāʾ*) in order to acquire their character traits (*tatakhallaq bi-akhlāqihim*) and reflect on their knowledge (*tanẓur fī ʿulūmihim*), and thus be saved together with them.[202] In other cases, the authors give council to the addressee on how to aid their brothers and friends, for example, by divulging the *Epistles* in order to awaken their desire (*targhīb*) for knowledge and to renounce this world for the sake of the afterlife.[203] Since the aim of such a brotherhood is the soul's salvation, the co-operation is characterized as "psychical" (*nafsānī*) and "spiritual" (*rūḥānī*), rather than corporeal;[204] the Brethren of Purity are, then, like "a single soul in many bodies."[205] In some passages, the metaphor of a vessel is employed for this spiritual brotherhood: "Strive, O brother, before your age passes and the appointed time draws near, to ride with your brothers on the ship of salvation (*safīnat al-najāt*) with God's mercy; do not be among those who get drowned, the brethren of demons."[206]

In consequence, the Brethren of Purity appear to be a community that seeks salvation through mutual aid and companionship. There is a utopian ring to these passages; the impression is not necessarily that this is a concrete society that exists in the real world. In most cases, the virtuous brothers are equated with friends and virtuous people in general who should join forces to save their souls. Thus, the passages in effect reiterate the commonplace notion of Islamic philosophy that human beings are political animals that cannot attain satisfactory existence, let alone happiness, in isolation.[207] That is, the authors encourage their readers to seek other virtuous people to perfect their souls and thus reach the afterlife; they do not clearly designate an actually existing group of such people.

---

201 Ibid, 117. See also *Rasāʾil*, vol. 2, 476; vol. 3, 29; vol. 4, 126, 134. 443; *On the Natural Sciences*, 378–381; *On Companionship and Belief*, 42–43, 51–52, 86–87, 94, 100–101, 110–111.

202 *On the Natural Sciences*, 62–63; *On Geography*, 61; *On Life, Death, and Languages*, 41–42.

203 *Rasāʾil*, vol. 4, 144; *On Magic*, 7; *On Companionship and Belief*, 36–38; *On Composition*, 92–93; *On Ethics*, 76.

204 *On the Natural Sciences*, 152–153; *On Companionship and Belief*, 112–113.

205 *Companionship and Belief*, 113.

206 *The Case of the Animals*, 10.

207 For this idea in the Brethren of Purity and other authors, see Janne Mattila, "Utopia," in *The Routledge Companion of Islamic Ethics*, ed. Mohammed Ghaly (New York: Routledge, forthcoming).

In the *Comprehensive Epistle*, again, one gets more the impression of a larger religious society than of a small group of the virtuous. However, many of the same themes are present as in the *Epistles*. The addressee is advised to pass knowledge only to trusted brothers,[208] and to join other brethren in sailing the "ship of salvation" to the afterlife.[209] But in the *Epistles*, often the context is the philosophical idea of the human being as a political animal; in the *Comprehensive Epistle*, this context seems to be absent. The emphasis is instead entirely on the afterlife, and in the identification of the Brethren of Purity as the group that enters paradise. Thus, for example, there are no evocations to adopt the character traits (*akhlāq*) of one's more morally advanced brethren; instead, one should follow the moral example of prophets, imāms, and sages: "Since this is God's custom (*sunna*) in His creation, and the custom of sages among His servants (*al-ḥukamāʾ min ʿibādihi*) and the chosen ones among His prophets (*wa-l-muṣṭafīna min anbiyāʾihi*), we must also follow their way of life (*nasīr bi-sīratihim*) and adopt their character traits (*natakhallaq bi-akhlāqihim*), so that they provide a good example (*uswa ḥasana*) for us."[210]

Yet, the idea of a religious community is obviously also present in the *Epistles*. In the sixth meaning of the term Brethren of Purity, they are defined as those who follow the right way designated by the prophets; their opponents, by contrast, have strayed from the true prophetic teachings. Besides the more generic connotations of virtue and vice, the term "brethren of demons," discussed above, clearly also carries this meaning of religious perdition. In the *Epistles*, one group of passages suggests this idea implicitly by referring to the alternation of fortunes in this world; the Brethren of Purity have suffered misfortunes and remain hidden during the current evil times,[211] but their time will come.[212] A further passage contrasts knowledge of bodies with knowledge of religions (*adyān*), and relates the latter kind of knowledge to prophets, caliphs, and saints (*awliyāʾ*) as "physicians of the soul" (*aṭibbāʾ al-nufūs*); the authors then state that this is the "doctrine" (*madhhab*) of "our noble brothers" (*ikhwāninā al-kirām*) to which "we call the rest of our brothers" (*nadʿū ilayhi ikhwānanā al-bāqīna*).[213] The authors also affirm that the Brethren of Purity's doctrine is not an illegitimate innovation (*raʾy mustaḥdath/madhhab muḥ-*

208 *Al-Jāmiʿa*, ep. 22; ed. Ghālib, 177; ed. Kacimi, 221.

209 Ibid, ep. 22; ed. Ghālib, 176; ed. Kacimi, 219–220.

210 Ibid, ep. 14; ed. Ghālib, 138; ed. Kacimi, 156. See also Ibid, eps. 1, 31; ed. Ghālib, 91, 324; ed. Kacimi, 92, 429.

211 *The Call to God*, 174–176; *On Life, Death, and Languages*, 31.

212 *On God and the World*, 244–246.

213 *On Companionship and Belief*, 35–36.

dath) but instead treads the path followed by prophets, caliphs, and imāms.[214] The reference to the "society" (*jamāʿa*) of the Brethren of Purity as sleeping "in the cave of our father" similarly depicts their group as the prophets' followers.[215]

A number of passages binds the Brethren of Purity's identity to a more clearly Shīʿīte viewpoint. The "brethren of the truthful, divine, and lordly people" (*ikhwān al-ṣiddīqīna al-mutaʾallihīna al-rabbāniyyīna*) are stated to be the "inheritors of the prophets" (*warathat al-nabiyyīna*), again contrasted with those demonic brothers who have strayed away from the prophetic path.[216] One passage reprimands people filled with hatred who declare each other infidels (*yukaffir baʿḍuhum baʿḍan*), citing imām Ḥusayn in its support; these people are also "opposed" (*muḍādd*) to the Brethren of Purity.[217] The authors also contrast those "demonic brothers" who pursue worldly goods with the Brethren of Purity, and exhort them to emulate the model of the Prophet's family when they submitted their bodies to death at the Battle of Karbalāʾ.[218] However, this is presented within a lengthy list of similar testimonies by Brahmins (*barāhima*), Moses, Jesus, Abraham, Socrates, Plato, Aristotle, and Pythagoras.

All these passages taken together affirm that the Brethren of Purity take themselves to be the prophets' followers both in their beliefs and in their virtuous way of life. The latter passages further attest to the authors' Shīʿīte tendencies. At the same time, the references appear too vague to clearly attest to either a specific doctrinal allegiance or a clearly defined group. In one of the mentioned passages, Karbalāʾ is presented as one among many religious and philosophical instances of sacrificing the body for the sake of the soul and is not seen as representing the authors' primary religious identity. Other passages speak against their exclusively Shīʿīte identity. One passage polemizes against a "group" (*ṭāʾifa*) that has turned Shīʿism (*tashayyuʿ*) into "merchandise" (*maksab*);[219] some of them claim that imāms do not hear or respond to prayers; others say that the expected imām is concealed due to fear of his opponents; but "on the contrary, he is manifest (*ẓāhir*) in their midst; he knows them but they are concealed (*maknūn*) from him."[220] All of these people allege that their imāms (*khulafāʾuhum al-aʾimma*) are "inheritors of the prophets' knowledge"

214 *Rasāʾil*, vol. 4, 126.
215 *On Companionship and Belief*, 38.
216 *Rasāʾil*, vol. 4, 82.
217 *Rasāʾil*, vol. 3, 537.
218 *On Companionship and Belief*, 74.
219 *The Call to God*, 164–168.
220 Ibid, 166.

(*wārithū ʿilm al-nubuwwāt*); but "they do not know the truth of what they attest" (*lā yadrūna ḥaqīqat mā yuqirrūna bihi*).[221] The addressee is, then, warned of joining them: "I implore you, O brother, to not become like them; but be instead rightly guided (*hādin*) and righteous (*rashīd*), a physician (*ṭabīb*) and companion (*rafīq*) to your brethren, friends, and neighbors, so that you may guide the lost, heal the blind, and revive the dead by God's permission."[222] The admonition is clearly directed against a specific Shīʿite group that upholds the doctrine of a hidden imām. But the authors' alternative does not seem to be another Shīʿite sect but a group of virtuous brothers.

Similarly, Epistle 50 (*Governance*) designates the Brethren of Purity's society as the "most truthful of people" (*aḥaqq al-nās*) in their observation of religious worship (*al-ʿibāda al-sharʿiyya*).[223] At the same time, the epistle invokes a higher form of philosophical worship modeled after the Sabian religion of Ḥarrān, as well as a third form of worship that pertains to the Brethren of Purity alone.[224] In Epistle 47 (*Divine Law*), the authors state that every society (*jamāʿa*) requires a leader to arbitrate on what constitutes their welfare (*ṣalāḥ*); for the society of the Brethren of Purity this leader is the intellect, which God created to be the leader for the "eminent among His creation" (*fuḍalāʾ min khalqihi*).[225]

All of this relates to the question of the Brethren of Purity's doctrinal allegiance, which has remained unresolved in scholarship. As discussed in the introduction, there are doctrinal arguments both for and against the claim that the authors of the *Epistles* were Ismāʿīlīs. The argument here is that when one only looks at the explicit definitions of the society of the Brethren of Purity, they are too vague to imply more than the claim that their doctrine accords with the prophets. This claim agrees with al-Tawḥīdī's account about their society's nature: they combine Greek philosophy with Islamic religion.

Some definitions in the *Comprehensive Epistle* are similarly vague. For example, the author defines the Brethren of Purity by their wisdom, grace (*niʿma*), and "form of perfection" (*ṣūrat al-tamām*), as well as by their knowledge of the "religion of Islam" (*sharīʿat al-islām*).[226] Another passage equates the Brethren of Purity with the "savants of the religious discipline" (*ʿulamāʾ al-ṣināʿa al-*

221 Ibid, 166–167.

222 Ibid, 167–168.

223 *On God and the World*, 280.

224 For these two forms of worship, see Mattila, "The Philosophical Worship." For the relationship between the *Epistles* and the Sabians, see also Mattila, "Sabians."

225 *Rasāʾil*, vol. 4, 127.

226 *Al-Jāmiʿa*, ep. 1; ed. Ghālib, 79; ed. Kacimi, 78.

dīniyya).[227] The *Epistles* make no use of these expressions to define the Brethren of Purity; but they also do not contradict the idea that the society constitutes the most truthful among the Prophet's followers.

Many other references diverge from the attributes given to the Brethren of Purity in the *Epistles*. The first instance is the address of the reader with the attached phrase of "our brethren before you" (*man qablaka min ikhwāninā*).[228] This expression never occurs in the *Epistles*. It emphasizes the Brethren of Purity's nature as a group whose existence goes back in time. This synchronic aspect is stated in more specific terms in reference to the Prophet's companions as the "first Brethren of Purity in the Muḥammadan law and Ḥashimite religion" (*awwal Ikhwān al-Ṣafāʾ fī al-sharīʿa al-muḥammadiyya wa-l-milla al-hāshimiyya*); "to them arrived revelation (*waḥy*) from heaven by Gabriel's tongue."[229] The upshot is that the Brethren of Purity designates not so much a concrete group at the present time as the people that since the birth of Islam have followed the Prophet. Every Muslim denomination may define itself in these terms. But a second group of references associates the Brethren of Purity with Ismāʿīlī doctrines. Thus, the appearance (*ẓuhūr*) of the Brethren of Purity's "call" (*daʿwa*) is attached to that of the "pure soul" (*al-nafs al-zakiyya*), that is, the seventh speaker-prophet (*nāṭiq*).[230] The "reign of the Brethren of Purity" (*dawlat Ikhwān al-Ṣafāʾ*) is also related to the final days of the human world.[231] Another passage states that the "seventh leader appearing at the end of time" is the "lord of the Brethren of Purity" (*sayyid Ikhwān al-Ṣafāʾ*); he comprises the knowledge of the previous seven leaders.[232] A further passage refers to the speaker-prophets as leaders of the Brethren of Purity.[233]

The implication is that the Brethren of Purity's leaders are prophets and imāms; their society is, then, in effect a codename for an Ismāʿīlī denomination. In the *Epistles*, the expression "lord of the Brethren of Purity" never occurs. Only Epistle 50 (*Governance*) makes a vague reference to "their lords" (*sayyiduhum*) in relation to the four Ikhwānian feasts.[234] They are not seven in number; nor are they connected with eschatology. In the *Epistles*, the general idea is rather

227 Ibid, ep. 39; ed. Ghālib, 472; ed. Kacimi, 616–617.
228 Ibid, ep. 22; ed. Ghālib, 199, 203; ed. Kacimi, 252, 257.
229 Ibid, ep. 25; ed. Ghālib, 245; ed. Kacimi, 317.
230 Ibid, ep. 22; ed. Ghālib, 203; ed. Kacimi, 257.
231 Ibid, ep. 36; ed. Ghālib, 399; ed. Kacimi, 527. See also Ibid, ep. 38; ed. Ghālib, 452; ed. Kacimi, 591.
232 Ibid, ep. 26; ed. Ghālib, 286; ed. Kacimi, 381.
233 Ibid, ep. 22; ed. Ghālib, 177; ed. Kacimi, 221.
234 *On God and the World*, 282.

that the Brethren of Purity constitute an elect society that best follows the precedent of past prophets and philosophers. The difference becomes even more evident in what the *Comprehensive Epistle* omits; it does not reiterate the claim that the intellect is the real leader of the Brethren of Purity.

This contrast is reinforced by the final category of definitions in the *Epistles*, which describe the Brethren of Purity as an actually existing society. These are restricted to a few epistles of the fourth part, in which the Brethren of Purity's community assumes more concrete forms. The subject of Epistle 45 (*Companionship of the Brethren of Purity*), in particular, is precisely the nature of that community. Much of the epistle, nevertheless, addresses friendship at an abstract level, resembling the various discussions of the subject in Arabic literature.[235] The theme of friendship is, however, complemented with concrete instructions for recruiting new members to the Brethren of Purity.[236] The treatise starts with an injunction to arrange regular meetings (*majlis*) for the study of philosophical and religious texts, including a generic list of the subjects to be studied:

> Know, O pious and compassionate brother (*al-akh al-bārr al-raḥīm*), may God support you and us with His spirit, that our brothers must, may God support them in the lands where they reside, have designated sessions (*majlis khāṣṣ*) where they gather in appointed times, which none but they attend. In them, they study together (*yatadhākarūna*) their sciences and discuss (*yataḥāwarūna*) their secrets. Most of their studies (*mudhākarāt*) must be on the science of the soul (*ʿilm al-nafs*), the senses and sensibles (*al-ḥiss wa-l-maḥsūs*), the intellect and intelligibles (*al-ʿaql wa-l-maʿqūl*), the investigation (*baḥth*) and inspection (*naẓar*) of the secrets of divine books (*al-kutub al-ilāhiyya*), prophetic revelations, and meanings comprised in religion (*maʿānī mā tataḍammanuhā al-sharīʿa*). They must also study the four mathematical sciences, I mean arithmetic, geometry, astronomy, and harmony (*taʾlīf*). But their greatest concern (*ʿināya*) and objective must be the investigation of theology (*al-ʿulūm al-ilāhiyya*), which is the ultimate objective. On the whole, our brothers, may God support them, must not oppose any science, renounce any book among the books of sages, or be zealous of any creed (*yataʿaṣṣabū ʿalā madhhab min al-madhāhib*). Because our view (*raʾy*) and creed embraces all creeds

235 See Nuha A. Alshaar, *Ethics in Islam: Friendship in the Political Thought of al-Tawḥīdī and his Contemporaries* (New York: Routledge, 2015).

236 See, for example, *On Companionship and Belief*, 100–101.

> and combines all sciences. This is because it consists in the inspection (*naẓar*) of the totality of existents, sensible and intelligible, from the first to the last, outer (*ẓāhir*) and inner (*bāṭin*), apparent (*jaliyy*) and hidden (*khafiyy*), by means of the eye of reality (*bi-ʿayn al-ḥaqīqa*) insofar as all of them come from one principle (*mabdaʾ wāḥid*), one cause, one knower (*ʿālim*), and one soul that comprises their different substances, diverse genera, wide-ranging species, and varied particularities (*juzʾiyyāt*).[237]

This sounds much like al-Tawḥīdī's account of the Brethren of Purity as a group of secretaries in Baṣra with study sessions on philosophical and religious sciences. The passage agrees with that account's claim that the Brethren of Purity's ultimate objective is to attain a synthesis between Greek philosophy and Islamic religion. The passage is also in agreement with the central characteristic of the *Epistles*; the famous phrase about the Brethren of Purity's creed comprising all sciences and doctrines perfectly summarizes the eclectic qualities of their encyclopedia. The passage, then, describes a group with both philosophical and religious interests. It seems difficult to reconcile the admonition against zealous adherence with the claim of the Brethren of Purity's exclusively Ismāʿīlī identity.

Perhaps the most interesting aspect of Epistle 45 is the introduction of a four-fold hierarchy of the Brethren of Purity: (1) those with pure soul characterized by a good intellectual faculty, (2) political leaders with a faculty of practical judgments (*al-quwwa al-ḥukmiyya*), (3) kings characterized with a legislative faculty (*al-quwwa al-nāmūsiyya*), and (4) those who have submitted to divine inspiration (*taʾyīd*) and a vision of the truth (*mushāhadat al-ḥaqq ʿiyānan*), characterized by an angelic faculty (*al-quwwa al-malakiyya*).[238] The four faculties are associated with the ages fifteen, thirty, forty, and fifty years, respectively; presumably, the theoretical capabilities of the first stage provide a foundation for the deliberative and political understanding gained by experience, while the final stage represents a profound vision in preparation for the afterlife. This classification is repeated in Epistle 48 (*The Call to God*).[239]

Epistle 48 also presents the Brethren of Purity as a larger secret society. They have supporters scattered in many lands and among all classes of people from princes and ministers to workers and secretaries.[240] As in Epistle 45, the

237 Ibid, 96–97.

238 Ibid, 131–133.

239 *The Call to God*, 69–72.

240 Ibid, 32–33.

addressee is given instructions for recruiting new members.[241] In other parts of the epistle, the authors offer perhaps their most explicitly utopian aspirations in calling the reader to join their version of the "virtuous city" (*madīnā fāḍila*), which is quite different from the one described by al-Fārābī.[242] The society's aims do not reside in worldly power, but they seek "heavenly rule" (*al-mulk al-samāwī*), that is, the soul's salvation in the spiritual world.[243] Through mutual aid, they intend to build a "spiritual virtuous city" (*madīna fāḍila rūḥāniyya*) whose members are all virtuous and wise; its foundation must lie in neither land nor water but in air so that the smoke of the "despotic cities" (*al-mudun al-jāʾira*) cannot reach it.[244]

A comparison with the corresponding commentaries in the *Comprehensive Epistle* is revealing. In most manuscripts of the *Comprehensive Epistle*, Epistle 45 is shifted into the position of the final epistle. This accords it particular prominence:

> Let us finish it [the *Comprehensive Epistle*] with a commentary on the epistle of the companionship of the Brethren of Purity and Friends of Loyalty; what is their mutual co-operation (*muʿāwana*) like; how must they be in their gatherings (*ijtimāʿāt*), feasts (*aʿyād*), and associations (*jamāʿāt*), places of worship (*mawāḍiʿ ʿibādāt*), and prayers (*ṣalawāt*); and in their study (*mudhakārāt*) of their sciences and wisdom. We inform you, O brother, of its nature so that you may arrange them in accordance with these grades (*li-turattibahum ʿalā hādhihi al-marātib*) and govern them by means of this governance (*siyāsa*) in order for them to be distinguished by the grade that pertains to them and not others .... This is the epistle composed on the quality of companionship (*kayfiyyat ʿishra*). It contains many lofty points (*mawāḍiʿ jalīla*) and valuable wisdoms (*ḥikām nafīsa*); for that reason, we ended [the *Comprehensive Epistle*] with a discussion and commentary on this epistle.[245]

241 Ibid, 32–38.

242 See Carmela Baffioni, "Al-Madīnah al-Fāḍilah in al-Fārābī and in the Ikhwān al-Ṣafāʾ: A Comparison," in *Studies in Arabic and Islam: Proceedings of the 19th Congress, Union Européenne des Arabisants et Islamisants, Halle 1998*, ed. S. Leder et al. (Leuven: Peeters, 2002), 3–12; Idem, "Temporal and Religious Connotations of the 'Regal Policy' in the Ikhwān al-Ṣafāʾ," *Mélanges de l'Université Saint-Joseph* 57 (2004).

243 *The Call to God*, 39–41.

244 Ibid, 58–62.

245 *Al-Jāmiʿa*, ep. 45; ed. Ghālib, 537; ed. Kacimi, 708–709.

Even this initial characterization does not entirely correspond to the original epistle's contents. The cited passage is correct in that Epistle 45 discusses the Brethren of Purity's mutual co-operation and study sessions; it also provides a scheme of four grades within their community. But the epistle says nothing about feasts, places of worship, or prayers. Nor does it provide instructions for governance. All these subjects are discussed in Epistle 50 (*Governance*). By highlighting these subjects at the end of the treatise, the passage, then, once again suggests a connection between Epistle 50 and the genesis of the *Comprehensive Epistle*. It also indicates that the author's aim is to read Epistle 45 as a description of a religious denomination. The original context of a philosophical account of friendship is completely absent; in fact, the commentary hardly mentions the concept of friendship at all.

Instead, the *Comprehensive Epistle*'s author gives instructions on how to organize a meeting of an esoteric religious denomination:

> Know, O brother, may God support you and us with His spirit, that what we must advise, deliver, and inform you about, and confide to you, is how to take care (*murāʿāt*) of your brothers, companions that precede you (*man qablaka min aṣḥāb*), and those that answer to your call (*man istajāba ilayka*), and they will answer, God willing. You must arrange them sessions that gather together their community (*tajmaʿ jamāʿatahum*) on one day every twelve days, during which they convene in whatever place or locality they happen to be in and where they feel safe. Their gathering must be founded on fear of God and His surveillance (*taqwā min Allāh wa-khīfat murāqabatihi*); they must clean (*yataṭahharūna*) and purify (*yatanaẓẓafūna*) themselves before their arrival; and they must adopt the seemliest appearance they can (*yaʾkhudhūna zaynatahum bi-aḥsan mā yaqdirūna ʿalayhi*). And when they gather at your presence and you observe them, none of them may sit down, unless some excuse prevents them from approaching and coming before you. Then come out and appear before them in your attire and garment, your beautiful and venerable appearance, just as the Universal Soul appears before particular souls. For they are to you like children, and you are their father; they are like bodies to you, you are like their soul; they are like houses to you, you are like their dweller. Because your wisdom will be deposited in them, and your spirit descends upon them. You must come to them with gravity (*sakīna*) and dignity (*waqār*), whether it is night or day … Then recite to them your wisdom and admonish them with your reminders (*tadhkira*) in accordance with their capabilities and what their minds allow. Teach and inform them by the admonitions (*mawʿiẓa*) and reminders you provide

> them; urge them to seek knowledge and [emphasize] that their greatest concern and aim, utmost intention and endeavor, is to investigate (*baḥth*) divine things and intellectual secrets (*al-asrār al-ʿaqliyya*). These are the ultimate objective in pleasure and utmost bliss of the people of paradise (*ghāyat naʿīm ahl al-janna*); through it is attained human perfection (*istikmāl*), ascent from the world of senses, liberation from the darkness of bodies, and salvation from the captivity of nature, desire, and the sea of matter ... Inform them that the most proper act and revered deed is to seek knowledge (*tafaqqud*) about your brothers and govern their affairs (*tadbīr umūrihim*), and [about] knowledge of religious and worldly forms of governance (*al-siyāsāt al-dīniyya wa-l-dunyawiyya*) ... and that they must love each other, seek together union, guidance, love, and mutual aid, and not oppose and be separated from each other, or quarrel and harbor enmity. Inform them about the conduct (*ādāb*) of the prophets, the qualities of the sages, and the character traits of believers. Recite to them these epistles from the first to the last, epistle-by-epistle, chapter-by-chapter, and explain to them by means of clear indications (*bi-awḍaḥ al-dalāla*) ... And [for the students who passed the test] recite to them the guarded books (*al-kutub al-maṣnūna*), preserved secrets (*al-asrār al-makhzūna*), and hidden sciences (*al-ʿulūm al-maknūna*) by explaining (*sharḥ*) what is contained in the *Comprehensive Epistle* and other books that we have presented and deposited to you, which are the four teachings (*al-madāris al-arbaʿa*), seven books (*al-kutub al-sabʿa*), twenty-five epistles (*rasāʾil*), fifty-two epistles, and the *Comprehensive Epistle*.[246]

This gathering seems very different from the study sessions described in Epistle 45, nor is this how "companionship" was characterized in the original epistle. In Epistle 45, the starting point was a philosophical ideal of a mutually beneficial friendship. The members of the Brethren of Purity should, then, convene to discuss both philosophical and religious sciences with an open mind for "every book" and "all creeds." The explicitly mentioned subjects included the "science of the soul," knowledge about sensible and intelligible reality, the four mathematical sciences, and the theological sciences. The *Comprehensive Epistle* instead gives instructions on how a religious instructor should address a congregation. No philosophical sciences are mentioned; the curriculum instead includes prophetic stories, "divine things," and "intellectual secrets," religious and worldly governance, the *Epistles*, and, for those students who have reached

246 Ibid, ep. 45; ed. Ghālib, 538–539; ed. Kacimi, 710–713.

the appropriate stage, the *Comprehensive Epistle* and further esoteric writings. Most conspicuously, Epistle 45 paints a picture of a study session with peers of equal standing; the *Comprehensive Epistle* instead tells of a religious authority preaching to a congregation. The *Comprehensive Epistle* also makes no mention of the four categories of the Brethren of Purity that correspond to distinct age groups. If, as the two cited passages claim, the epistle's aim is to give instructions about how to govern the community of the Brethren of Purity, it is strange that these stages of development within the brotherhood are not mentioned. Once again, then, the *Epistles* characterize the Brethren of Purity as a community of virtuous, wisdom-seeking brothers; the Comprehensive *Epistle* depicts them instead as a hierarchic religious denomination.

As for the utopian community of Epistle 48, the *Comprehensive Epistle* translates it into purely Ismāʿīlī terms. The *Comprehensive Epistle*'s author starts by an enigmatic reference to books that can only be read by them, and whose truths are only known by the ones that preceded them.[247] Before introducing the virtuous city, the author states that "we have established (*aqamnā*) for each faction (*ṭāʾifa*) among nations (*umam*), to whom the call of prophets (*daʿwat al-anbiyāʾ*) has been directed, people who call toward us, indicate our presence, inform of our coming … and of the appearance of our *mahdī* and the rise of our *qāʾim*."[248] The author then calls for the founding of a "city that joins us together" (*madīna taḍumm shamlanā*),[249] and proceeds to explain the real meaning of the "spiritual and noble virtuous city" (*madīna rūḥāniyya fāḍila sharīfa*).[250] The foundation of that city is the "kingdom of the master of the greatest law" (*mamlakat ṣāḥib al-nāmūs al-akbar*), the "seventh one" (*al-amr al-sābiʿ*) who governs souls during the "kingdom of the sixth one" (*mamlakat al-sādis*).[251] That city's foundation is in the "fear of God" (*taqwā min Allāh*) and "truthful speech" (*al-ṣidq fī al-aqāwīl*); it is built on the four "columns" (*arkān*) of (1) fidelity, (2) purity (*ṣafāʾ*), (3) veracity (*ṣiḥḥa*), and (4) certainty (*yaqīn*), as well as the "pillars" (*daʿāʾim*) of the knowledge of the (1) Intellect, (2) Soul, (3) nature, and (4) matter.[252] Its inhabitans excel in reading the Qurʾān (*al-kitāb al-mubīn*), and their ultimate objective consists in seeking the eternal bliss of paradise.[253]

247 Ibid, ep. 48; ed. Ghālib, 526; ed. Kacimi, 691.
248 Ibid, ep. 48; ed. Ghālib, 526; ed. Kacimi, 692.
249 Ibid.
250 Ibid, ep. 48; ed. Ghālib, 526–527; ed. Kacimi, 692.
251 Ibid, ep. 48; ed. Ghālib, 527; ed. Kacimi, 693.
252 Ibid, ep. 48; ed. Ghālib, 527–528; ed. Kacimi, 694.
253 Ibid, ep. 48; ed. Ghālib, 528; ed. Kacimi, 694.

Both treatises, then, depict the society of the Brethren of Purity as aiming for a utopian "virtuous city," a spiritual kingdom whose goal is salvation in the afterlife. In the *Epistles*, however, the utopia's emphasis is on friendship and brotherhood, an egalitarian society seeking salvation in co-operation. In the *Comprehensive Epistle*, in contrast, the utopian society clearly refers to an Ismāʿīlī community. This is especially evident in the fact that the treatise binds this utopia to the eschatological figure of the seventh speaker-prophet.

One final detail remains to be mentioned, a detail that provides a further contrast between the two works. This concerns the classification of the Brethren of Purity into four classes in Epistles 45 and 48. In these epistles, the classification is presented as a temporal progression from theoretical aptitude to political knowledge, and finally preparedness for the afterlife. Thus, it bears little resemblance to the hierarchical arrangement of the Ismāʿīlī *daʿwa*. An interesting aspect of the classification is that the first three classes are given distinct names: (1) pious and compassionate brethren (*ikhwānunā al-abrār al-ruḥamāʾ*), (2) virtuous and good brethren (*ikhwānunā al-fuḍalāʾ al-akhyār*), and (3) virtuous and noble brethren (*ikhwānunā al-fuḍalāʾ al-kirām*).[254] The classification itself is not repeated outside the two epistles. However, in the *Epistles*, the standard invocation employed for the addressee is "O pious and compassionate brother" (*ayyuhā al-akh al-bārr al-raḥīm*).[255] This seems to imply that the addressees lie at the first stage of the four categories, that is, they have the intellectual disposition to join the society of the Brethren of Purity. Parts of the other two titles are employed in three invocations,[256] but their precise forms are not. Instead, the epithet "good and virtuous" (*al-akhyār al-fuḍalāʾ*) is applied to those savants and sages that follow the prophets,[257] and the epithet "noble and virtuous" (*al-kirām al-fuḍalāʾ*) to the brothers that the addressee should follow in either knowledge or moral qualities.[258] In one passage, the souls of the "believing, truthful, good, virtuous, and pious" (*al-muʾminīna al-muṣaddiqīna al-akhyār al-fuḍalāʾ al-abrār*) are presented as a third stage corresponding to prophets and saints (*ṣiddīqīna*).[259] In the *Comprehensive Epistle*, the epithet "pious and compassionate" is used in many invocations, and that of

254 *On Companionship and Belief*, 131–135.

255 In the Beirut edition, it is used in the singular a total of 75 times.

256 Epistle 27 (*Rasāʾil*, vol. 3, 8) employs the phrase "noble, pious, and compassionate brother" *al-akh al-karīm al-bārr al-raḥīm*; Epistles 28 and 48 (*Rasāʾil*, vol. 3, 25; *The Call to God*, 149) use the expression "virtuous brother" (*al-akh al-fāḍil*).

257 *The Case of the Animals*, 277; *On Composition*, 96.

258 *Rasāʾil*, vol. 3, 29, 302, 452; vol. 4, 64; *On Companionship and Belief*, 51, 100.

259 *Rasāʾil*, vol. 4, 117.

the "virtuous brother" once.[260] The other two epithets are never attributed to either the Brethren of Purity, prophets, or sages. Thus, the *Comprehensive Epistle* seemingly employs the common epithet it takes from the *Epistles*, but it is ignorant of the coded meaning of the other two epithets.

In this section, I have presented the meanings assigned to the term Brethren of Purity as the final piece of evidence against the theory of common authorship. It seems that the two works assign a fundamentally different overall sense to that community. In the *Epistles*, the Brethren of Purity's community is a group of individuals striving for knowledge and virtue; in the *Comprehensive Epistle*, it is an organized religious association often defined in explicitly Ismāʿīlī terms. In providing this contrast, I may have attributed a false unity to the *Epistles*, which goes against the argument for the work's gradual redaction that I presented in the previous part of this book. In fact, the previous analysis of the term's employment also provides additional support for that conclusion. There does indeed seem to be a dissonance between the epistles of the first three sections and a few epistles of the fourth section of the *Epistles*. The former relate the term "Brethren of Purity" either to a specific doctrinal view or to moral purity, or else to the necessity of co-operation for their attainment. None of the references clearly indicate that such a society exists; the references instead take the form of a call to pursue knowledge, moral purity, and mutual aid. It is only in very few epistles in the fourth section that the Brethren of Purity's society assumes a concrete form. The *Comprehensive Epistle*, finally, transforms that concrete philosophical-religious society into an organized religion.

4 From the *Epistles* to the *Comprehensive Epistle*

The *Epistles* and the *Comprehensive Epistle* make an identical claim about their mutual relationship: the latter is the final epistle that provides explicit explanations for the meanings left implicit in the former. Thus, the *Epistles* act as an introduction to the *Comprehensive Epistle*. The implication is that both works form a part of an original plan of composition. This second part of the book has argued that this claim is not in fact true; it is highly unlikely that the *Epistles* and the *Comprehensive Epistle* were composed by the same authors. This is because textual evidence does not support the claim.

The first class of counter-evidence, discussed in Chapter 4, is technical in nature. First, the claim of common authorship for the two works has in part

260 *Al-Jāmiʿa*, ep. 1; ed. Ghālib, 65; ed. Kacimi, 61.

been based on the fact that the *Epistles* contain references to the *Comprehensive Epistle*. Chapter 4 shows that most of these references do not occur in the most ancient manuscripts. Only the four references in Epistle 52b (*Magic*) are unambiguously present not only in the Beirut edition but also in most manuscripts. Given the likelihood that Epistle 52b postdates the *Epistles*, the conclusion is that the references in the *Epistles* do not support the theory of common authorship. On the contrary, the fact that most references are later interpolations suggests that the *Comprehensive Epistle* was finished well after the *Epistles*, and, only at that point, were references to the *Comprehensive Epistle* added to some manuscripts to update the expanded nature of the Ikhwānian corpus.

Second, it was also shown in Chapter 4 that the beginning of every epistle commentary in the *Comprehensive Epistle* is textually based on the corresponding entry in variant c of the *Fihrist*. The *Fihrist*'s manuscript variants and status within the *Epistles* were discussed in Chapter 1. The conclusion was that the *Fihrist* in general is likely to have been composed after the *Epistles*; this is almost certainly the case for its latest variant, *Fihrist c*. Since the *Comprehensive Epistle* depends on *Fihrist c*, the same conclusion obviously applies to its dating.

A third, somewhat related, technical argument may be added at this point. Compared to the major manuscript variation that the *Epistles* manifest, the *Comprehensive Epistle*'s text appears surprisingly stable. The work also gives no indications of a gradual redaction of the *Epistles*; instead, it seems to be based on the fixed variant of the text that we have in the Beirut edition. This is also what the text's dependence on *Fihrist c* tells us; *Fihrist c* was clearly composed at a point when the *Epistles* had assumed their final form. As a result, the *Comprehensive Epistle* comments on (1) fifty-two, rather than fifty-one, epistles; (2) in the form and (3) order in which they appear in the latest iteration of the *Epistles*. Thus, for example, in most manuscripts of the *Epistles*, Epistles 12 (*On Interpretation*) and 13 (*Prior Analytics*) constitute a single epistle, instead of two separate epistles; all *Comprehensive Epistle* manuscripts treat them as two separate epistles.

A second class of counter-evidence, discussed in Chapters 5 and 6, is doctrinal. The doctrinal differences between the two works are subtle enough for the theory of common authorship to have appeared plausible. But the doctrines of the two works also involve contradictions. Thus, unlike the *Epistles*, the *Comprehensive Epistle* relates the story of Adam and Eve's fall, first, to the lapse in the cosmological order of creation and, second, to the cycles of manifestation and occultation. The *Epistles* uphold God's transcendence in a more lenient sense; at times, the authors allow human knowledge and predication of attributes of God; at other times, their position appears close to the negative theology in the Muʿtazilite sense. The *Comprehensive Epistle* is more categoric in its denial of the attributes of God, even suggesting that the real creator of the world is the Intellect rather than God. The concept of the cycles of manifestation and occultation is almost non-existent in the *Epistles*; it is central in the *Comprehensive Epistle*. The number seven is of little significance in the *Epistles*; when a list of prophets is given, it does not correspond to the seven speaker-prophets of Ismāʿīlī doctrine. The *Comprehensive Epistle* is adamant that there are precisely seven prophetic cycles, and that the final eschatological era is initiated upon the appearance of the seventh speaker-prophet. The *Epistles* depict the society of the Brethren of Purity as an egalitarian group of individuals pursuing knowledge and virtue; in the *Comprehensive Epistle*, it is instead a hierarchic religious denomination.

All these differences attest to an Ismāʿīlī adoption of the *Epistles*. This is also supported by the third class of terminological evidence. In all of the subjects mentioned above, the doctrinal contrast is made more apparent by the *Comprehensive Epistle*'s employment of technical Ismāʿīlī terms. In general, many Ismāʿīlī terms that are omnipresent in the *Comprehensive Epistle* occur rarely or not at all in the *Epistles*. Beyond this, in some cases, the use of these terms in the two works involves a contradiction in meaning. Thus, for example, the *Epistles* only employ the term "speaking" (*nāṭiq*) with reference to prophets in the sense that Adam was the first human being with the faculty of speech. The authors seem to be completely unaware that the term has a technical meaning in reference to the seven speaker-prophets initiating a new religious era. In contrast, the *Comprehensive Epistle* consistently employs the term in precisely the sense of a speaker-prophet. In the *Epistles*, "virtuous individual" (*shakhṣ fāḍil*) refers to a planetary sphere; in the *Comprehensive Epistle*, it instead refers to a speaker-prophet. In the former work, Throne and Footstool refer to the two outermost astronomical spheres; in the latter work, they refer to the hypostases of the Intellect and the Soul.

One could explain these differences by the *Comprehensive Epistle*'s declaration of its objective: it explicates the meanings left implicit in the *Epistles*. Thus, the fact that it attributes explicitly Ismāʿīlī explanations to the *Epistles* is due to the original authors' effort to veil these meanings under concealed references. This might be a plausible explanation why Ismāʿīlī terms are so much more prominent in the *Comprehensive Epistle*. But it does not explain actual contradictions regarding the doctrine or meaning assigned to specific terms. Moreover, with regard to the contents of the doctrine or the employed terms, in most cases, there seems to be no clear rationale for the original authors to have concealed them.

The results of the comparison, then, suggest that the *Comprehensive Epistle* is an Ismāʿīlī adaptation of the *Epistles*. By implication, this means that the *Epis-*

tles are not an exclusively Ismāʿīlī treatise, even if the work manifests generic Shīʿīte and, to a lesser extent, specifically Ismāʿīlī influences. Thus, the history of the Ikhwānian corpus is one of a gradual Ismāʿīlī adoption of the Brethren of Purity. Besides the *Comprehensive Epistle*, traces of this adoption are also visible in the manuscripts of the *Epistles*. We have seen instances of this adoption in the two parts of this book. In the first part, we saw that some of the readings that *Fihrist c*, in particular, attributes to individual epistles are of an Ismāʿīlī character. In the second part, we saw that some manuscripts tend to introduce Shīʿīte terms to Epistle 48 (*The Call to God*); the interpolation of references to the *Comprehensive Epistle* obviously also constitutes an effort to incorporate the original epistles to its doctrinal framework.

The manuscripts provide abundant further evidence for an effort to give the *Epistles* a more explicitly Ismāʿīlī character. In the appendices of her critical editions and a series of articles, Baffioni has documented various additions of this nature present in individual manuscripts.[261] Some of these additions also occur in a close to identical form in the *Comprehensive Epistle*, as well as in the *Super-Comprehensive Epistle*.[262] Thus, two manuscripts include at the end of Epistle 16 (*Heaven and World*) a lengthy additional section on cosmology, prophetic cycles, and Judgment Day.[263] The addition conforms to Ismāʿīlī doctrine and terminology, such as the idea of seven "speaker-prophets" (*nāṭiq*). In the *Comprehensive Epistle*, an identical passage is divided between the commentaries on Epistles 34 (*The World is a Big Human Being*) and 47 (*Divine Law*).[264] A single manuscript (MS أ) contains an addition at the end of Epistle 50 (*Governance*) that discusses letters in the context of the cycles of manifestation and occultation.[265] This passage is not contained in the *Comprehensive Epistle*, but the sections on Adam agree doctrinally with similar discussions in that work.[266] An addition in a single manuscript of Epistle 10 (*Eisagōgē*), as well as a concise alternative variant of Epistle 50, also present in only one manuscript, depict the hierarchy of creation in Ismāʿīlī terms.[267] Neither passage occurs in the *Comprehensive Epistle*.

261 Baffioni, "Esoteric Shiʿism"; Idem, "Embryology"; Idem, "Ismaili Onto-Cosmological Doctrines"; Idem, "Onto-cosmology"; Idem, "The Figure of Iblīs"; Idem, "The Esoteric Meaning."

262 For the textual relation between the *Comprehensive Epistle* and the *Super-Comprehensive Epistle*, see the next chapter.

263 The addition is edited and translated in Baffioni, "Esoteric Shiʿism."

264 *Al-Jāmiʿa*, eps. 34, 47; ed. Ghālib, 351–357, 519–525; ed. Kacimi, 466–476, 683–690.

265 Baffioni, "Embryology," 177–179. For this passage, see also Idem, "The Figure of Iblīs."

266 Baffioni, "The Esoteric Meaning."

267 Baffioni, "Ismaili Onto-Cosmological Doctrines," 38–39, 43–44. For the short variant of Epistle 50, see also Idem, "Onto-cosmology."

Baffioni provides two possible interpretations for these additions: a stronger thesis takes them as evidence for an Ismāʿīlī commitment of the authors of the *Epistles*; a weaker thesis interprets them as showing that the *Epistles* were adopted by Ismāʿīlī authors earlier than is often assumed.[268] I concur with the latter alternative. For me, these additions, once again, show that the *Epistles* did not sufficiently correspond to Ismāʿīlī doctrine for some authors. Thus, there was an effort to incorporate them more firmly into an Ismāʿīlī framework. The fact that some of the added passages also appear in the *Comprehensive Epistle* shows that this process was at least in part connected with the composition of that work. The oldest known manuscript of the *Comprehensive Epistle* is dated at 1200 CE; it is, then, only slightly younger than the most ancient manuscript of the *Epistles* (dated 1182 CE). These manuscripts postdate the final stages of the redaction of the *Epistles* by at least two centuries. It is, then, clear that the process of Ismāʿīlī adoption was initiated during the intervening period.

As a result, we have the *a parte ante* and *a parte post* dates for the composition of the *Comprehensive Epistle*: it was composed after the final redaction of the *Epistles*, that is, the late tenth century, if we accept al-Tawḥīdī's account; the dating of the first manuscript at 1200 CE provides the last possible date. This leaves an interval of roughly two centuries. Since the *Comprehensive Epistle* is textually based on *Fihrist c*, which clearly postdates the final redaction of the *Epistles*, the work is more likely to have been written toward the end of these two centuries than immediately after the *Epistles*.

The results of this second part of the book also provide some indications of the concrete methods employed in the composition of the *Comprehensive Epistle*. First, it was shown in Chapter 4 that the beginning of each epistle commentary draws on the corresponding epistle entry of the *Fihrist*. The most plausible explanation for this textual relation is that the author employed the *Fihrist* as a primary source for the composition of the *Comprehensive Epistle*. An alternative explanation is that the same author composed both texts. The latter alternative seems unlikely since the *Comprehensive Epistle* regularly expands on *Fihrist c* with more explicitly Ismāʿīlī readings of the corresponding epistle.

Second, the evidence discussed in Chapter 4 suggests that, in the *Epistles*, Epistle 50 (*Governance*) was considered the original *Comprehensive Epistle*. In Chapter 6, it was mentioned that the *Comprehensive Epistle* justifies the trans-

---

268 Baffioni, "Esoteric Shiʿism," 326; Idem, "Embryology," 179; Idem, "Onto-cosmology," 217–218; Idem, "The Figure of Iblīs," 138–141.

position of Epistle 45 (*Companionship of the Brethren of Purity*) to the position of a final epistle by attributing to this epistle subjects that are actually only discussed in Epistle 50. The subject matter of Epistle 50, therefore, assumes particular significance for the *Comprehensive Epistle*. As a result, there is clearly a particular relation between Epistle 50 and the *Comprehensive Epistle*. Against this background, it is highly interesting that a single manuscript provides an alternative version of Epistle 50, which is in effect a summary of much of the *Epistles*.[269] This variant presents an account of the cosmological order of creation employing the Qurʾānic terms employed in similar Ismāʿīlī accounts; kinds of governance (*siyāsa*), the second of which is related to the imāmate; the degrees of existence in our sublunar world; the human being as a microcosm; eternity and the ultimate fate of the human soul; and talismans and forms of administration.[270] Most of these themes are also present in the *Comprehensive Epistle*, others are not. It is, then, not clear that this is a proto-*Comprehensive Epistle*. But at least this shows that Epistle 50 was perceived as occupying a central role in interpreting the *Epistles* in the manuscript tradition, even in the sense of summarizing the central contents of the *Epistles*.

The *Comprehensive Epistle* is, then, the most prominent testimony for the Ismāʿīlī adoption of the *Epistles*. As shown in this part of the book, the adoption process consists of several components. The first component is, as shown in Chapter 6, the naturalization of the *Epistles* into an Ismāʿīlī framework in the sense of reading the original work by means of Ismāʿīlī doctrines and technical terms. Second, as shown in Chapter 5, it consists of reducing the contents of the *Epistles* to a number of key theological themes, such as the cosmological order of creation, particular souls' afterlife, eschatology, prophecy and imāmate, and an esoteric interpretation of the story of Adam and Eve. These themes are, to varying extents, already present in the *Epistles*. But the *Comprehensive Epistle* transforms them into almost the complete contents of the *Epistles*. The author attains this by two main methods. First, the *Comprehensive Epistle*'s focus is on a relatively restricted number of epistles; the commentaries on most epistles are concise and are often limited to expanding the corresponding epistle entry in the *Fihrist*. The epistles related to the original work's philosophical-scientific stratum, in particular, are given little space. Second, even in cases where the commentary on an epistle with scientific subject matter is more extended, the *Comprehensive Epistle*'s author omits the original epistle's technical contents. As a result, the thematic emphasis of the epistles is reversed: the original epis-

269 This variant is edited and translated by Baffioni in *On God and the World*, 279–319.

270 See Baffioni's summary in Ibid, 279–283.

tle mainly consists of an overview of a specific subject, such as number theory, to which esoteric concerns are attached almost incidentally. In the *Comprehensive Epistle*, that specific subject, if mentioned at all, becomes a footnote in the midst of the work's major theological themes. Thus, an encyclopedia of philosophical sciences with theological aims is transformed into a treatise of philosophical theology.

Epilogue: The *Super-Comprehensive Epistle*

This book's thesis is that the textual corpus attributed to the Brethren of Purity underwent a gradual evolution from a mainly philosophical work to a largely theological work. The first stage involves the emergence of the *Epistles* of the Brethren of Purity, the seminal text in the corpus, as a result of gradual redaction. At a later stage, the contents of the final version of the *Epistles* were filtered through an Ismāʿīlī lens in the *Comprehensive Epistle*. This progression reaches its culmination in the *Super-Comprehensive Epistle* (*Risālat Jāmiʿat al-jāmiʿa*). In contrast to the *Comprehensive Epistle*, the *Super-Comprehensive Epistle* makes no effort to provide a systematic overview of the contents of the *Epistles* in the sense of summarizing each of its epistles. Instead, the work claims to present the doctrinal core of the original epistles; at the beginning and end, the author identifies the treatise's aim with presenting the "prime part" or "quintessence" (*zubda*) of the *Epistles* of the Brethren of Purity.[1] The result is a dogmatic presentation of a number of key doctrines where the textual relation with the original work is more obscure than was the case with the *Comprehensive Epistle*.

There have been very few studies on the *Super-Comprehensive Epistle*. Like the *Comprehensive Epistle*, the work presents itself as a work of the Brethren of Purity in the sense of employing the first person plural and Ikhwānian stock phrases. Tamir, the treatise's editor, subscribes to the Ismāʿīlī theory of authorship with regard to the *Epistles* and seems to accept that both the *Comprehensive Epistle* and *Super-Comprehensive Epistle* were composed by the same authors as the original epistles.[2] In contrast to the *Comprehensive Epistle*, however, the theory of common authorship seems less likely for the *Super-Comprehensive Epistle* at the outset. Kacimi accepts the common authorship of the *Epistles* and the *Comprehensive Epistle* but situates the *Super-Comprehensive Epistle* in a later Ismāʿīlī context.[3] This is supported by the manuscript tradition. All four known manuscripts originate in the Syrian Salamiyya region with Ismāʿīlī connections; one manuscript relates the text explicitly to "Ismāʿīlī doctrine" (*ʿaqāʾid al-Ismāʿīliyya*).[4] Since the oldest known manuscript is dated at 1241 AH/1825 CE, and, unlike the *Comprehensive Epistle*, the work is apparently not cited by medieval Ismāʿīlī authors, it also seems likely that the work con-

1 *Jāmiʿat al-jāmiʿa*, 62, 229.
2 See Tamir's introduction in Ibid, 5–61.
3 Kacimi, "Reflexiones."
4 Ibid, 331–332.

siderably postdates the *Comprehensive Epistle*. Kacimi therefore concludes that the treatise was written by a resident of Salamiyya with the aim of providing an Ismāʿīlī résumé of the *Comprehensive Epistle*, with the focus on key themes rather than the entire text. Baffioni's two studies note the *Super-Comprehensive Epistle*'s attribution to the Ismāʿīlī imām Aḥmad Ibn ʿAbd Allāh, and the Ismāʿīlī nature of its interpretation of the story of Adam.[5] Baffioni does not take a stance on its textual relation with either the *Epistles* or the *Comprehensive Epistle*.

Many questions concerning the *Super-Comprehensive Epistle* remain unanswered. Who is the author of the treatise? Does the author share the ideological commitment of the authors of the *Comprehensive Epistle* or the *Epistles*? Or does the author rather rework the material to suit doctrinal commitments pertaining to a different context? Most importantly, what is the treatise's textual relation with the other two texts? The work's title suggests that its author depends primarily on the *Comprehensive Epistle*. That is, the *Super-Comprehensive Epistle* is a summary of the *Comprehensive Epistle*, which is itself a summary of the *Epistles*. If this is the case, we have three clearly defined textual steps in the evolution of the Ikhwānian corpus; the *Super-Comprehensive Epistle* is in effect a super-commentary of the *Epistles*. In the absence of systematic studies, a definite answer to questions regarding the authorship and dating of the *Super-Comprehensive Epistle* is not yet possible. As regards the work's textual relation with the two other works, however, I am able to provide a definite answer.

A major question concerning the *Super-Comprehensive Epistle* is why it was written in the first place. If we have the *Epistles* and their Ismāʿīlī reading in the *Comprehensive Epistle*, what need is there for a further Ismāʿīlī summary? Kacimi's hypothesis of a late Ismāʿīlī treatise seems plausible. The author's intention is then to "Ismāʿīlize" the *Epistles* by reducing its contents to a few doctrines. In contrast to the many explicit references that the *Comprehensive Epistle* makes to its esoteric objectives, the *Super-Comprehensive Epistle* does not provide as systematic a disclaimer. A few passages, however, indicate motives for the composition of the treatise. In one passage, the author states that the treatise provides "what is necessary" (*bi-mā yuḥtāj ilayhi*) so that it may "substitute" the epistles in case they are not available.[6] Another passage promises to provide "clarity and proof" (*al-bayān wa-l-tawḍīḥ fī al-burhān*), as has been "the stipulation (*sharṭ*) for this *Super-Comprehensive Epistle*, where we

5 Baffioni, "Eléments"; Idem, "La figure d'Adam."

6 *Jāmiʿat al-jāmiʿa*, 81.

gather the goals (*aghrāḍ*) that befit its eminent name and perfect position (*mā yalīq bi-ismihā al-fāḍil wa-maḥallihā al-kāmil*) so that it may be sufficient for the one who attains it … and understands its contents with clarity."[7] The main point of both passages is the same: the *Super-Comprehensive Epistle* conveys the indispensable doctrines of the *Epistles*; one who reads it no longer really needs the original epistles. This is the relation that the *Comprehensive Epistle* claims to have with the *Epistles*. Thus, like the former work, the author claims to have left "open and clear speech" (*al-qawl bi-l-taṣrīḥ al-wāḍiḥ*) to this epistle.[8]

This brings up the question of the *Super-Comprehensive Epistle*'s relation with the *Comprehensive Epistle*. The two works supposedly occupy a largely identical function: both explicate the essential doctrines that were left implicit in the *Epistles*. If the *Comprehensive Epistle* provides the final explication, does the *Super-Comprehensive Epistle* provide an even more final one? Given that the author of the *Super-Comprehensive Epistle* approaches this task by summarizing the central themes of the *Epistles*, as opposed to providing a systematic commentary, this seems to be the case. The *Super-Comprehensive Epistle*'s role is then to provide a more accessible summary of an Ismāʿīlī reading of the *Epistles* that dispenses with the structural division of the original work. It is remarkable, however, that the work never mentions the *Comprehensive Epistle*. Thus, the impression is that the author intends the work to replace the *Comprehensive Epistle* rather than to complement it with a more concise summary.

This impression is confirmed by the work's structure. It is unlikely to be accidental that the *Super-Comprehensive Epistle* consists of 52 chapters, which is the number of epistles in later manuscripts and modern editions of the *Epistles*, as well as the number of epistle commentaries in the *Comprehensive Epistle*. This is the case despite the fact that the order of the subjects that the chapters address is entirely different from that of the original epistles. It seems that the author agrees with the authors of the *Comprehensive Epistle* and the final variant of the *Epistles* that this particular number is of special significance. Thus, the number of chapters confirms that the *Super-Comprehensive Epistle* indeed provides a comprehensive summary of the doctrinal contents of the *Epistles*.

A textual comparison with the *Comprehensive Epistle* largely resolves the enigma of the *Super-Comprehensive Epistle*. It turns out that the latter work is not only a summary of, but indeed a digest of passages cited from, the former work (see Table 19). This relationship may be illustrated by two instances. The subject matter of Chapter 1 of the *Super-Comprehensive Epistle* corresponds to

7 Ibid.
8 Ibid, 112.

that of Epistle 29 (*Life and Death*). When the text of the *Super-Comprehensive Epistle* is compared with the corresponding epistle commentary in the *Comprehensive Epistle*, it is found that the former simply reproduces a passage in the latter.[9] There are some differences between the wordings of the two texts, but these are minor. This is the case until the end of the passage where both texts cite Qurʾānic verses. The *Super-Comprehensive Epistle* expands the Qurʾānic citations in the *Comprehensive Epistle* and inserts new verses. The concluding phrase of Chapter 1 is again cited verbatim from the *Comprehensive Epistle*.

Chapter 4 in the *Super-Comprehensive Epistle* corresponds to Epistle 34 (*The World is a Big Human Being*). In this case, the *Super-Comprehensive Epistle* reproduces the beginning of the commentary on Epistle 34 in the *Comprehensive Epistle*.[10] The variation in wordings is slightly greater than in the previous case; the Qurʾānic citations are now almost identical. The *Super-Comprehensive Epistle* even reproduces the *Comprehensive Epistle*'s self-reference in a modified form. The latter refers to "this chapter in this *Comprehensive Epistle*"; the former amends the reference to "this chapter of this *Super-Comprehensive Epistle*."[11]

More than half of the chapters (1–2, 4–26, 30, 43–44, 48, 50–52) of the *Super-Comprehensive Epistle* have this simple textual relation with the *Comprehensive Epistle*: each chapter consists of a single passage taken word for word from the *Comprehensive Epistle*. Moreover, groups of chapters in the former treatise consist of a more or less continuous citation; Chapters 4–12 and 14–24 both cite a continuous section corresponding to Epistles 34 and 38, respectively.

For the rest of the chapters, the textual relation between the two works is more complex; each chapter is patched together from two or more cited passages. A few chapters in the *Super-Comprehensive Epistle* are composed of several passages within a single epistle commentary of the *Comprehensive Epistle*. Chapter 3, for example, consists of two separate passages taken from the commentary of Epistle 32 (*Intellectual Principles according to Pythagoreans*).[12] The chapter follows the order in which the passages appear in the *Comprehensive Epistle*; a lengthy section between the two passages is simply left out. Other chapters are composites of passages from distinct epistle commentaries. Chapter 33, thus, consists of six passages taken from sections of the *Comprehensive*

9 *Al-Jāmiʿa*, ep. 29; ed. Ghālib, 309–310; ed. Kacimi, 410–411; *Jāmiʿat al-jāmiʿa*, 65–66.

10 *Al-Jāmiʿa*, ep. 34; ed. Ghālib, 346–349; ed. Kacimi, 461–464; *Jāmiʿat al-jāmiʿa*, 72–75.

11 *Al-Jāmiʿa*, ep. 34; ed. Ghālib, 347; ed. Kacimi, 462; *Jāmiʿat al-jāmiʿa*, 73.

12 *Al-Jāmiʿa*, ep. 32; ed. Ghālib, 339–340, 344–345; ed. Kacimi, 453–454, 458–459; *Jāmiʿat al-jāmiʿa*, 70–71.

Epistle that correspond to Epistles 25, 1, and 5, respectively.[13] Even in these more complex cases, the chapters of the *Super-Comprehensive Epistle* simply reproduce the text of the *Comprehensive Epistle*. Here, as in many other cases, some of the cited passages only consist of a few lines. Nor is there always a clear rationale for why these particular passages have been combined to form a chapter.

As a result, each chapter in the *Super-Comprehensive Epistle* corresponds to one or more passages in the *Comprehensive Epistle*, which in turn correspond to an epistle in the original work (the correspondences are indicated in Table 19). Only two chapters (43 and 49) contain lengthy sections for which I have not been able to identify a corresponding passage in the *Comprehensive Epistle*. Doctrinally, the most interesting of these is the classification of degrees of paradise and hell, as well as of the distinct meanings of resurrection, in Chapter 49,[14] which apparently is not conveyed by the *Comprehensive Epistle*.

Otherwise, the author's independent contribution to the text is restricted to minor editorial interventions. There is some textual variation between the two texts concerning Qur'ānic citations. In most cases, the two texts cite the same Qur'ānic verses in the same location of a passage; on a few occasions, the *Super-Comprehensive Epistle* expands the citations of a particular verse, or even inserts new verses. The author of the *Super-Comprehensive Epistle* also consciously edits the source text in a few instances. The author inserts transition phrases of varying length to introduce a cited passage or, in chapters that cite more than one passage, to connect the passages. Often these include some variant of the Ikhwānian stock phrase of "know (O brother)" (*iʿlam yā akhī*). At the end of Chapter 26, the author helpfully explains that "this epistle (*risāla*) has been completed previously at the beginning of the book."[15] Indeed, the citation of the commentary of Epistle 32 of the *Comprehensive Epistle* in Chapter 26 is continued in Chapters 2 and 3 of the *Super-Comprehensive Epistle*. Perhaps the most revealing instance concerns the occasional references to epistles or chapters. References to the *Comprehensive Epistle* are consistently edited so that they refer to the *Super-Comprehensive Epistle* instead, as in the case of Chapter 4 mentioned above.[16] As a result, the *Super-Comprehensive Epistle* con-

13 *Al-Jāmiʿa*, eps. 25, 1, 5; ed. Ghālib, 251, 90, 91–92, 64, 83, 103; ed. Kacimi, 326–327, 91, 93, 60, 82, 111; *Jāmiʿat al-jāmiʿa*, 166–169.

14 *Jāmiʿat al-jāmiʿa*, 219–220.

15 *Jāmiʿat al-jāmiʿa*, 145.

16 For another example, see *al-Jāmiʿa*, ep. 34; ed. Ghālib, 352; ed. Kacimi, 468; *Jāmiʿat al-jāmiʿa*, 81. In another passage (*al-Jāmiʿa*, ep. 34; ed. Ghālib, 363; ed. Kacimi, 484; *Jāmiʿat al-jāmiʿa*, 95), the reference to the *Comprehensive Epistle* is simply dropped in the *Super-Comprehensive epistle*: "once our discussion [in the *Comprehensive Epistle*] reaches the *epistle on resurrection and Judgment Day*."

tains no references to the *Comprehensive Epistle*. In a similar vein, the reference in the *Comprehensive Epistle* to the "ultimate goal of this epistle" (*al-gharaḍ al-maqṣūd ilayhi min hādhihi al-risāla*) is edited to the "ultimate goal of this chapter" (*al-gharaḍ al-maqṣūd min hādhā al-faṣl*).[17]

The prominence of Qur'ānic citations in the *Super-Comprehensive Epistle* reinforces the impression of a theological treatise. It is true that most of the verses are present in the passages cited from the *Comprehensive Epistle*; in a few cases, they are expanded in the *Super-Comprehensive Epistle*. However, the author's selection in part contributes to an impression that the work's emphasis is on the Qur'ānic basis of the presented doctrine. In particular, Chapters 15, 22, 23, and 24 consist almost entirely of citations of Qur'ānic verses on Judgment Day. There is little difference between the two works; in both, the passages mostly contain a lengthy chain of Qur'ānic verses interspersed by succinct exegesis.[18] These chapters reinforce the idea that the treatise's doctrinal claims on these questions are grounded firmly in the Qur'ān. The authors of the original epistles would undoubtedly agree. The difference is, however, that the *Epistles* constitute an elaborate argument for the soul's eternity and afterlife that engages with ancient and contemporary philosophers and theologians. For the *Super-Comprehensive Treatise*'s author, in contrast, the sole argument appears to be the identification of certain dogmatic claims with passages in the Qur'ān.

As a result, the *Super-Comprehensive Epistle* is in essence a reader's digest of the *Comprehensive Epistle*. The treatise is not an independent commentary on either the *Epistles* or the *Comprehensive Epistle*, in the way the *Comprehensive Epistle* presents a novel interpretation of the *Epistles*. Instead, the author's main contribution consists of the selection and ordering of the cited passages. It has been shown that the *Comprehensive Epistle* presents a reading of the *Epistles* in an Ismāʿīlī doctrinal context. Clearly, this is equally true of the *Super-Comprehensive Epistle*. As suggested by Kacimi, the treatise's reason for existence is, then, to present a more concise Ismāʿīlī reading of the *Epistles* than was the case for the *Comprehensive Epistle*.

As regards its nature as a digest, the *Super-Comprehensive Epistle* is not entirely an anomaly in the history of the Ikhwānian corpus. In the manuscript tradition of the *Epistles*, most manuscripts portray the entire work; some convey a single epistle. But there are also manuscripts that contain a selection of the *Epistles*. In most cases, these carry a number of epistles in their entirety. This

17 *Al-Jāmiʿa*, ep. 34; ed. Ghālib, 348; ed. Kacimi, 463; *Jāmiʿat al-jāmiʿa*, 75.

18 *Al-Jāmiʿa*, ep. 38; ed. Ghālib, 429–435, 446–453; ed. Kacimi, 564–569, 583–592; *Jāmiʿat al-jāmiʿa*, 102–110, 128–140.

is true, for example, of a manuscript that contains epistles 1–14, 33, 25, and 41.[19] There are also manuscripts that carry selected parts of a number of epistles, such as the Tübingen manuscript.[20] Perhaps the editors of these manuscripts perceived that their selection conveyed a sufficient digest of the *Epistles* for a reader. The *Super-Comprehensive Epistle* obviously differs from these digests in that it summarizes the *Comprehensive Epistle* rather than the *Epistles*. It is thus a digest for an Ismāʿīlī audience.

Given all this, the value of the *Super-Comprehensive Treatise* is obviously diminished as an independent treatise. But the treatise is still of great interest for the question of the evolution of the Ikhwānian corpus: it provides a testimony of how the *Epistles* were read by a later Ismāʿīlī audience. The important questions now concern the author's method in selecting and ordering the passages. Clearly, for the author, the selected passages constitute the "core" of the *Epistles*.

As regards the selection of epistles, it is evident that the author of the *Super-Comprehensive Epistle* does not follow the same criteria as the author of the *Comprehensive Epistle*. Many of the lengthiest epistle commentaries in the latter treatise, such as those on Epistles 1 and 22, are not given much prominence in the former treatise. Instead, the *Super-Comprehensive Epistle*'s focus is primarily on the third and secondarily on the second section of the *Epistles*. The first thirty-two chapters of the *Super-Comprehensive Epistle* are devoted to epistles in the third section with the exception of three chapters corresponding to epistles in the second section. Chapters 33–52 draw primarily on epistles in the second section and secondarily on epistles in the third section. As regards the first and forth sections, only Epistle 1 (*Arithmetic*) is accorded a whole chapter; even then, the cited passages relate to the spiritual interpretation of the story of Adam rather than the original epistle's subject matter.

Beyond this, the *Super-Comprehensive Epistle* lays emphasis on certain epistles. Eleven chapters are devoted exclusively to Epistle 38 (*Resurrection*) and nine chapters to Epistle 34 (*The World is a Big Human Being*); Epistle 25 (*Fall of the Sperm*) is cited in seven chapters. Since most chapters cite only a small part of the epistle commentary in the *Comprehensive Epistle*, the author's editorial role is obviously also present in the selection of passages within each commentary.

The result is a very select reading of the doctrinal contents of the *Epistles*. Judgment Day and the soul's afterlife emerge as the single most important

19 MS Amcazade Hüseyn Paşa 452 (1090/1680).

20 Tübingen, MS MaVI86 (612/1215).

theme for the treatise. This is already evident in the treatise's beginning with the theme of life and death, which is related to the eternal life in the celestial world. It is also evident in the fact that Epistle 38, devoted to the doctrine of resurrection, is the most cited epistle in the *Super-Comprehensive Epistle*. A second central theme concerns the cosmological order of creation, including the microcosmic correspondences between the terrestrial and the celestial worlds. As a result, the treatise's main message perhaps concerns the subject of beginning and return: the descent of existence from God through the hypostases of the Intellect and the Soul down to the material world; and the particular souls' return to the supernal world. This is fairly similar to the *Comprehensive Epistle*'s thematic emphasis. The difference is that a reader of the *Super-Comprehensive Epistle* is given the impression that this is the whole doctrinal content of the *Epistles*. The *Super-Comprehensive Epistle* thus represents the final stage in the *Epistles*' reduction to Ismāʿīlī doctrine.

The story of the Ikhwānian corpus may, then, be the following. At the beginning, there was an encyclopedia of philosophical sciences. Its aim was to convey the main tenets of every philosophical science from mathematics, through logic, to natural philosophy and theology. As in al-Kindī's school, the authors of the encyclopedia subscribed to the late ancient classification of the parts of philosophy. Historically, this was the first attempt to present the learning of ancients within a single volume. Even at the beginning, the project did not serve an exclusively philosophical purpose; the philosophical themes of creation and the soul's afterlife were, after all, already related to the Islamic context by al-Kindī. But the idea of a larger synthesis of philosophy and religion emerged later. Thus, and perhaps by a different generation of authors, the original project was expanded to a comprehensive philosophical as well as religious encyclopedia. As a result, a fourth part with religious epistles was added, and the original epistles were in part restructured to fit the new purpose. Ismāʿīlī thought probably constituted one line of influence on the eclectically inclined authors of the final redaction of the *Epistles*. But it was only in the *Comprehensive Epistle* that the doctrinal contents of the *Epistles* were given a fully Ismāʿīlī garb. In that work, and even more in its abridgement in the *Super-Comprehensive Epistle*, the initially Kindīan philosophical encyclopedia was transformed to an Ismāʿīlī treatise of philosophical theology.

Tables

TABLE 1 The *Epistles* of the Brethren of Purity

No	Variant nos[1]	Title	Variation in MSS[2]	Length (BE)	Length (MS ع)[3]
		Fihrist	Versions a, b, c[4]	27	8
		Propaedeutic Sciences		**404**	**256**
1		Arithmetic	Minor	30	19
2		Geometry	Minor[5]	36	16
3		Astronomy	Minor[6]	44	28
4	5, 6[7]	Geography	Minor[8]	25	13
5	4	Music	Minor	59	41
6	2, 4[9]	Arithmetical and Geometrical Proportions	Minor[10]	16	11
7	8[11]	Theoretical Sciences[12]	Minor[13]	18	11
8	7	Practical sciences	Minor[14]	20	12
9		Ethics	Minor[15]	94	65
10		Eisagōgē	Major[16]	14	11
11		Categories	Major	10	7
12	13[17]	On Interpretation	Major	6	7
13		Prior Analytics[18]	Major	9	
14		Posterior Analytics	Major	23	15
		Physical Sciences		**647**	**343**
15		Matter and Form	Major[19]	19	13
16		Heaven and World	Major	28	17
17		Generation and Corruption	Major	10	6
18		Meteorology	Major	25	17
19		Minerals	Major	45	27
20	13[20]	Essence of Nature	Major	18	12
21		Plants	Major	28	17
22		Animals	Minor[21]	200	115
23		Constitution of the Human Body		18	11
24		Senses and Sensibles	Minor[22]	21	13
25		Fall of the Sperm		39	24
26		The Human Being is a Microcosm		23	15
27		Spread of Particular Souls into Physical Bodies		13	7
28		Human Capability for Knowledge	Minor	16	9
29		Death and Life	Minor	18	10
30		Pleasure and Pain	Minor[23]	32	18
31		Differences of Languages	Versions a, b[24]	94	12

TABLE 1 *Epistles* of the Brethren of Purity (*cont.*)

No	Variant nos	Title	Variation in MSS	Length (BE)	Length (MS ع)
		Psychical-Intelligible Sciences		**223**	**128**
32		Intellectual Principles according to the Pythagoreans	Versions a, b[25]	21	12
33		Intellectual Principles according to the Brethren of Purity		13	7
34		The World is a Big Human Being	Minor[26]	19	6
35		Intellect and the Intelligibles	Minor	18	11
36		Cycles and Revolutions	Minor	20	12
37		Love		18	11
38		Resurrection		34	20
39		Movements	Minor	23	14
40		Causes and Effects	Minor[27]	40	26
41		Definitions and Descriptions	Major[28]	17	9
		Nomic-Divine Sciences		**595**	**349**
42		Opinions and Doctrines		137	89
43		The Way to God	Minor	9	6
44		Beliefs of the Brethren of Purity	Minor	27	17
45	52[29]	Companionship of the Brethren of Purity	Minor	20	12
46		Belief		63	38
47		Divine Law		21	17
48		The Call to God	Major[30]	53	34
49		Spiritual Beings	Versions a, b[31]	52	26
50		Governance[32]	Versions a, b[33]	23	13
51		Constitution of the World	Versions a, b[34]	10	8
52		Magic	Versions a, b, c[35]	180	89 (52b)

1 These refer to the positions of epistles in some manuscripts of the *Epistles* that differ from the positions they occupy in the Beirut and the Institute of Ismaili Studies/Oxford University Press editions. Variant numbers exclude the cases where all subsequent epistles are one unit smaller or higher than they would be otherwise due to the fact that, (1) in most manuscripts, Epistles 12 and 13 form a single epistle; and, (2) in MSS ز and ك, Epistle 48 constitutes two epistles and Epistles 49 and 50 form a single epistle.

2 The assessment of the extent of manuscript variation for each epistle, summarized in these endnotes, is based mostly on the technical introductions provided by each editor of the Institute of Ismaili Studies/Oxford University Press editions. I have, in some cases, supplemented them by my personal observations but have not systematically studied the manuscripts for this purpose. Within the nineteen manuscripts employed in these editions, the editors have often used different subsets of a smaller number of manuscripts; some editors have also used additional manuscripts outside the nineteen manuscripts. These conclusions should, then, be viewed as a provisional overview of the conclusions made by the editors of the Institute of Ismaili Studies/Oxford University Press editions

complemented by some additional findings. A systematic study of a larger sample of manuscripts, including those not used in the IIS/OUP editions, would be a desideratum in the future.

3 The length is given in pages, rather than folios, excluding the title page.

4 The manuscripts carry various variants of the *Fihrist*. For the table of contents section, (1) most manuscripts contain a short variant (*Fihrist a*); (2) D contains a slightly expanded variant (*Fihrist b*); and (3) the Beirut edition contains the longest variant (*Fihrist c*). For the final part of the *Fihrist*, (1) ع and F contain a very concise variant; (2) most manuscripts contain a longer variant; and (3) the Beirut edition carries a much more elaborate variant which includes an additional cosmological section.

5 ط carries a long additional passage (edited in Appendix 2 of the edition); ف contains additional passages, partly shared with غ, ل, ن, at the end of Chapter 21; the end of س is followed by a passage celebrating the benefits of knowledge not present in other manuscripts. ل contains a text corresponding to Epistle 6 between Epistles 1 and 2. The Beirut edition contains an additional section in Chapter 26.

6 أ and ف contain thirteen additional chapters of mainly technical nature (edited in Appendix A); ع contains an additional explanation at the end of Chapter 16.

7 The variant positions are 5 (ج, خ, ر, ع, غ, ن, F, G) and 6 (أ, ف).

8 ع contains significant omissions, including most images and tables, which, however, also vary between other manuscripts; خ also contains significant omissions.

9 The variant positions are 2 (ل, B) and 4 (أ, ف).

10 أ, ع, ف, ل contain an additional short appendix on the extraction of cubic root; the final passage of the Beirut edition is omitted in أ, ف, ك, ل, ن.

11 Epistles 7 and 8 occupy the reverse positions in ل, ن, B, G, H, I.

12 Shaded epistles occupy an arbitrary position within the *Epistles*.

13 أ, ح, د, ر, ط, ف, ك, ل contain an additional chapter on the importance of self-knowledge at the end (edited in Appendix A of the edition).

14 أ, ف, ل, and the Beirut edition contain an additional passage with Qurʾānic and *ḥadīth* citations.

15 ف has a short addition at the end (Appendix 1 of the edition) with quotes on ascetic virtues and two citations of poems, as well as a concise passage on genera and species which probably should form a part of the logical epistles that follow.

16 For the logical epistles (10–14), manuscripts manifest major additions and omissions, sometimes concerning entire epistles: ف contains only Epistles 10 and 11, even though its *Fihrist* lists thirteen epistles for the first part. For Epistle 10, ع contains a cosmological addition with Ismāʿīlī-like terminology; خ and غ carry an alternative short, but more technical, version of the epistle (edited in Appendices A and B of the edition). أ contains an additional passage at the end of Epistle 11. In أ, خ, ع, غ, Epistles 12 and 13 constitute a single epistle, so that epistle 13 forms a chapter of Epistle 12. أ contains a lengthy addition for Epistle 12 (Appendix C of the edition), possibly paraphrasing chapters 13–14 of Aristotle's *On Interpretation*.

17 In ك, *On Interpretation* is numbered as the thirteenth and the *Prior Analytics* as the twelfth epistle, but their actual positions are inverse, and thereby coincide with those of the other manuscripts.

18 The *Prior Analytics* is a separate epistle in ك, ل, B, F, I, and the Beirut edition; it is a chapter of Epistle 12 in the rest of the manuscripts.

19 For the physical epistles (15–21), many manuscripts contain an incomplete text; أ omits Epistle 20 and ط ends in Epistle 19. There are also short additions.

20 In ك, Epistle 20 is incorrectly numbered as the thirteenth epistle, despite occupying the correct position and being correctly numbered as the sixth of the physical section.

21 ب contains an incomplete text ending at chapter 21.

22 ع contains an additional chapter at the end.

23 MSS ح, ف, ل, and ن contain an additional passage at the end on animals' aversion for death and love of life.

24 The manuscripts carry a shorter and a longer version of this epistle. In the Beirut edition and أ, it is the fourth longest epistle in the entire work; in ع, the epistle is fairly short. In ف, there is a shorter and a longer version of the epistle one after the other, each with separate epistle title and both numbered as the seventeenth of the second part. For the most part, the text of the shorter version corresponds to passages in the longer version.

25 Epistles 32 and 33 are intertwined in the manuscripts, which manifest great variation. This implies the existence of two or more versions for these epistles. All manuscripts contain mostly the same material which is arranged in different orders. The editor posits the arrangement in ع as correct, where the variants 32a and 32b are followed by 33. The corresponding epistles in the Beirut edition follow the arrangement 32a, 33, 32b. In addition, the Beirut edition and some manuscripts contain additional material not shared by other manuscripts.

26 غ and the Beirut edition include a lengthy addition at the end, replicated from other epistles, in particular Epistle 21.

27 أ includes a list of letters (*ḥurūf muʿjama*) at the end of the epistle; the text in ج is incomplete. The list of nine philosophical questions differs in the manuscripts.

28 The manuscripts vary greatly regarding the order of sections, definenda, and the wordings of definitions. There are frequent omissions: غ omits three of the eight sections and خ two sections present in ع. ق adds a lengthy additional passage from another manuscript at the end of the epistle. Nāṣir-i Khosrow's Persian translation of the epistle in *Jāmiʿ-i ḥikmatayn*, predating the oldest manuscript by a century, appears to have employed a shorter version of the treatise.

29 In most manuscripts of the *Comprehensive Epistle*, this is the final epistle.

30 The manuscripts vary greatly regarding the order of chapters and contain lengthy additions and omissions. Of the twelve manuscripts employed by the editor, only seven (ح, د, ر, ز, ع, ل, ن) contain the complete text. The editor postulates two main manuscript groups for the arrangement of the chapters; the order in أ forms a special case. ز and ك divide the epistle into two separate epistles. Chapters 5–7 reproduce verbatim a section of Epistle 45. Since these chapters are missing in some manuscripts, although present in the Beirut edition and ع, and contain a reference to Epistle 48 as a different epistle, these could be a later addition. Another section appears to be derived from Epistle 15.

31 The manuscripts carry a shorter version of the epistle (a) in ح, ل, ن, and a longer version (b) in أ, خ, د, ز, ع, ق, ك. Variant b is an elaboration on a; the editor thus believes variant a to be earlier. The Beirut edition represents a hybrid of both variants, although it does not contain the entirety of variant a. ر and ك contain an incomplete text with an abrupt transition to Epistle 50.

32 Epistles 49 and 50 form a single epistle in ز and ك.

33 ز and ك only contain a partial version of the epistle, presented as a part of Epistle 49. خ and غ carry completely different texts (variants b and c); the former, with the title *Description of the World in Its Entirety* (edited in Appendix D of the edition), corresponds to Epistle 32a, and the latter (Appendix B of the edition) appears to be a summary of much of the *Epistles*, while it also relates to variant b. أ contains a lengthy addition (Appendix A of the edition).

34 The editor posits three different variants for the epistle: a) أ, خ, ع, غ, ق; b) ح, د, ز, ك, ل, ن, the Beirut edition; c) ر. Variant a draws on Epistle 32b and variant b on Epistle 34. Variant c combines variants a and b.

35 Manuscripts carry three different variants of the epistle: a) ل and ن, b) most manuscripts, and c) أ. The three variants are completely different and bear no textual relation with each other. أ includes variant c after variant b; the copyist introduces it as a version he found in another, unidentified, manuscript. In the Beirut edition, variant a is followed by variant b with no indication that they are separate treatises.

TABLE 2 Theories of redaction chronology

Marquet	**Hamdani**	**Madelung**
IV	**I**	**Version I (899–937)**
48		7
45		50 eps following the 4-division in ep. 7
50		52c
44		49a
47		32a
43		
46, 49		
52		
42[1]		
51		
III	**II**	**Version II (ca. 937)**
32–40		*Fihrist* a
41		51 eps following a new 4-division
		52a
		50a
		42a
		32b
		12 = 12 + 13
I	**III**	
3		
1–2		**Version III (937–964)**
5–6 → 5, 6		*Fihrist* b
10–14 → 10, 11, 12, 13, 14	the *Fihrist*	52b
7		49b
8		50b
9		42b
4		32c
II	**IV**	12, 13 (separated)
15, 16, 17, 18, 19, 21, 22, 24	42–50	RJ
25, 26, 27, 28, 29, 30, 31	52	
23	51	
22 (animal fable)	RJ	
	the *Fihrist* (updated)	
		Version IV (after 964)
		4 and 5 shift places

1 Bolded epistles either changed position or were inserted later into the middle of existing epistles.

TABLE 3 Classification of sciences by al-Kindī and al-Fārābī

Al-Kindī	Al-Fārābī
1. Mathematics	1. Linguistics
a) arithmetic, b) geometry, c) astronomy, d) harmonics (music)	a) single expressions, b) composed expressions, c) rules of single expressions, d) rules of composed expressions, e) rules of correct writing, f) rules of correct reading, g) rules of poems
2. Logic	2. Logic
a) Categories, b) On Interpretation, c) Prior Analytics, d) Posterior Analytics, e) Topics, f) Sophistics, g) Rhetoric, h) Poetics	a) Categories, b) On Interpretation, c) Prior Analytics, d) Posterior Analytics, e) Topics, f) Sophistics, g) Rhetoric, h) Poetics
3. Physics	3. Mathematics
a) Physics, b) On the Heaven, c) Generation and Corruption, d) Meteorology, e) On Minerals, f) On Plants, g) On Animals	a) arithmetic, b) geometry, c) optics, d) astronomy, e) music, g) science of weights, h) mechanics
4. Psychical existents	4. Physics
a) On the Soul, b) On Senses and the Sensibles, c) On Sleep and Waking, d) On the Length and Shortness of Life	a) Physics, b) On the Heaven, c) Generation and Corruption, d) Meteorology, I-III, e) Meteorology, IV, f) Book of Minerals, g) Book of Plants, h) Book of Animals and On the Soul
5. Metaphysics	5. Metaphysics
a) Metaphysics	a) existents as existents, b) premises of the theoretical sciences, c) incorporeal existents
6. Practical philosophy	6. Practical philosophy
a) Nicomachean Ethics b) Politics	a) political science, b) jurisprudence, c) theology

TABLE 4 Classification of sciences in Epistle 7

I. PROPAEDEUTIC SCIENCES	
1.	Writing and reading
2.	Language and grammar
3.	Calculation and transactions
4.	Poetry and prosody
5.	Auguries (*zajr*) and auspicies (*fa'l*)
6.	Magic, incantations, alchemy, and artifices (*ḥiyal*)
7.	Professions and crafts
8.	Sale and purchase, trade, cultivation, and breeding
9.	Tales (*siyar*) and history (*akhbār*)
II. RELIGIOUS SCIENCES	
1.	Revelation
2.	Interpretation (*ta'wīl*)
3.	Transmission (*riwāyāt*) and reports (*akhbār*)
4.	Jurisprudence, laws (*sunan*), and stipulations (*aḥkām*)
5.	Remembrance (*tadhākir*), exhortations (*mawāʿiẓ*), asceticism (*zuhd*), and Sufism (*taṣawwuf*)
6.	Interpretation of dreams
III. PHILOSOPHICAL SCIENCES	
i.	**Mathematics**
1.	Arithmetic
2.	Geometry
3.	Astronomy
4.	Music
ii.	**Logic**
1.	Poetics
2.	Rhetoric
3.	Topics
4.	Analytics
5.	Sophistry

TABLE 4 Classification of sciences in Epistle 7 (*cont.*)

III. PHILOSOPHICAL SCIENCES	
iii.	**Natural sciences**
1.	Corporeal principles (*al-mabādiʾ al-jismāniyya*)
2.	Heaven and the world
3.	Generation and corruption
4.	Meteorology
5.	Minerals
6.	Plants
7.	Animals
iv.	**Theology**
1.	Creator
2.	Spiritual beings (*rūḥāniyyāt*)
3.	Psychical beings
4.	Governance: a) prophetic, b) royal, c) common (*ʿāmmiyya*), d) particular (*khāṣṣiyya*), e) individual (*dhātiyya*)
5.	Return

TABLE 5 Classification of sciences in Epistles 1, 2, and 15

Epistle 15		Epistles 1 (10–11), 2 (73–76)	
I.	**Preparatory** (*riyāḍiyyāt*)	I.	**Mathematics** (*riyāḍīyyāt*)
1.	Arithmetic	1.	Arithmetic
2.	Geometry	2.	Geometry
3.	Astronomy	3.	Astronomy
4.	Arithmetical, geometrical, and harmonious proportions	4.	Music
5.	Logic	II.	**Logic**
II.	**Physics**	III.	**Physics**
1.	Matter, form, movement, space, time		
2.	Senses and sensibles		
3.	Heaven and the Earth		
4.	Generation and corruption		
5.	Meteorology		
6.	Minerals		
7.	Plants		
8.	Animals		
		IV.	**Theology** (*ilāhiyyāt*)

TABLE 6 Contents of the epistles in the *Fihrist* and the *Comprehensive Epistle*

Epistle	*Fihrist a*	*Fihrist c*	the *Comprehensive Epistle*
1	Number, its essence and quantity	Manner of its properties	
2	Geometry, its essence, and the quantity of its kinds	Manner of its substrates	
3	Stars, spheres, zodiacs, planets	Their influence on the lower world, on elements, and on generation and corruption	Signs written on the "spheres and souls"
4	Geography, form of the Earth and the climes, and all that it contains	The world as spherical; the world as a living being that resembles the form of an animal worshipping God with all of its parts	The savants' differing opinions on the Earth lying in the middle of the air; their agreement on God's decree of the proper position of the Earth and the spheres; world as a living being as the most important part of this science
5	Music; effects of balanced melodies on the souls as resembling medicine on the bodies; melodies caused by the movements of the spheres		Balanced melodies discovered by the philosophers and the savants; their connection with the supernal world; properties of the supernal world and the paradise
6	Arithmetical, geometrical, and harmonious relations; the number of their kinds; the manner of their order		
7	Theoretical disciplines	Their parts and arrangement, and ways and doctrines	Their aims
8	Practical disciplines	Their kinds	Their parts and arrangement
9	Differences of character traits	Causes for the differences in character traits; customs, ways of life, and character traits of the prophets and sages	Differences of character traits as proportional to the climate and the four qualities; its divine wisdom and providence; actions emerging in souls in accordance with the faculties; union of the souls with bodies; examples of caliphs and imāms; knowledge of the spiritual beings and their revelation to the prophets; knowledge of Iblīs, the demons, and their followers; microcosm as an image of the macrocosm

TABLE 6 Contents of the epistles in the *Fihrist* and the *Comprehensive Epistle* (*cont.*)

Epistle	*Fihrist a*	*Fihrist c*	the *Comprehensive Epistle*
10	*Eisagōgē*, i.e., the six predicables employed by philosophers in logic, speech, and books	Employed also in their arguments and demonstrations	The six predicables: genus, species, individual, differentia, idion, and accident
11	Categories; each denotes a genus of existents		Names of the ten categories
12	*Peri hermēneias* [and Analytics]	Expressions; their true meanings and their explication	
13	First Analytics, i.e., syllogism		
14	Second Analytics	Demonstration	
15	Matter, form, time, place, movement; sages' sayings on them		
16	Heaven and the world; sphere, Throne, and Footstool		
17	Generation and corruption		Individuals, bodies, spread (*nashr*), and destruction
18	Meteorology		
19	Generation of minerals; quantity of minerals; causes for their differences		Generation and qualities of minerals as resembling plants and animals; mercury and sulfur; their special qualities
20	Essence of nature, its qualities and actions; four elements and composed bodies	Difference between voluntary and natural actions	
21	Kinds and qualities of plants		Development of plants toward their perfection
22	Kinds of animals; their compositions and states		
23	Composition of the body; the human being as microcosm; the human being as a virtuous city and soul as its king		
24	Senses and sensibles		
25	Fall of the sperm; attachment of the soul to the body; transformation of the body in the womb; effects of the spheres on bodily composition		Perfection of the soul; ascent to the spiritual world; the world of generation and corruption; descent of particular souls
26	The human being as a microcosm; the human form as a summary of the spiritual and corporeal worlds	The human being as intermediate between the material and spiritual worlds	God, the Intellect, the Universal Soul, and first matter; macrocosm, microcosm, and the Universal Soul
27	Spread of particular souls in human bodies		Ascent to the grade of angels by good actions; descent to the grade of demons by bad actions

TABLE 6 Contents of the epistles in the *Fihrist* and the *Comprehensive Epistle* (*cont.*)

Epistle	*Fihrist a*	*Fihrist c*	the *Comprehensive Epistle*
28	Human capability for knowledge and its limits		
29	Essence of death and life	Afterlife	
30	Corporeal and spiritual pleasure and pain; cause for animal repugnance of death; pleasure and pain in the afterlife		Ascent of the godly to a heavenly grade; descent of the ungodly to a demonic form; contact of pure souls with angelic substances and God
31	Differences of languages	Scripts; principles of religions and doctrines; transformation of doctrines from one nation to another	Celestial causes of the transformation of doctrines
32	Intelligible principles according to Pythagoras		Doctrine of those who have purified their souls to reach the higher world
33	Intelligible principles according to the Brethren of Purity		
34	The world as a big human being with soul and spirit; no void or plenitude in the world	The world not being in a place; everything in the world having its appointed place	
35	The intellect and the intelligible	Material, potential, actual, acquired, and active intellects	
36	Cycles and revolutions	Changes of times	
37	Love; divine illness	Desire of souls for union	Emanation of the moving faculties
38	Resurrection; ultimate end of all the epistles		
39	Kinds of movement		
40	Causes and effects		
41	Definitions and descriptions		
42	Doctrines and religions; disagreements of scholars	Proper knowledge for common people versus the elect	Principles of beliefs
43	The way to God		God's writing in the macrocosm and the microcosm; God's signs in "horizons and souls"
44	Beliefs of the Brethren of Purity and doctrines of the lordly		Sincere love
45	Companionship and mutual co-operation of the Brethren of Purity	Their meetings	
46	Essence of belief and properties of believers		Believers contemplating the signs in "horizons and souls"

TABLE 6 Contents of the epistles in the *Fihrist* and the *Comprehensive Epistle* (*cont.*)

Epistle	*Fihrist a*	*Fihrist c*	the *Comprehensive Epistle*
47	Essence of divine law; conditions of prophecy; doctrines of the lordly		
48	The call to God; pure brotherhood and love; classes of the called		Qualities of the caller and the called
49	Acts of spiritual beings	Proximate angels and demons	Followers of angels; powers flowing from angels; followers of demons
50	Kinds of governance; degrees of the governed		
51	Order of the world; order of existents; world being like a city, an animal, or a human being		
52	Magic, spells, divination, auguries, imagination, incantations, talismans; earthly beings, *jinn*s, demons, and angels and their effects	Evil eye	

TABLE 7 Goals of the epistles in the *Fihrist* and the *Comprehensive Epistle*

Epistle	*Fihrist a*	*Fihrist c*	*Comprehensive Epistle*
1	Training of students of philosophy; correspondence between the forms of numbers and existents; arithmetic as the root of knowledge	Arithmetic as a stepping stone to other philosophical sciences	Arithmetic as the first elixir and highest alchemy
2	Guidance from sensibles to intelligibles; vision of immaterial forms	Guidance from corporeal to spiritual beings; properties of the spiritual beings; ascent to and union with spiritual beings	Causes of time and generation
3	Arousing desire for ascent toward the world of spheres	Ascent to spiritual beings; arrival at the holy spirit	Reaching the degree of perfection; indifference to death and the material world; certainty on the superiority of the spiritual world; reaching the prophets, believers, gnostics, and martyrs; proximity with the Lord in paradise
4	Cause for the soul's arrival in the material world; instigation on contemplation of the signs on the "horizons"; reminder for the ignorant on preparing for the afterlife	The way and cause of the soul's union with the body; its employment of senses and reason; its release from bodies	Release from the bonds of matter and nature; ignorance of the souls not desiring the spiritual world; the soul's suffering in successive bodies; the Earth as a prison of the souls; bodies as "bridges"; sacrificial animals as the ultimate debasement; the human form as leading toward paradise; incitement toward the approach of and trust in God; paradise and hell
5	Arousing desire for ascent toward the spiritual world; abode of prophets and martyrs after separation from the body		Souls purified by the philosophical sciences as prepared for the ascent; the spiritual world as the abode of "people of secrets and insights"
6	Guidance to the inner meanings of sciences; existents of balanced and unbalanced proportions	The persistence of unbalanced composition	Things occupying their proper positions and arrangements as being their well-being
7	The kinds of sciences and their goals; guidance toward the way of knowledge		

TABLE 7 Goals of the epistles in the *Fihrist* and the *Comprehensive Epistle* (*cont.*)

Epistle	*Fihrist a*	*Fihrist c*	*Comprehensive Epistle*
8	Classification of the kinds of knowledge and crafts; reminder for ignorant souls of their substances being active and their bodies being the instruments of souls; the faculties of particular souls as supported by the Universal Soul		Principles of the productive sciences; their extraction by pure minds; the perfection of crafts implying the perfection of their Creator; spheres, planets, and elements as instruments of the Universal Soul; the Universal Soul as guiding all beings toward their perfection
9	Purification of the soul and rectification of character traits	Means to attain eternity, perfection, and happiness in this life and the next	
10	Philosophical and linguistic logic/speech and their difference	Reminder of what constitutes and completes the essence of the soul; the soul's preparation for the afterlife; knowledge of what is required for the guidance of the intellect toward the truth and away from error; the parallel between logic and grammar	
11	Meanings of all existents as combined in the ten categories; each category being a genus of genera; division of genera into species and individuals; categories as the gardens of knowledge and fruits of souls	Division of individuals into elements	The virtuous as distinguished from the non-virtuous
12		Simple assertive expressions; composition of syllogistic premises; positive and negative statements; classification of propositions	
13	Kinds of syllogisms employed by logicians, sages, and theologians; balance introduced by philosophers and sages to distinguish true from false and good from evil	Corrupt opinions	Balance for signs of the "horizons and souls"; mysteries of creation
14	Manner of true syllogism; balance of sages for distinguishing the truth from falsehood	Balance of justice employed by the "divine treasurers"	Examination of "horizons and souls;" mysteries of creation; resurrection, second creation, and return; following of a guide leading to knowledge and instruction

TABLE 7 Goals of the epistles in the *Fihrist* and the *Comprehensive Epistle* (*cont.*)

Epistle	*Fihrist a*	*Fihrist c*	*Comprehensive Epistle*
15	Essence of body; its necessary accidents and constituent forms; physics		Matter as the locus of form; form as perfecting matter; form released from matter and reverting to its essence; soul as united with individuals; matter and form as reminders for the souls of the second creation
16	Movements of spheres and planets; the Universal Soul as their mover	Holy spirit	Throne and Footstool; universal faculties of the supernal world; supernal faculties flowing to the lower world; sin leading to the fall from paradise; return to the supernal world; revelation and prophecy
17	Forms of the four elements; bodies as composed of the elements; transformation of elements due to the spheres; Nature as a faculty of the Universal Soul	Nature as moving bodies toward their perfection; Nature as an angel	Nature as a means of divine will and providence; prophets as supported by spirits from the Universal Soul; release of particular souls from matter; return to the supernal world; pure soul and the cycle of revelation
18	Events and transformations in the air; light, dark, hot, cold, winds, etc.		
19	Minerals as the first sublunar acts of Nature and the Universal Soul; beginning of the ascent of particular souls up to the supernal world	Nature as a means of the Creator; the supernal world as the abode of the pious and the prophets	Derivation of the knowledge of the sages as resembling the generation of minerals by Nature
20	Psychical activities; angels and spiritual beings	Spiritual beings as moving sublunar bodies toward their perfection	Ascent to the supernal world; mysteries and signs of nature; divine providence; the human being as a microcosm; paradise and hell
21	Kinds of plants; their generation and differences; plants between minerals and animals		Special properties of plants; their faculties; pure philosophers' and "virtuous individuals'" knowledge of mysteries

TABLE 7 Goals of the epistles in the *Fihrist* and the *Comprehensive Epistle* (*cont.*)

Epistle	*Fihrist a*	*Fihrist c*	*Comprehensive Epistle*
22	Kinds of animals and their differences; their generation; animals between plants and humans; angels as being above humans; the human beings as vice-regents on Earth; good humans as angels, bad humans as demons		Animal bodies as punishment for prior sins
23	Knowledge of the human body; the human body as a summary of the world in the Tablet; human body between paradise and hell; the human soul as God's vice-regent; knowledge of the self as leading to knowledge of God; ascent toward God	The human body as the finest among animals; ascent to the supernal world	Human beings as the highest in degree in the sublunar world
24	Sense perception; imaginative faculty; thinking faculty; memory; speaking faculty; productive faculty; writing and transmission of knowledge	Common sense; rational faculty as the human essence	Ascent to the spiritual world as requiring the body
25	The state of souls before their connection with particular bodies; bodies' stay in the womb to perfect their forms		The Sun's role in the flowing of spiritual powers to particular bodies and souls
26	Knowledge of true human nature; humans containing the ideas of existents	State of human souls before their connection with bodies; reason for the connection; the separation of souls from bodies	Human nature as a summary of the Tablet
27	The manner in which human beings attain the rank of angels	Gradual ascent to human perfection through the human substance	Ascent to knowledge of God
28	Knowledge of and ascent to God		Three kinds of human knowledge; prophets as leading to knowledge of God; the seventh prophet
29	Cause for the union of rational souls with human bodies; abolition of fear of death	Preparation for death; insignificance of death; necessity of death for ascent to the spiritual world	Desire for death
30	Paradise and hell		

TABLE 7 Goals of the epistles in the *Fihrist* and the *Comprehensive Epistle* (*cont.*)

Epistle	*Fihrist a*	*Fihrist c*	*Comprehensive Epistle*
31		Acts of the soul lying in its nature; the search for hidden things lying in the human substance; the soul having knowledge in potency and forming opinions through it; words and writing expressing ideas; human capability of choosing beliefs	Revelation, the spirituality of the Sun, and emanation of the "second degree" (*al-ḥadd al-thānī*); effects of the spheres and the "virtuous soul" (*al-nafs al-fāḍila*); succession of religions; beneficence of doctrinal disagreements
32	God ordering creation as corresponding to numbers	The Intellect as the first created being; intermediacy of the Soul in creation	
33	Generation of the world; universal and particular causes of generated beings; order of existents corresponding to numbers		
34	None	None	None
35	True nature of the soul; union of forms of knowledge in the soul	Generation of existents from non-existence; ascent to immaterial existence	Eternity of the soul in the spiritual world; particular souls' acquisition of knowledge
36	Generation and destruction of the world	Order and final causes of the world	
37	God as the true object of love	Existents' movement toward their perfection	
38	None	None	None
39	Generation of the world from the Creator; cause for the destruction of the corporeal world	Movement of natures toward their perfection	
40	Principles and rules of sciences		
41	True natures of things		Answer to the question: Who is the creator of existents?
42	The aim of religions and doctrines being the souls' pursuit of happiness in the afterlife; salvation from the world of generation and corruption; ascent to the world of the spheres; most religions having deviated from the way to salvation	Religions and doctrines as remedies for the diseases of the soul; liberation from the "sea of matter and the captivity of nature"	Prophets as physicians of the soul
43	Purification of the soul; afterlife, resurrection, reward and punishment		Awakening from the slumber of ignorance

TABLE 7 Goals of the epistles in the *Fihrist* and the *Comprehensive Epistle* (*cont.*)

Epistle	*Fihrist a*	*Fihrist c*	*Comprehensive Epistle*
44	The soul's survival of the body		Agreement of prophets and philosophers on the soul's eternity
45	Harmony of the hearts	Co-operation as the cause for salvation	
46	Nature of revelation and demonic whisper	Spiritual sublimity; success, guidance, and perdition	True knowledge of belief as assent to revelation; heavenly degree
47	Mysteries of prophetic books; guidance to and unveiling of mysteries; expected imām	Expected *mahdī*	Differences of commands and prohibitions; differences of languages of revelation; true nature of rites of worship
48	Rule of the good as beginning with an agreement on a single doctrine	The good following a just way of life	The good contemplating the signs of "horizons and souls"; the salvation of the good
49	Proof of the existence of incorporeal beings	Spiritual beings being beyond time, space, and sensation; knowledge about them through their effects	
50	God as the governor of the All; the best governor as closest to God	The best governor as the one most familiar with God's wisdom and decree	
51	Knowledge of the true nature of things; God as the beginning and end of everything	Principles and consequences of things; God as uncreated pure creation; First Existent	Mysteries and signs of existents
52	Proof of the existence of spiritual beings	Actions of spiritual beings by the intermediacy of Nature, the Soul, and the Intellect	Mysteries of nature; the effects of magic

TABLE 8 Variation of epistle titles in manuscript title pages

No	Main part(s) of the title in most MSS	Significant MS additions	Variant elements
1	Fī al-ʿadad	Arithmāṭīqī (ز, D)	Wa-l-maʿrūf bi-l-nisab al-ʿadadiyya (F)
2	Fī al-handasa	Jūmuṭriyā (BE, ع, ك, C), madkhal ilā ʿilm al-handasa (ع, ز, د, C)	Fī al-nisab al-ʿadadiyya wa-l-handasiyya wa-bi-l-jūmuṭriyā wa-huwa al-ʿulūm al-handasiyya (ل)
3	Fī al-nujūm	Asṭrūnūmiyā (BE, ع, ط, أ, ك, ل, ر, ز, F); (shibh) al-madkhal (ط, ف, ك, ل, ن, C, F, G, K)	
4	Fī jughrāfiyā, ṣūrat al-arḍ wa-l-aqālīm	Wa-l-bayān fa-innahā kurriyyat al-shakl (ل)	
5	Fī (ʿilm) al-mūsīqā	Al-madkhal ilā ʿilm ṣināʿat al-taʾlīf (ل)	Al-nisab al-ʿadadiyya wa-l-handasiyya
6	Fī al-nisba al-ʿadadiyya wa-l-handasiyya	Wa-l-taʾlīfiyya (ع, ن, أ, ف)	Wa-bi-l-jūmuṭriyā wa-huwa al-ʿulūm al-handasiyya (ل)
7	Fī al-ṣanāʾiʿ al-ʿilmiyya	Kammiyyat ajnās al-ʿulūm (ع, أ, ر)	
8	Fī al-ṣanāʾiʿ al-ʿamaliyya	Man al-fāʿil bi-l-ḥaqīqa (ر)	
9	Fī bayān ikhtilāf al-akhlāq	Akhlāq al-nafs (ف); ādāb al-anbiyāʾ, akhlāq al-ḥukamāʾ (BE)	Bayān ikhtilāf al-lughāt (ل), wa-l-alwān wa-l-ajsām fī al-akwār wa-l-adwār (ل)
10	Fī Īsāghūjī, al-alfāẓ al-sitta	Al-Madkhal ilā ʿilm al-manṭiq (ع, ل, J); al-madkhal min Rasāʾil Ikhwān al-Ṣafāʾ wa-Khullān al-Wafāʾ (I)	Wa-yatlūhā fī ākhirihā Qāṭīghuriyās (ل)
11	Fī (maʿnā) Qāṭīghūriyās	Al-maʿqūlāt al-ʿashara (BE, I); tafsīr maʿānī al-ʿashara al-alfāẓ allatī fī Qāṭīghūriyās (أ, ك)	Fī ʿilm al-manṭiq (I)
12	Fī (maʿnā) Bārī Armīniyās		Wa-anūlūṭīqā (ع, ر)
13	Fī (maʿnā) Anūlūṭīqā (al-ūlā)	Qiyās (Fihrist c)	Fī al-manṭiq (ك)
14	Fī Anūlūṭīqā al-thāniya	Afūdiqṭīqā (Fihrist a, ع); burhān (Fihrist b, I)	
15	Fī (bayān) al-hayūlā wa-l-ṣūra	Al-ḥaraka wa-l-zamān (wa-l-makān) (أ, ف); al-samʿ al-kiyānī (ر, أ, د, ز, ط, ف, ك, ل, ن)	
16	Fī al-samāʾ wa-l-ʿālam	Bi-asrihi (ر); bayān kammiyyat aṭbāq al-samawāt wa-kayfiyyat tarkīb al-aflāk (ل)	Wa-mā huwa al-ʿirsh al-ʿaẓīm wa-l-kursiyy al-wāsiʿ (ل)
17	Fī (bayān) al-kawn wa-l-fasād		
18	Fī (bayān) al-āthār al-ʿulwiyya	Wa-ḥawādith al-jaww (ل)	Wa-l-qirānat wa-l-adwār wa-l-akwār (RJ MSS)
19	Fī (bayān takwīn) al-maʿādin	Kammiyyat al-jawāhir al-maʿdaniyya (أ, ل, ف)	

TABLE 8 Variation of epistle titles in manuscript title pages (*cont.*)

No	Main part(s) of the title in most MSS	Significant MS additions	Variant elements
20	Fī māhiyyat al-ṭabīʿa, kayfiyyat ṣanāʾiʿihā/afʿālihā	Wa-kayfiyyat afʿālihā fī al-arkān al-arbaʿa wa-kayfiyyat mawālīdihā allatī hiya al-ḥayawān wa-l-nabāt wa-l-maʿādin (ك, ل, ف)	
21	Fī (ajnās) al-nabāt	Wa-kayfiyyat sarayān al-quwā al-nāmiya fīhā (ل); wa-kayfiyyat sarayān al-quwā al-nabātiyya fīhā (ف)	
22	Fī (aṣnaf) al-ḥayāwān(āt)	Kayfiyyat takwīn al-ḥayawān (BE); wa-ʿajāʾib hayākilihā wa-gharāʾib aḥwālihā (ل, ف)	
23	Fī tarkīb al-jasad	Ḥālāt al-nafs (fīhi) (ر, أ, ع); sarayān quwāhā fī aʿḍāʾ al-jasad (أ)	Inna al-insān ʿālam ṣaghīr (Fihrist)
24	Fī al-ḥāss wa-l-maḥsūs		
25	Fī masqaṭ al-nuṭfa	Kayfiyyat ribāṭ al-nafs bi-l-jasad (ل); kayfiyyat riyāḍat al-nafs (ف); ḥāl jinīn wa-taʾthīrāt al-kawākib (ع); kayfyyat taʾthīrāt al-kawākib fī al-mawlūd (I)	
26	Fī (maʿnā qawl- al-ḥukamāʾ inna) al-insān ʿālam ṣaghīr		
27	Fī kayfiyyat nashr/nushūʾ al-anfus al-juzʾiyya fī al-ajsād (al-bashariyya al-ṭabīʿiyya)	Kayfa inbiʿāthuhā min al-ajsād al-bashariyya ilā rutbat al-malāʾika (ر, ع)	
28	Fī bayān ṭāqat al-insān fī al-maʿārif	Tanāhī ʿilm al-insān (ع, ك); ilā ayy ḥadd mablūghuhu min al-ʿulūm wa-ilā ayy ḥadd yantahī (ر, د, أ, ل, ن, ف)	
29	Fī (māhiyyat) ḥikmat al-mawt (wa-l-ḥayāt)		
30	Fī māhiyyat al-ladhdhāt wa-l-ālām (al-jismāniyyā wa-l-rūḥāniyya)	ʿIllat karāhiyyat al-ḥayawān (al-alam wa-) al-mawt (wa-maḥabbatihā li-l-ḥayāt) (F, ع); fī khāṣṣiyyāt al-ladhdhāt (BE); fī māhiyyat al-ālām (أ, J)	Wa-fī ḥikmat al-ḥayāt wa-l-mawt (BE)
31	Fī ʿilal ikhtilāf al-lughāt	Wa-rusūm al-khuṭūṭ wa-l-ʿibārāt (BE); kayfiyyat wujūdihā (ع); maʿrifat uṣūlihā (ر, J)	
32	Fī (al-)mabādiʾ (al-mawjūdāt) al-ʿaqliyya ʿalā raʾy Fīthāghū-ras/al-fīthāghūriyyīn	Fī mabādiʾ al-mawjūdāt ʿalā raʾy Fīthāghūras (ح)	Fī al-mabādiʾ al-ʿaqliyya wa-l-jismāniyya ʿalā raʾy Fīthāghūras min al-maʿārif al-ṣūfiyya (ز); fī al-mabādiʾ al-jismāniyya (غ); fī al-mabādiʾ al-ʿaqliyya wa-l-jismāniyya (د)

TABLE 8 Variation of epistle titles in manuscript title pages (*cont.*)

No	Main part(s) of the title in most MSS	Significant MS additions	Variant elements
33	Fī al-mabādiʾ al-ʿaqliyya ʿalā raʾy Ikhwān al-Ṣafāʾ		Fī al-mabādiʾ al-ʿaqliyya ʿalā raʾy al-aḥdāth (ع, ج, ح, ر); fī mabādiʾ al-mawjūdāt ʿalā raʾy Fīthāghūras (أ); maqāla fī al-mabādiʾ wa-marātibihā (د); maqāla fī al-mabādiʾ wa-marātibihā ʿalā al-qawm razaqanā Allāh al-luḥūq lahum (ل); faṣl al-risāla al-thāniya fī al-mabādiʾ al-ʿaqliyya (خ, ك, J); kitāb al-mabādiʾ al-ʿaqliyya ʿalā al-ārāʾ al-falsafiyya wa-l-ishārāt al-naẓariyya (غ); al-risāla al-ūlā fī qaṣd al-ʿālam wa-marātib al-mawjūdāt (I)
34	Fī (maʿnā qawl al-ḥukamāʾ) inna al-ʿālam insān kabīr	Dhū nafs wa jism wa-maʿnā man qāla bi-ghayr al-jamʿ (أ); dhū nafs wa-rūḥ ḥayy ʿālim ṭāʿī bi-rabbihi (ن)	Fī anna al-insān ʿālam ṣaghīr (د); fī anna al-insān ʿālam kabīr (J)
35	Fī al-ʿaql wa-l-maʿqūl		
36	Fī al-adwār wa-l-akwār		Faṣl dhikr al-risāla al-khāmisa min al-qism al-thālith al-mawsūma bi-l-akwār wa-l-adwār (خ)
37	Fī māhiyyat al-ʿishq	Wa-maḥabbat al-nafs (أ); wa-l-maraḍ al-ilāhī (أ)	
38	Fī (māhiyyat) al-baʿth wa-l-qiyāma	Wa-kayfiyyat al-miʿrāj (أ, ر); wa-l-nushūr (ع); wa-l-dār al-ākhira (ر)	
39	Fī kammiyyat (ajnās) al-ḥarakāt	Fī māhiyyat al-ḥarakāt (ع); fī kayfiyyat ajnās al-ḥarakāt (ح); wa-l-jihāt allatī taḥarraka ilayhā al-mutaḥarrikāt (أ, ل)	
40	Fi al-ʿilla/ʿilal wa-l-maʿlūl(āt)	Wa-kayfa yaḥkī awākhiruhā awāʾilāhā wa-tarjiʿ awākhiruhā ilā awāʾilihā (أ)	
41	Fī al-ḥudūd wa-l-rusūm		
42	Fī al-ārāʾ wa-l-diyānāt (wa-l-madhāhib al-falsafiyya wa-l-nabawiyya)		
43	Fī māhiyyat al-ṭarīq ilā Allāh	Wa-kayfiyyat al-wuṣūl ilayhi (ك, ق, ل)	Wa-tahdhīb al-nafs wa-iṣlāḥ al-akhlāq min kalām al-ṣūfiyya (A)

TABLE 8 Variation of epistle titles in manuscript title pages (*cont.*)

No	Main part(s) of the title in most MSS	Significant MS additions	Variant elements
44	Fī (bayān) iʿtiqād Ikhwān al-Ṣafāʾ (fī amr baqāʾ al-nafs baʿd mufāraqatihā al-jasad)	Wa-madhhab al-rabbāniyyīn (BE, ق, ن); wa-ḥāl man sabaqathu al-maniyya qabla ṣāḥibihi (ر, ل, ك); wa-wuḍūḥ al-ḥujja ʿalā baqāʾ al-nafs baʿd mufāraqatihā al-jasad bi-ṭarīq al-iqnāʿ lā ṭarīq al-burhān (إ)	
45	Fī kayfiyyat ʿishrat/muʿāsharat Ikhwān al-Ṣafāʾ	Wa-taʿāwun/muʿāwanat baʿḍihim baʿḍan wa/li-ṣidq al-mawadda (wa-l-shafaqa wa-l-taḥannun wa-l-raḥma) (BE, ل, ك, ق, إ)	
46	Fī māhiyyat al-īmān wa-khiṣāl al-muʾminīn (al-muḥaqqiqīn)		
47	Fī māhiyyat al-nāmūs al-ilāhī wa-sharāʾiṭ/khiṣāl al-nubuwwa	Wa-madhāhib al-rabbāniyyīn (إ, ل); wa-marmūzātihim al-mawḍūʿa al-nāmūsiyya (إ)	Wa-kayfiyyat ṣifat al-ākhira wa-l-daʿwa al-ilāhiyya (إ, ع)
48	Fī kayfiyyat al-daʿwa ilā Allāh	Wa-bayān jawāhir al-nufūs wa-ṭabaqāt al-nās (ع, A); wa-khiṭāb al-mutafalsifīn al-shākkīn fī amr al-sharīʿa al-ghāfilīn min asrār kutub al-nubuwwa (ع, إ); wa-khiṭāb ṭabaqāt al-madʿuwwīna min al-kidhb (ل)	Wa-ilā/bi ṣafwat al-ukhuwwa wa-ṣidq al-wafāʾ (F, ع, ر); wa-ilā ṣafwat al-ikhwān wa-taṣdīq al-mawadda (ل); wa-ilā ṣafwat al-ukhuwwa (A)
49	Fī (kayfiyyat) afʿāl al-rūḥāniyyīn	Fī kayfiyyat aḥwāl al-rūḥāniyyīn (ل, ق, ك, خ, ز)	Wa-kayfiyyat al-siyāsa wa-l-riyāsa (ك, ز); al-bāb al-thāmin fī kayfiyyat afʿāl al-rūḥāniyyīn (غ); wa-l-gharaḍ minhā huwa al-bayān annā fī al-ʿālam fāʿilīn nafsāniyyīn rūḥāniyyīn ghayr jismāniyyīn (Fihrist c)
50	Fī kammiyyat anwāʿ al-siyāsāt wa-kayfiyyatihā	Wa-marātib al-musawwasīn wa-ṣifāt al-mudabbarīn lahā (fī al-ʿālam) (ع, ل)	Risālat al-sharḥ wa-l-bayān wa-l-burhān (dhāt al-faṣl al-jāmiʿ li-funūn al-manāfiʿ) (إ, ق, A)
51	Fī (kayfiyyat) naḍd al-ʿālam (bi-asrihi)	Fī qaṣd al-ʿālam bi-asrihi (ق, غ, A); niṣf al-ʿālam bi-asrihi (خ),	Fī al-mabādiʾ al-ṭabīʿiyya wa-l-jismāniyya ʿalā nushūʾ al-marātib al-ʿadadiyya (ع, إ); fī bayān marātib al-mawjūdāt wa-niẓām al-makhlūqāt (ر); fī (marātib) al-mawjūdāt wa-niẓām al-kāʾināt (ق, ن, ل, ح); fī al-mabādiʾ al-ʿaqliyya wa-l-jismāniyya ʿalā marātib al-ʿadadiyya (إ); fī al-mabādiʾ al-jismāniyya (A)

TABLE 8 Variation of epistle titles in manuscript title pages (*cont.*)

No	Main part(s) of the title in most MSS	Significant MS additions	Variant elements
52a	Fī māhiyyat al-siḥr wa-l-ʿazāʾim wa-l-ʿiyan wa-l-ruqā (wa-l-zajar wa-l-wahm) wa-kayfiyyat afʿāl al-ṭilismāt (wa-mā ʿummār al-arḍ) wa-mā al-jinn wa-mā al-shayāṭīn wa-mā al-malāʾika wa-kayfa afʿāluhum wa-taʾthīrāt baʿḍihim fī baʿḍ		
52b	Fī (māhiyyat) al-siḥr wa-l-ʿazāʾim wa-l-ruqā (wa-l-kihāna wa-l-fāl wa-l-zijr wa-l-ṭilismāt wa-l-tijāma wa-l-ʿiyan wa-l-zajar wa-l-wahm)	Kayfiyyat aʿmāl al-ṭilismāt (ق); mā ʿummār al-arḍ wa-mā al-jinn wa-mā al-shayāṭīn wa-mā al-malāʾika wa-kayfa afʿāluhum wa-taʾthīrāt baʿḍihim fī baʿḍ (ق, غ)	Bayān anna fī al-ʿālam rūḥāniyyīn (ق, غ)

TABLE 9 Variation of epistle titles in cross-references

Epistle	Most common title	Alternative titles
1	Risālat al-ʿadad	Khawāṣṣ al-ʿadad; al-Arithmāṭīqī; Tarkīb al-ʿadad
2	Risālat al-handasa	Jūmuṭriyā; Shibh al-madkhal ilā ṣināʿat al-handasa; al-Handasa al-ḥissiyya
3	Risālat al-nujūm	Asṭrunūmiyā; al-Tanjīm; (Kitāb) madkhal al-nujūm; ʿIlm al-nujūm; al-Madkhal ʿalā ʿulūm al-nujūm
4	Risālat jughrāfiyā	
5	Risālat al-mūsīqā	Ta'līf al-luḥūn; Nisab al-mūsīqā
6	Risālat al-nisba al-ʿadadiyya wa-l-handasiyya	Al-nisab al-ʿadadiyya; al-Nisba al-ʿadadiyya wa-l-handasiyya wa-l-ta'līfiyya
7–8		Al-Ṣanā'iʿ al-bashariyya; al-Ṣanā'iʿ; Ajnās al-ʿulūm
7	Risāla fī al-ṣanā'iʿ al-ʿilmiyya	Ajnās al-ʿulūm wa-anwāʿuhā; Ajnās al-ʿulūm; al-Ajnās min al-ʿulūm
8	Risāla fī al-ṣanā'iʿ al-ʿamaliyya	Al-Ṣanā'iʿ al-bashariyya; al-Ṣanā'iʿ; al-Ṣanā'iʿ al-ʿaqliyya
9	Risālat al-akhlāq	
10–14	Risālat al-manṭiq	Kitāb al-manṭiq; Kutub al-manṭiq
10	Risālat Īsāghūjī	
11	Risālat Qāṭīghuriyās	Sharḥ al-maqūlāt al-ʿashar
12	None	
13	None	
14	Risālat al-burhān	Al-Barāhīn
15	Risālat al-hayūlā	Al-Hayūlā wa-l-ṣūra; Samʿ al-kiyān; al-Jism al-muṭlaq
16	Risālat al-samā' wa-l-ʿālam	
17	Risālat al-kawn wa-l-fasād	
18	Risālat al-āthār al-ʿulwiyya	
19	Risālat al-maʿādin	Jawāhir al-maʿādin; al-Jawāhir al-maʿdaniyya; Takwīn al-maʿādin
20	Risālat māhiyyat al-ṭabīʿa	Al-Ṭabīʿa
21	Risālat al-nabāt	
22	Risālat al-ḥayawān	Al-Ḥayawānāt; Ajnās al-ḥayawānāt; Khawāṣṣ al-ḥayawānāt
21–22	Risalat al-ḥayawān wa-l-nabāt	
23	Risālat tarkīb al-jasad	Kitāb al-tashrīḥ (tarkīb jasad al-insān); al-Jasad
24	Risālat al-ḥāss wa-l-maḥsūs	Al-Ḥāss wa-l-maḥsūsāt; al-Ḥāss; al-Ḥiss wa-l-maḥsūs; Kayfiyyat idrāk al-ḥawāss li-maḥsūsātihā
25	Risālat masqaṭ al-nuṭfa	

TABLE 9 Variation of epistle titles in cross-references (*cont.*)

Epistle	Most common title	Alternative titles
26	Risālat al-insān ʿālam ṣaghīr	Inna al-insān ʿālam ṣaghīr; Qawl al-ḥukamāʾ al-insān ʿālam kabīr
26+34	Risālat maʿnā qawl al-ḥukamāʾ inna al-insān ʿālam ṣaghīr wa-l-ʿālam insān kabīr	
27	Risālat nushūʾ al-anfus al-juzʾiyya	
28	Risālat al-ʿishq	
29	Risālat ḥikmat al-mawt	Karāhiyyat al-ḥayāt wa-l-mawt; Risālat karāhiyyat al-ḥayawān li-l-mawt
30	Risālat al-ālām wa-l-ladhdhāt	Al-Ālām; al-Ladhdhāt
31	Risālat ikhtilāf al-lughāt	
32	None	
33	None	
32–33	Risālat al-mabādiʾ al-ʿaqliyya	Al-Mabādiʾ; al-Mabādiʾ al-ʿaqliyya wa-l-afʿāl al-rūḥāniyya
34	Risālat anna al-ʿālam insān kabīr	Maʿnā qawl al-ḥukamāʾ inna al-ʿālam insān kabīr
35	Risālat al-ʿaql wa-l-maʿqūl	Al-ʿAql wa-l-maʿqūlāt; al-ʿAqliyyāt
36	Risālat al-adwār wa-l-akwār	Al-Akwār wa-l-adwār; al-Adwār; Taʾthīrāt al-nujūm fī al-mawālīd
37	None	
38	Risālat al-baʿth wa-l-qiyāma	Al-Baʿth; al-Maʿād
39	Risālat al-ḥarakāt	
40	Risālat al-ʿilal wa-l-maʿlūlāt	Al-ʿIlal wa-l-maʿlūl
41	Al-Ḥudūd	
42	Risālat al-ārāʾ wa-l-madhāhib	Al-Ārāʾ wa-l-madhāhib wa-l-diyānāt
43	Risālat kayfiyyat al-ṭarīq ilā Allāh	
44	Risālat iʿtiqād Ikhwān al-Ṣafāʾ	
45	Risālat ʿishrat Ikhwān al-Ṣafāʾ	
46	Risālat al-īmān	Al-Īmān wa-māhiyyatuhu wa-khiṣāl al-muʾminīn; al-Muʾminīn; Māhiyyat al-īmān wa-khiṣāl al-muʾminīn; Ṣifāt al-muʾminīn al-muḥaqqiqīn; Khiṣāl al-muʾminīn wa-sharāʾiṭ al-īmān

TABLE 9 Variation of epistle titles in cross-references (*cont.*)

Epistle	Most common title	Alternative titles
47	Risālat al-nāmūs	Al-Nāmūs al-ilāhī; al-Nawāmīs; al-Nawāmīs al-ilāhiyya; Al-Nawāmīs al-ilāhiyya wa-l-madhāhib al-rabbāniyya
48	Risālat al-daʿwa ilā Allāh	
49	Risālat afʿāl al-rūḥāniyyāt	Al-Afʿāl al-rūḥāniyya; al-Rūḥāniyyāt; al-Rūḥāniyyūn
3/52	Risālat ʿilm al-nujūm wa-l-siḥr wa-l-ṭilismāt	
50	Risālat al-siyāsāt	Al-Siyāsiyyāt; Siyāsāt al-nubuwwa wa-l-mulk
51	Risāla fī naḍd al-ʿālam	
52	Risālat al-siḥr wa-l-ṭilismāt	Al-Ṭilismāt wa-l-ʿazāʾim; al-Zajar; al-Zajar wa-l-firāsa

TABLE 10 Orphan epistles in cross-references

Epistle title	Alternative titles	Referred by
Kitāb al-ʿurūḍ		5
Kitāb al-anbiyāʾ		5
Nisbat al-ḥurūf		6
Risālat al-taʿālīm	Risālat al-taʿlīm	10
Māhiyyat nafs al-ʿālam		16
Risālat al-ṣifāt wa-l-mawṣūfāt		16
Kitāb ṭabāʾiʿ al-aghdhiya wa-darajāt quwāhā		25
Risālat faḍīlat al-nisab		26

TABLE 11 Epistle transitions

Epistle	Incipit (BE)[1]	Incipit (ع)	Variant nos
1	Correct	Correct	No
2	Correct	Correct	No
3	Correct	Correct	No
4	Incomplete	Incomplete	No
5	Incorrect	Incorrect	7–8
6	Incomplete	None	No
7	Correct	Incorrect	7, 8
8	Incorrect	Incorrect	15–31
9	Incorrect	Incorrect	25
10	None	None	No
11	Correct	Correct	No
12	Correct	None	No
13	None	None	No
14	Correct	Correct	No
15	Correct	Incomplete	No
16	Correct	Correct	No
17	Correct	Correct	No
18	Correct	Correct	No
19	Incorrect	Incorrect	32/33, 38, 42
20	Incorrect	Incorrect	8
21	Incorrect	Incorrect	19
22	Correct	Correct	No
23	Correct	None	No
24	Correct	None	No
25	None	None	No
26	Correct	Incorrect	34
27	Correct	None	No
28	Correct	None	No
29	Correct	None	No
30	Correct	Correct	No
31	Correct	None	No
32	Correct	None	No
33	None	None	No
34	Correct	None	No
35	Correct	Incorrect	24
36	Correct	Incorrect	20

TABLE 11 Epistle transitions (*cont.*)

Epistle	Incipit (BE)	Incipit (ع)	Variant nos
37	Correct	None	No
38	Correct	None	No
39	Correct	Incorrect	II, III
40	Correct	None	No
41	Correct	None	No
42	Correct	None	No
43	None	None	No
44	Correct	None	No
45	None	None	No
46	None	None	No
47	None	None	No
48	None	None	No
49	None	None	No
50	None	None	No
51	Incomplete	None	No
52a	Correct	Correct	No
52b	Incomplete	Incomplete	No

1 The incipits are classified into four categories: (1) correct: correctly indicates the preceding epistle; (2) incomplete: introduces the epistle but does not indicate the preceding epistle; (3) none: the epistle contains no proper introduction; and (4) incorrect: the epistle indicates the preceding epistle incorrectly. In the last case, the incorrect number is given in the next column. The cases where the reference differs from the Beirut and Oxford editions because the epistles in a manuscript are ordered differently, but the reference is correct within that manuscript, are not included.

TABLE 12 Cross-references between epistles

Epistle	Refers to (most MSS)[1]	Referred by
Fihrist	All	42, 52a[2]
1	2, 32–33[3]	2 (5), 3, 5 (2), 7, 10 (2), 11, 15 (2), 32 (2), 33, 39, 40 (4), 41 (2), 50
2	1 (5), 5, 52[4]	1, 3, 5 (4), 7, 15, 16, 18, 34, 42, 51
3	1, 2, 5, 18, 24, 42[5]	5 (2), 8, 15, 20, 26 (2), 34, 49, 51, 52a (2)
4	None	17, 18, 19, 20, 28
5	1 (2), 2 (4), 3 (2), 6, 7, 8, 9, 10–14 (2), 15 (2), 18, 24, 25 (2), 29, 32–33 (2), 38, TBD (2)[6]	2, 3, 6, 15, 20 (2), 25, 31 (6), 33, 35 (2), 37, 39, 42
6	5, TBD (3)[7]	5, 15, 26
7	1, 2, 10–14 (2), 11, 15 (3), 23, 32–33, 35[8]	5, 8, 11, 14, 29, 40 (3)
8	3, 7, 10–14, 16 (2), 17, 23, 24, 26, 35 (2), 47[9]	5, 9, 10 (2), 11, 18, 20 (2), 25, 26, 35, 39, 42, 49
9	8, 22, 24, 25 (2), 26 (2), 29, 30, 38 (2), 42, 45, 46 (3), 47, 48, 50[10]	5, 22, 30, 37 (2), 42 (3), 45, 46
10	1 (2), 8 (2), 15 (2), 17, 24, 25, 31, 32–33 (2), 34, TBD[11]	14
11	1, 7, 8, 17 (2), 38[12]	7, 29, 42
12	None[13]	None
13	None	None
14	7, 10, 15, 22, 24 (4), 32–33, 35, 38, 40, 41, TBD[14]	35, 48
10–14		5 (2), 7 (2), 8, 15, 26, 35 (2), 42, 46
15	1 (2), 2, 3, 5, 6, 10–14, 16, 17, 18, 19, 20, 21, 22, 24[15]	5 (2), 7 (3), 10 (2), 14, 16, 24 (3), 26 (2), 29, 31, 34, 35, 39, 42 (2), 51
16	2, 15, 20, 23, 24, 35, 39, 40, TBD[16]	8 (2), 15, 17 (2), 18 (2), 20 (2), 24, 26, 34 (2), 36, 39 (3), 49, 51(2)
17	4, 16 (2), 19, 21, 22[17]	8, 10, 11 (2), 15, 18, 19, 20 (2), 24, 26, 30, 34, 35, 39, 51
18	2, 4, 8, 16 (2), 17, 22, 23, 24 (3), 36, 39, 52	3, 5, 15, 19 (2), 20 (3), 25, 31 (2), 34, 39 (2), 41 (2), 51
19	4, 17, 18 (2), 20, 21 (2), 22, 32–33 (3), 34, 36, 38, 40, 42 (2)[18]	15, 17, 20 (2), 21 (3), 25, 31, 34 (2), 36 (3), 39 (3), 49, 51

TABLE 12 Cross-references between epistles (*cont.*)

Epistle	Refers to (most MSS)	Referred by
20	3, 4, 5 (2), 8 (2), 16 (2), 17 (2), 18 (3), 19 (2), 21 (2), 22 (2), 23 (2), 34 (2), 38, 49[19]	15, 16, 19, 25 (2), 36, 37, 39
21	19 (3), 22[20]	15, 17, 19 (2), 20 (2), 22, 25 (2), 26, 34, 36 (3), 39, 49, 51
22	9, 21, 25 (3), 32–33, 40, 49, TBD[21]	9, 14, 15, 17, 18, 19, 20 (2), 21, 24, 25 (2), 30, 31, 36 (3), 39, 42 (2), 46, 49
23	TBD[22]	7, 8, 16, 18, 20 (2), 26 (3), 29 (3), 30 (2), 35, 40, 42 (3), 46, 48, 49
24	15 (3), 16, 17, 22, 32–33, 35 (2)	3, 5, 8, 9, 10, 14 (4), 15, 16, 18 (3), 26, 28, 29 (2), 30 (2), 31, 35 (6), 39, 42 (4), 46 (4)
25	5, 8, 18, 19, 20 (2), 21 (2), 22 (2), 27, 34, 36 (4), 38 (2), 40, 49 (2), TBD (2)[23]	5 (2), 9 (2), 10, 22 (3), 29, 30, 31, 32, 37 (2), 48, 52b
26	3 (2), 6, 8, 10–14, 15 (2), 16, 17, 21, 23 (3), 24, 27, 29, 32–33 (2), 35 (2), 38, TBD[24]	8, 9 (2), 29 (2), 30, 35 (2), 42, 46, 48, 49 (3)
27	None	25, 26, 30
28	4, 24, 32–33 (3)	42 (2), 46
29	7, 11, 15, 23 (3), 24 (2), 25, 26 (2), 32–33, 34, 35, 38	5, 9, 26, 30 (4), 42
30	9, 17, 22, 23 (2), 24 (2), 25, 26, 27, 29 (4), 32–33 (2), 38 (4), 40, 44, 46 (3), 47 (2)[25]	9, 38, 39, 42, 46, 52b
31	5 (6), 15, 18 (2), 19, 22, 24, 25, 32–33	10, 35
32	1 (2), 25[26]	None
33	1, 5	None
32–33		1, 5 (2), 7, 10 (2), 14, 19 (3), 22, 24, 26 (2), 28 (3), 29, 30 (2), 31, 35 (3), 36, 37, 40 (5), 42 (5)
34	2, 3, 15, 16 (2), 17, 18, 19 (2), 21, 36, 49, TBD[27]	10, 19, 20 (2), 25, 29, 48
35	5 (2), 8, 10–14 (2), 14, 15, 17, 23, 24 (6), 26 (2), 31, 32–33 (3), 47[28]	7, 8 (2), 14, 16, 24 (2), 26 (2), 29, 42, 46

TABLE 12 Cross-references between epistles (*cont.*)

Epistle	Refers to (most MSS)	Referred by
36	16, 19 (3), 20, 21 (3), 22 (3), 32–33, 38[29]	18, 19, 25 (4), 34, 37, 46, 48, 51
37	5, 9 (2), 20, 25 (2), 32–33, 36, 38 (2), 49	52c (3)
38	30, 42, 49	5, 9 (2), 11, 14, 19, 20, 25 (2), 26, 29, 30 (4), 36, 37 (2), 39, 42 (4), 48
39	1, 5, 8, 15, 16 (3), 17, 18 (2), 19 (2), 20, 21, 22, 24, 30, 38, 42, 46[30]	16, 18
40	1 (4), 7 (3), 23, 32–33 (5), 51[31]	14, 16, 19, 22, 25, 30, 42 (3), 52b
41	1 (2), 18 (2)[32]	14
42	F, 2, 5, 8, 9 (3), 10–14, 11, 15 (2), 22 (2), 23 (3), 24 (4), 26, 28 (2), 29, 30, 32–33 (5), 35, 38 (4), 40 (3), 46, 47 (3), 52[33]	3, 9, 19 (2), 38, 39, 52a
43	TBD (2)[34]	46
44	TBD[35]	30, 46, 48 (2), 52b
45	9, TBD (2)[36]	9, 46, 48
46	9, 10–14, 22, 23, 24 (4), 26, 28, 30, 35, 36, 43, 44, 45, 47 (5), 50, TBD[37]	9 (3), 30 (3), 39, 42, 48
47	TBD (2)[38]	8, 9, 30 (2), 35, 42 (3), 46 (5), 48
48	14, 23, 25, 26, 34, 36, 38, 44 (2), 45, 46, 47, TBD (3)[39]	9
49	3, 8, 16, 19, 21, 22, 23, 26 (3), 50, 52 (3)[40]	20, 22, 25 (2), 34, 37, 38, 51, 52a, 52b (2), 52c (5)
50	1	9, 46, 49
51	2, 3, 15, 16 (2), 17, 18, 19, 21, 36, 49[41]	40
52a	F, 3 (2), 42, 49[42]	
52b	25, 30, 40, 44, 49 (2)[43]	
52c	37 (3), 49 (5)[44]	
52		2, 18, 42, 49 (3)

1 In the table, the number designates the ordinal number of the referred epistle; the number in parentheses designates the number of references to that epistle, in case there are more than one. References to unidentified epistles are indicated by TBD ("to be determined"). The references have been collected from the Beirut edition. Each reference has

been checked against the critical edition, or, if no critical edition was available, against MS ع. The manuscript variation in the cross-references is surprisingly minor; for each epistle, major variants are indicated in these endnotes. The number of references excludes (1) the references to the previous epistle at the beginning when it only occurs in the "corrected" variant of some manuscripts (the *incipit*s are discussed in Chapter 2 and summarized in Table 11); (2) the references to the next epistle at the end of most epistles in many manuscripts (which seem to come from the scribe's pen rather than the original text); (3) the references that an epistle makes to itself; and (4) the references to the *Comprehensive Epistle* (discussed in Chapter 4 and summarized in Table 17). The list of refences is not entirely definitive. First, some references are ambiguous; these are indicated in the following endnotes. Second, since the main method for seeking the references was a digital search of the Beirut edition, those references that occur in one or more manuscripts, but not in the Beirut edition, are mostly not present. However, given the small extent of manuscript variation in this case, there should not be many such references.

2 The beginning of Epistle 42 contains a reference to the *Fihrist* only in the Beirut edition and some manuscripts.

3 The end of MS ز (fol. 39[a]) contains a reference (probably added by the copyist) to Epistle 6.

4 In one of the five references to Epistle 1, the Beirut edition and MS ك omit the explicit reference and only refer vaguely to a previous passage. MSS ك and the margin of ل contain a sixth reference to Epistle 1, which explicates a vague reference in most manuscripts ("as we have shown previously (*qabla dhālika*)"). The Beirut edition and some manuscripts (including ع) contain a self-reference that gives the impression that the reference is to another epistle: "as we have shown in the epistle on geometry (*risālat jūmuṭriyā*)."

5 Most manuscripts refer to Epistles 1 and 2 at the beginning; the Beirut edition only refers to Epistle 1.

6 The references to Epistles 7 and 8 occur as a single reference to "two epistles of ours" (*fī risālatayni lanā*) addressing intellectual and corporeal disciplines. The first of the two references to an "epistle on logic" (*risālat al-manṭiq*), that is, Epistles 10–14, takes the form "book/books of logic" (*kitāb/kutub al-manṭiq*) in the Beirut edition and some manuscripts. Both references to Epistle 25 use the vague allusion "epistle of ours" (*risāla lanā*), although the referent epistle is clear from the context. The epistle also contains references to a "book of prosody" (*kitab al-ʿurūḍ*) and a "book of prophets" (*kitab al-anbiyāʾ*). Both are attributed to the authors by the use of the first person plural, although for the latter some manuscripts employ the passive voice suggesting that the work was not composed by the Brethren of Purity.

7 The reference to Epistle 5 is present in the Beirut edition and MS ك, but absent in many manuscripts, including ع. The epistle contains three unidentified references (pp. 31–32 of the edition) to (1) an epistle addressing the "mutual relations of letters" (*nisbat al-ḥurūf baʿḍuhā min baʿḍ*); (2) "another epistle" (*risāla ukhrā*) on the proportions of colors (*aṣbāgh*); and (3) a "previously mentioned epistle" (*al-risāla al-mutaqaddim dhikruhā*) addressing the proportions of forms (*ṣuwar*). Since the references occur close to each other, and all address proportions, they may all indicate the same epistle.

8 An addition to the epistle (present in MSS أ, ح, د, ز, ط, ف, ك, ل, and ن, and edited in Appendix A of the edition) also contains references to Epistles 23, 24, 25, 27, 29, 35, and 38.

9 The reference to Epistle 3 (*Astronomy*) is in the form "epistle of the effects of stars on generated beings" (*risālat taʾthīrāt al-nujūm fī al-mawālīd*). This expression could plausibly

designate several other epistles, such as Epistle 17 (*Generation and Corruption*), Epistle 18 (*Meteorology*), or Epistle 36 (*Cycles and Revolutions*). The Beirut edition also contains a second, vague, reference to Epistle 7: "and previously we discussed the scientific [disciplines]" (*wa-taqaddama al-qawl fī al-ʿilmiyya*), which is omitted in most manuscripts. The second reference to Epistle 35 (p. 123 of the edition) is ambiguous: "and we ascribed this epistle to the epistles of intellect and the intelligibles" (*wa-nasabnā hādhihi al-risāla ilā rasāʾil al-ʿaql wa-l-maʿqūl*). Given the plural form of the "epistles," this could perhaps also mean the third part of the *Epistles*. In addition, there is the following passage (p. 102): "the soul gains the forms of objects of knowledge through three ways: one is by way of the five senses, the other by way of demonstration, and the last by way of thought and deliberation. And we have composed an epistle for each of these." It is unclear to which specific epistles the passage refers.

10 The reference to Epistle 8, "epistle of human disciplines" (*risālat al-ṣanāʾiʿ al-bashariyya*), is ambiguous and could also designate Epistle 7. The reference to Epistle 42 speaks of "an epistle of ours" (*risāla lanā*) that addresses the subject of "corrupt views" (*al-ārāʾ al-fāsida*). The reference to Epistle 47 (*Divine Law*) speaks of both law (*nāmūs*) and governance (*siyāsa*), to which "we designated an epistle" (*afradnā lahu risāla*), and could also refer to Epistle 50 (*Governance*). The epistle contains a reference (p. 92 of the edition) to "our legal and intellectual epistles" (*rasāʾilinā al-nāmūsiyya wa-l-ʿaqliyya*), that is, the fourth and third sections of the *Epistles*, respectively.

11 In the reference to Epistle 25 (p. 26 of the edition), the title of that epistle is absent in many manuscripts. The reference is prefaced by a mention of the manner of particular souls' spread into matter, while some manuscripts contain the addition: "I gave it the title (*tarjamtuhā*) the human being is a microcosm." Thus, the reference could also be to either Epistle 26 (*The Human Being is a Microcosm*) or Epistle 27 (*Spread of Particular Souls into Physical Bodies*). A discussion of intelligible knowledge makes an unidentifiable reference to an "epistle of knowledge/mathematics" (*risālat al-taʿālīm*).

12 The reference to Epistle 38 (*Resurrection*) employs the form "epistle of afterlife" (*risālat al-maʿād*), which is ambiguous. The otherworldly pleasures, to which the passage refers, are also discussed in Epistle 30 (*Pleasure and Pain*). At least one manuscript (أ) adds "and resurrection" (*wa-l-baʿth wa-l-qiyāma*) to the epistle title, perhaps to explicate the reference.

13 In the transition passage at the beginning of the epistle, some manuscripts refer to Epistle 11 (*Categories*).

14 The epistle (p. 151 of the edition) also contains a reference, depending on the manuscript, to either an "epistle of theology" (*al-risāla al-ilāhiyya*) or "theological epistles" (*al-rasāʾil al-ilāhiyya*), which supposedly shows that the world's creation does not occur in space and time. This could be a reference to the fourth section, where only Epistles 49 and 50 deal with philosophical theology, or else to some unidentified epistle.

15 The reference (p. 53 of the edition) to Epistle 6 ("an epistle in which we discussed numerical, geometrical, and harmonious proportions") occurs in the context of a list of the four mathematical sciences. Thus, the reference might also designate Epistle 5 (*Music*). The reference to Epistle 20 (*Essence of Nature*) is ambiguous. The passage (p. 9 of the edition) refers to "another epistle" (*risāla ukhrā*) where the authors addressed nature as a faculty of the Universal Soul.

16 The reference (p. 70 of the edition) to Epistle 20 (*Essence of Nature*) is ambiguous: "then we describe ... in another epistle the essence of the world soul." There is an unidentified reference (p. 72) to an "epistle of the attributes and their substrates" (*risālat al-ṣifāt wa-l-mawṣūfāt*).

17 Epistles 19, 21, and 22 are referred to together (p. 175 of the edition); the authors state that "we have dedicated a separate epistle to each kind" (*wa-qad afradnā li-kull nawʿ minhā risāla mufrada*) of minerals, plants, and animals.

18 The first reference (p. 250 of the edition) to Epistle 21 (*Plants*) has, in many manuscripts, the ambiguous referent of "another epistle" (*risāla ukhrā*). In other manuscripts, the reference is explicitly to an "epistle of plants." The reference to Epistle 22 (p. 250) is similarly ambiguous in most manuscripts: "an epistle after the discussion of plants" (*risāla baʿd dhikr al-nabāt*); some manuscripts refer explicitly to an "epistle on animals." The references (p. 252) to Epistle 36 (*Cycles and Revolutions*) and Epistle 20 (*Essence of Nature*) are to an "epistle of ours" (*risāla lanā*) that is specified to discuss the subject in question.

19 The first reference to Epistle 34 (*The World is a Big Human Being*), and the second reference to Epistle 8 (*Practical Sciences*), are absent in many manuscripts. One passage (p. 391 of the edition) in many manuscripts refers to the authors' "five books" which are identified as Epistles 3, 16, 17, 18, and 34. In the Beirut edition and other manuscripts, the reference is instead to Epistle 16 only. The second reference (p. 396) to Epistles 19, 20, and 22 is in the form the "epistle of minerals, plants, and animals."

20 The third reference to Epistle 19 (*Minerals*) is ambiguous: the fact that the active and passive powers are not separate in some individuals is said to be shown in an "epistle of ours" (*risāla lanā*). Five manuscripts specify this to be the "epistle of minerals."

21 An unidentified reference (p. 9 of the edition) to "an epistle of ours" (*risāla lanā*) addresses the Universal Soul's faculties and their spread into celestial and earthly bodies. This could be either Epistle 20 (*Essence of Nature*) or Epistle 27 (*Spread of Particular Souls into Physical Bodies*).

22 The epistle contains a reference to the "first epistle" (*al-risāla al-ūlā*) where an organic body's constituent parts (flesh, blood, bones, etc.) were explained. The first epistle of the second part, Epistle 15 (*Matter and Form*), contains no such explanation. The parts of an animal body are addressed, for example, in Epistle 22 (*Animals*) (pp. 13–15 of the edition).

23 The epistle also contains a reference to what "was mentioned in the book of introduction to astronomy (*madhkūra fī kitāb madkhal al-nujūm*)." This is presumably not a reference to Epistle 3 (*Astronomy*), to which many manuscripts also refer as an "introduction," but to an external source. The reference to Epistle 19 and the second reference to Epistle 21 occur within a list of five epistles in MS ع. In the Beirut edition, the list only contains Epistles 18, 20, and 22, while it omits Epistles 19 and 21. The reference to Epistle 34 (*The World is a Big Human Being*) is clear in MS ع. In the Beirut edition, however, the reference is in the form "the epistle on the meaning of the sages' saying that the human being is a microcosm and the world is a big human being." This would also include Epistle 26 (*The Human Being is a Microcosm*) in the reference. The first reference to Epistle 36 is present in the Beirut edition but missing in MS ع. The epistle contains two unidentified references (*Rasāʾil*, vol. 2, 428): (1) the "book of natures of nourishments and degrees of their powers" (*kitāb ṭabāʾiʿ al-aghdhiya wa-darajāt quwāhā*) and (2) the "book of anatomy" (*kitāb al-tashrīḥ*). The authors refer to both as their own works. The subject matter of both might relate to Epistle 23 (*Constitution of the Human Body*).

24 The reference to Epistle 21 is missing in MS ع but present in the Beirut edition. The reference to Epistle 6 (*Arithmetical and Geometrical Proportions*) (*Rasāʾil*, vol. 2, 476) uses the ambiguous title "epistle of the virtue of proportions" (*risālat faḍīlat al-nisab*). The mentioned subject, harmonious proportions of animal bodies, is in fact addressed in Epistle 6 (35–36).

25 The first references to Epistles 23, 24, and 38, and the references to Epistles 9 and 26, are omitted in many manuscripts. Three of the four references to Epistle 29 refer to an "epistle" in the Beirut edition but to a "chapter" (*faṣl*) or a "previous passage" (*qablu*) in most manuscripts.

26 Epistles 32 and 33 are intertwined in the manuscripts. The references here follow the Beirut edition division of these epistles. Among the Oxford edition variants, Epistle 32a has two references to Epistle 1; 32b has one reference to Epistles 1 and 5 each; and Epistle 33 has one reference to Epistle 25. The reference to Epistle 25 refers to "another epistle" (*risāla ukhrā*) which addresses the subject of sperm.

27 The epistle contains a reference to "another epistle" (*risāla ukhrā*), which addresses the animal body as either the soul's prison or as a "bridge" (*barzakh*) to resurrection.

28 The "corrected" incipit of this epistle in the Beirut edition contains a fourth reference to Epistle 32/33, which is not included in the final count. The epistle contains a reference to "our physical epistles" (*rasāʾilinā al-ṭabīʿiyya*), that is, the second part of the *Epistles*.

29 The second references to Epistles 21 and 22 occur in the combined form of "epistle of animals and plants" (*risālat al-ḥayawān wa-l-nabāt*).

30 The beginning of the epistle contains a reference (p. 5 of the edition) to "our physical and theological epistles" (*rasāʾilinā al-ṭabīʿiyyāt wa-l-ilāhiyyāt*), that is, the second and, presumably, the fourth sections of the *Epistles*. At the end (p. 56), there is another reference to "our theological epistles."

31 The third reference (pp. 109–110 of the edition) to Epistle 32/33 (*Intellectual Principles*) occurs in the form "two epistles on the principles" (*risālatayni fī al-mabādiʾ*). This appears to be the only reference in the *Epistles* that treats these epistles as two separate epistles. However, even here the epistles' full titles are not mentioned. Moreover, in the same passage, the authors also refer to "two epistles on the constitution of the world" (*risālatayni fī naḍd al-ʿālam*). Only Epistle 50 carries this name. The other epistle could potentially be Epistle 49 (*Spiritual Beings*), which addresses similarly metaphysical themes. Since this is the only reference to Epistle 50 in the entire *Epistles*, and the term "constitution" (*naḍd*) varies considerably in the manuscripts, it could also be that the reference is to two epistles on the coming-to-be of the world in general, rather than to Epistle 50 in particular. The third reference (p. 110) to Epistle 7 (*Theoretical Sciences*) states: "as we have shown in the epistle on the kinds of sciences (*ajnās al-ʿulūm*), the nine questions and their philosophical answers." The Beirut edition and many manuscripts, however, omit the word "sciences." Hence, the passage then refers to the epistles or epistles on the nine questions. The epistle also contains a reference to "our physical epistles" (*rasāʾilinā al-ṭabīʿiyyāt*), that is, the second section of the *Epistles*.

32 The second reference to Epistle 18 is ambiguous. In most manuscripts, the reference is to "an epistle of ours" (*risāla lanā*) addressing the causes of the saltiness versus sweetness of sea and river water. In the Beirut edition and some manuscripts, this is specified to be the "previously mentioned epistle" (*risāla lanā qad taqaddama dhikruhā*), which would be Epistle 18.

33 The fifth reference to Epistle 32/33 is present in the Beirut edition but omitted in MS ع. The Beirut edition contains what seems to be a second reference to Epistles 10–14: "and we have discussed some of that in length in the books of logic (*kutub al-manṭiq*)." The reference is missing in MS ع. The two references to Epistle 28 (*Human Capability for Knowledge*) appear in the form "epistle of knowledge" (*risālat al-maʿārif*), and are thus ambiguous. The epistle contains references (*Rasāʾil*, vol. 3, 457, 464, 465, 510, 512) to "our preparatory (*al-riyāḍiyyāt*)" epistles, "physical" (*ṭabīʿiyya*) epistles, an "intellectual" epistle or epistles

(*risāla/rasāʾil al-ʿaqliyyāt*), and "our theological epistles" (*rasāʾilinā al-ilāhiyya*), which presumably refer to the four sections of the *Epistles*.

34 The epistle contains a reference (p. 20 of the edition) to "two first epistles that we composed to learners and disciples" (*al-risālatayni al-awwalatayni allatayni waḍaʿnāhumā li-l-mutaʿallimīna wa-l-murīdīna*) where the authors addressed the gradual way of instruction.

35 The epistle contains a reference (p. 38) to a "second epistle" (*risāla thāniya*) addressing "our spiritual city" (*madīnatunā al-rūḥāniyya*).

36 The epistle contains a reference (p. 20 of the edition) to "five epistles which we formed (*ṣawwarnā*) in the human form (*ʿalā ṣūrat al-insān*)". The epistle contains two references (pp. 98, 131 of the edition) to a "second epistle" (*al-risāla al-thāniya*). In the first, the authors tell that their sciences are derived from four books; the second refers to the four degrees of the Brethren of Purity and their city.

37 The reference to Epistles 10–14 (*Rasāʾil*, vol. 4, 65) occurs in the form "our logical epistles" (*fī rasāʾilinā al-manṭiqiyyāt*), as opposed to an "epistle of logic," as is usually the case. The reference to Epistle 28 (*Human Capability for Knowledge*) is ambiguous as it has the form "epistle of knowledge" (*risālat al-maʿārif*). The references to Epistles 44 and 45, and the third reference to Epistle 47, occur within a passage that is present in the Beirut edition but absent in MS ع. The second reference to Epistle 45 (*Companionship of the Brethren of Purity*) occurs in the Beirut edition in the form "epistles of the Brethren of Purity." The epistle contains an unidentified reference to "another epistle" (*risāla ukhrā*) addressing the "classes" (*ṭabaqāt*) of demons (*shayāṭīn*). The epistle also contains references (*Rasāʾil*, vol. 4, 83) to "our physical epistles" (*rasāʾilinā al-ṭabīʿiyya*) and our "intellectual epistles" (*rasāʾilinā al-ʿaqliyya*), that is, the second and third sections of the *Epistles*.

38 The epistle contains two unidentified references to (1) "our epistle after this one" (*risāla lanā baʿda hādhihi*), addressing the properties of prophecy, and to (2) "our epistle before this one" (*risāla lanā qabla hādhihi*), addressing either "pure brotherhood" (*ṣafwat al-ukhuwwa*) or the "attributes of brotherhood" (*ṣifat al-ukhuwwa*). The latter reference could be to Epistle 45 (*Companionship of the Brethren of Purity*).

39 In the first reference to Epistle 44 (*Beliefs of the Brethren of Purity*), only MS خ includes the full title in the reference. Other testimonies have either "epistle of the Brethren of Purity" (*risālat ikhwān al-ṣafāʾ*), "epistle by the Brethren of Purity" (*risāla li-ikhwān al-ṣafāʾ*), or "an epistle" (*al-risāla*). In the second reference to Epistle 44, the word "epistle" (*risāla*) is absent in the Beirut edition but present in most manuscripts. The epistle contains a reference (p. 222 of the edition) to "our legal epistles" (*rasāʾilinā al-nāmūsiyya*), that is, the fourth section of the *Epistles*. The epistle contains three unidentified references (pp. 69, 97, 113 of the edition) to (1) "another epistle" (*al-risāla al-thāniya*), which addresses the four degrees of the Brethren of Purity, (2) "another epistle" or a "third epistle," addressing the prophets' twelve characteristics, and (3) an "epistle of ours" (*risāla lanā*), which addresses the question of whether soul is a substance or an accident.

40 These represent the references of the Beirut edition variant of this epistle. The manuscripts carry a short and a long version; the Beirut edition combines both. All of the Beirut edition references are contained in one or another of the variants: Epistle 49a refers once to Epistles 3 and 50, and twice to Epistle 52; the rest of the references correspond to Epistle 49b. The reference (p. 44 of the edition) to Epistle 3 (*Astronomy*) and the second reference to Epistle 52 (*Magic*) occur in the form "epistle of astronomy, magic, and talismans," as if the two epistles constituted a single epistle.

41 These are the references in the Beirut edition variant, which corresponds to Epistle 51b in the critical edition. All the Beirut edition references are also present in the other manuscripts with that variant.

42 Since the three variants of Epistle 52 are textually entirely independent, their references are given separately. The Beirut edition version of this epistle combines Epistles 52a and 52b.

43 The reference to Epistle 30 is to the "commentary" (*sharḥ*) of this epistle in the *Comprehensive Epistle*. The references to Epistles 25 and 40 are absent in the Beirut edition but present in most manuscripts.

44 The references to Epistle 49 (*Spiritual Beings*) are in the form of either an "epistle on angels" (*risāla fī al-malāʾika*) or "epistle on tranquil angels" (*risāla fī al-sakīnāt*). There is no such epistle, but angels are discussed within Epistle 49 as one kind of spiritual beings.

TABLE 13 Cross-references between the four parts of the *Epistles*

	I	II	III	IV	Total
I	43	41	19	11	114
II	37	111	48	13	209
III	27	51	16	8	102
IV	19	45	19	30	113
Total	126	248	102	62	538

TABLE 14 Reference ratio of the epistles

Epistle	References	Length (MS ع)	References / 20 pages	Referents	Ratio
1	2	19	2.1	25	0.1
2	7	16	8,8	13	0.7
3	6	28	4.3	12	0.4
4	0	13	0	5	0
5	26	41	12.7	19	0.7
6	4	11	7.3	3	2.4
7	11	11	20	8	2.5
8	12	12	20.0	14	1.4
7–8	23	23	20.0	22	0.9
9	19	65	5.8	10	0.6
10	14	11	25.5	1	25.5
11	6	7	17.1	3	5.7
12	0	7	0	0	N/A
13	0	N/A	0	0	N/A
14	14	15	18.7	2	9.3
10–14	34	40	17.0	6	2.8
15	15	13	23.1	22	1.0
16	9	17	10.6	20	0.5
17	6	6	20.0	16	1.3
18	14	17	16.5	14	1.2
19	17	27	12.6	19	0.7
20	25	12	41.7	8	5.2
21	4	17	4.7	17	0.3
22	9	115	1.6	22	0.1
23	1	11	1.8	22	0.1
24	9	13	13.8	36	0.4
25	23	24	19.2	16	1.2
26	22	15	29.3	14	2.1
27	0	7	0	3	0
28	5	9	11.1	3	3.7
29	15	10	30.0	8	3.8
30	27	18	30.0	6	5.0
31	14	94 (BE)	3.0	2	1.5
32	3	12	5.0	0	N/A

TABLE 14 Reference ratio of the epistles (*cont.*)

Epistle	References	Length (MS ع)	References / 20 pages	Referents	Ratio
33	2	7	5.7	0	N/A
32–33	5	19	5.3	36	0.1
34	13	6	43.3	7	6.2
35	22	11	40.0	12	3.3
36	13	12	21.7	11	2.0
37	11	11	20.0	3	6.7
38	3	20	3.0	24	0.1
39	20	14	28.6	2	14.3
40	14	26	10.8	10	1.1
41	4	9	8.9	1	8.9
42	43	89	9.7	7	1.4
43	2	6	6.7	1	6.7
44	1	17	1.2	5	0.2
45	3	12	5.0	3	1.7
46	23	38	12.1	9	1.3
47	2	17	2.4	14	0.2
48	15	34	8.8	1	8.8
49	14	26	10.8	16	0.7
50	1	13	1.5	3	0.5
51	11	8	27.5	1	27.5
52b	6	89	1.3	6	0.2

TABLE 15 Chronology of epistles by ratio

Very early epistles	1, 3, 4, 21, 22, 23, 24, 27, 32/33, 38, 44, 47, 52
Early epistles	2, 5, 7–8, 9, 16, 19, 49, 50
Middle epistles	8, 15, 17, 18, 25, 31, 36, 40, 42, 45, 46
Late epistles	6, 7, 10–14, 26, 28, 29, 35, 43
Very late epistles	10, 11, 12, 13, 14, 20, 30, 34, 37, 39, 41, 48, 51

TABLE 16 Hypothesis of epistle chronology

I:	Mathematical/Pythagorean	1–6, 32–33
II:	Aristotelian physical	15–19, 21–24
III:	Non-Aristotelian physical	28, 31, 34–36, 38–40
IV:	Curricular	7–8
V:	Logical	10–14, 41
VI:	Non-Aristotelian physical II	20, 25–27
VII:	Cosmological/metaphysical	49, 51
VIII:	Ethical/religious	43–45, 47, 50
IX:	Ethical/religious II	9, 29–30, 37, 42, 46, 48
X:	Very late/post-RIS	52a, 52b, 52c, *Fihrist*, RJ

TABLE 17 References from the *Epistles* to the *Comprehensive Epistle*

Epistle	Beirut edition	Most manuscripts
Fihrist	4	–
I	2	–
7	1	–
9	1	–
II	–	–
III	–	–
IV	8	6
42	1	–
49	1	1
50	2	1
52b	4	4
Total	14	6

TABLE 18 Epistle lengths in the *Comprehensive Epistle*

Epistle	Ed. Ghālib	Ed. Kacimi	RJ *Fihrist*	RJ NON-*Fihrist*
Intro	3	3	–	
1	75	88		
2	2	4	1	3
3	1	1	1	–
5	3	4	2	2
4	4	4	4	–
6	4	5	1	4
7	9	12	1	11
8	2	2	2	–
9	11	13	2	11
10	1	1	1	–
11	1	1	1	–
12	1	1	1	–
13	1	1	1	–
14	12	15	1	14
Transition I–II				
15	2	2	2	–
16	2	2	2	–
17	1	2	2	–
18	1	1 / 9[1]	1	– / 8
19	2	3	3	–
20	2	2	2	–
21	3	4	3	1
22	70	99	1	98
23	4	6	1	5
24	2	3	3	–
25	23	35	5	30
26	35	56	8	48
27	9	16	2	14
28	2	3	3	–
29	2	2	1	1
30	11	14	1	13
31	13	17	8	9
32	11	14	1	13
33	1	1	0,5	0,5

TABLE 18 Epistle lengths in the *Comprehensive Epistle* (*cont.*)

Epistle	Ed. Ghālib	Ed. Kacimi	RJ *Fihrist*	RJ NON-*Fihrist*
Transition II–III				
34	18	23	1	22
35	18	24	2	22
36	32	36	0,5	35,5
37	15	18	2	16
38	25	30	0,5	29,5
39	19	24	0,5	23,5
40	17	21	0,5	20,5
41	3	5	0,5	4,5
Transition III–IV				
42	5	6	2	4
43	3	4	1	3
44	4	6	1	5
46	1	1	1	–
47	7	9	1	8
48	5	7	1	6
49	2	3	1	2
50	2	3	1	2
51	2	2	2	–
52	2	3	3	–
45	6	10	1	9

1 This epistle commentary has a short and a long version in the manuscripts.

TABLE 19 The relation between the *Super-Comprehensive Epistle* and the *Comprehensive Epistle*

RJJ	Subject	RIS[1]	RJ[2]
Intro	Death and life	29	G308–309/K409
1	Death and life	29	G309–310/K410–411
2	Soul and Intellect	32	G337–339/K450–452
3	Particular souls	32	G339–340/K453–454 G344–345/K458–459
4	Macrocosm	34	G346–349/K461–464
5	Soul	34	G349–350/K464–466
6	Hierarchy of creation	34	G351–352/K466–468
7	Macrocosm	34	G352–354/K468–471
8	Particular souls	34	G354–356/K471–474
9	Judgment Day	34	G356–358/K475–477
10	Judgment Day	34	G358–359/K477–478
11	Judgment Day	34	G359–360/K478–480
12	Judgment Day	34	G360–363/K480–484
13	Macrocosm; prophets and imāms	26	G287–290/K383–389
14	Judgment Day	38	G428–429/K563
15	Judgment Day	38	G429–435/K564–569
16	Judgment Day	38	G435–436/K569–571
17	Judgment Day	38	G436–437/K572–573
18	Judgment Day	38	G437–439/K573–574
19	Judgment Day	38	G439–441/K575–577
20	Judgment Day	38	G441–443/K577–579
21	Judgment Day	38	G443–445/K579–583
22	Judgment Day	38	G446–448/K583–587
23	Judgment Day	38	G448–451/K587–590
24	Judgment Day	38	G451–453/K590–592
25	Hierarchy of creation	32	G334–336/K446–449
26	Two primary principles	32	G336–337/K449–450
27	Spiritual father and mother; microcosm	25, 26	G253–254/K330–332 G260–263/K342–346
28	Intellect and Soul	44, 48, 40	G517/K680 G527/K693 G484–485/K632–3
29	Intellect	35, 40	G363–384/K484–485 G485/K633–634

TABLE 19 The *Super-Comprehensive Epistle* (*cont.*)

RJJ	Subject	RIS	RJ
30	Soul	40	G485–489/K634–637
31	Soul	40, 39	G489/K637–638 G461–463/K602–604
32	Knowledge of God; belief	35, 34, 46	G370/K493 G370/K492–493 G353/K469 G517–518/K681–682
33	Adam's sin; good and bad beliefs	25, 1, 5	G251/K326–327 G90/K91 G91–92/K93 G64/K60 G83/K82 G103/K111
34	Iblīs	1	G75–77/K73–76 G69/K66–67
35	Iblīs	28, 9	G306/K406–407 G307–308/K408 G124–126/K139–140
36	Iblīs; love	9, 37	G128/K142–143 G427–428/K561–562
37	Particular souls; Iblīs	37, 49, 7, 22	G416/K547 G532–534/K702–703 G112/K123 G171–172/K213–214
38	False imāms; microcosm	22, 23	G206/K262 G226/K289–290
39	Microcosm	26	G264–265/K347–349 G272–273/K361–362
40	Paradise and hell	30	G313/K414–415 G316–318/K418–420
41	Paradise and hell	30	G319–321/K422–425
42	Human form; God, Intellect, Soul	26, 41	G257/K336–337 G490–492/K639–643
43	Souls and intellects; spiritual and corporeal existents	35	G373–374/K497

TABLE 19 The *Super-Comprehensive Epistle* (*cont.*)

RJJ	Subject	RIS	RJ
44	Causes of existents	35	G374–375/K498–500
45	Macrocosm	26	G281–282/K374–375 G283/K376–377
46	Intellect and Soul	25, 41	G240–241/K309–312 G490–491/K639–640
47	Intellect, Soul, matter	41, 25	G491/K641 G241–242/K312 G242–243/K312–315 G246–247/K318–320
48	Macrocosm	25	G235–236/K301–304
49	Embryo; end of movement; grades of heaven and hell; kinds of resurrection	36	G383–384/K510
50	Intellect and Soul	25	G248–251/K320–325
51	Embryo	25	G251–252/K326–327
52	Jinns	22	G194–196/K245–248

1 The number refers to the epistle(s) of the *Rasāʾil Ikhwān al-Ṣafāʾ* (RIS) to which the cited section of *al-Risāla al-jāmiʿa* corresponds.

2 The page numbers give the sections in the Ghālib and Kacimi editions of *al-Risāla al-jāmiʿa* that are cited in the chapter of *Risālat Jāmiʿat al-jāmiʿa*.

Bibliography

Manuscripts

MS Amcazade Hüseyn Paşa 452 (1090/1680) [I]
MS Atıf Efendi 1681 (578 /1182) [ع]
Bibliothèque nationale de France
MS 2303 (1020/1611) [ر]
MS 2304 (1065/1654) [ز]
MS 2305 (1153/1740) [D]
MS 6000 (undated?) [K]
MS 6647–6648 (675/1275) [د]
MS Carullah 982 (undated) [J]
MS Damat Ibrahim 808 (undated) [E]
MS Esad Efendi 3637 (ca. seventh/thirteenth cent.) [ن]
MS Esad Efendi 3638 (686/1287) [أ]
MS Feyzullah Efendi 2130 (704/1304) [ف]
MS Feyzullah Efendi 2131 (704/1304) [ق]
MS Köprülü 871 (820/1417) [ل]
MS Köprülü 870 (ca. 857–886/1453–1481) [ك]
Princeton University Library, Garrett Collection MS 4263 (956/1549) [C]
MS Raghıp Paşa, 839 (undated) [F]
MS Raghıp Paşa, 840 (undated) [G]
Universitätsbibliothek Tübingen, MS MaVI86 (612/1215) [H]
MS Yeni Camı 1199 (887/1482) [B]
A private manuscript sold at the Sotheby auction on April 26th, 2017 (683/1284) [A]

Primary Sources

ʿAbd al-Jabbār. *Tathbīt dalāʾil al-nubuwwa*. Edited by ʿAbd al-Karīm ʿUthmān. 2 vols. Beirut: Dār al-ʿarabiyya, 1962.

Avicenna. "Aqsām al-ʿulūm al-ʿaqliyya." In *Tisʿ rasāʾil fī al-ḥikma wa-l-ṭabīʿiyyāt*, 104–118. Cairo: Dār al-ʿarab li-l-bustānī, 1326 AH.

Avicenna. *Al-Mabdaʾ wa-l-maʿād*. Wisdom of Persia. Edited by ʿAbd Allāh Nūrānī. Tehran: Institute of Islamic Studies and Tehran University, 1984.

Avicenna. *The Metaphysics of the Healing: A Parallel English-Arabic Text*. Edited and translated by Michael E. Marmura. Provo: Brigham Young University Press, 2004.

Al-Bayhaqī. *Tārikh ḥukamāʾ al-islām*. Edited by Muḥammad Kurd ʿAlī. Damascus: Maṭbaʿat al-taraqqī, 1946.

Al-Fārābī. *Al-Farabi on the Perfect State: Abū Naṣr al-Fārābī's Mabādi' ārā' ahl al-madīna al-fāḍila*. Edited and translated by Richard Walzer. Oxford: Clarendon Press, 1985.

Al-Fārābī. *Iḥṣā' al-ʿulūm li-Abī Naṣr al-Fārābī*. Edited by ʿUthmān Amīn. Cairo: Maṭbaʿat al-saʿāda, 1931.

Al-Fārābī. *Kitāb Taḥṣīl al-saʿāda*. Edited by Jaʿfar Āl Yāsīn. Beirut: Dār al-Andalus, 1981.

Al-Ghazālī. *Al-Munqidh min al-ḍalāl*. Edited by Jamīl Ṣalībā and Kāmil ʿAyyār. Beirut: Dār al-Andalus, 1967.

Ibn al-Qifṭī. *Tārīkh al-ḥukamā'*. Edited by Julius Lippert. Leipzig: Dieterichische Verlagsbuchhandlung, 1903.

Ikhwān al-Ṣafā'. *The Call to God: An Arabic Critical Edition and English Translation of Epistle 48*. Epistles of the Brethren of Purity. Edited and translated by Abbas Hamdani and Abdallah Soufan. Oxford: Oxford University Press and the Institute of Ismaili Studies, 2019.

Ikhwān al-Ṣafā'. *The Case of the Animals versus Man Before the King of the Jinn: An Arabic Critical Edition and English Translation of Epistle 22*. Epistles of the Brethren of Purity. Edited and translated by Lenn E. Goodman and Richard McGregor. Oxford: Oxford University Press and the Institute of Ismaili Studies, 2010.

Ikhwān al-Ṣafā'. "Edición crítica árabe y estudio de la *Risālat al-ŷāmiʿa ḏāt al-fawā'id al-nāfiʿa, tāŷ Rasā'il Ijwān al-Ṣafā'* (finales del s. IV H./X E.C., principios del s. V H./XI E.C.): Tomo II." Edited by Mourad Kacimi. Doctoral thesis, University of Alicante, 2015.

Ikhwān al-Ṣafā'. *On Arithmetic & Geometry: An Arabic Critical Edition and English Translation of Epistles 1–2*. Epistles of the Brethren of Purity. Edited and translated by Nader El-Bizri. Oxford: Oxford University Press and the Institute of Ismaili Studies, 2013.

Ikhwān al-Ṣafā'. *On Astronomia: An Arabic Critical Edition and English Translation of Epistle 3*. Epistles of the Brethren of Purity. Edited and translated by Jamil F. Ragep and Taro Mimura. Oxford: Oxford University Press and the Institute of Ismaili Studies, 2015.

Ikhwān al-Ṣafā'. *On Companionship and Belief: An Arabic Critical Edition and English Translation of Epistles 43–45*. Epistles of the Brethren of Purity. Edited and translated by Toby Mayer, Ian Richard Netton, and Samer F. Traboulsi. Oxford: Oxford University Press and the Institute of Ismaili Studies, 2017.

Ikhwān al-Ṣafā'. *On Composition and the Arts: An Arabic Critical Edition and English Translation of Epistles 6–8*. Epistles of the Brethren of Purity. Edited and translated by Nader El-Bizri and Godefroid de Callataÿ. Oxford: Oxford University Press and the Institute of Ismaili Studies, 2018.

Ikhwān al-Ṣafā'. *On Ethics and Character Traits: An Arabic Critical Edition and English Translation of Epistle 9*. Epistles of the Brethren of Purity. Edited and translated by Omar Alí-de-Unzaga. Oxford: Oxford University Press and the Institute of Ismaili Studies, 2024.

Ikhwān al-Ṣafāʾ. *On Geography: An Arabic Critical Edition and English Translation of Epistle 4*. Epistles of the Brethren of Purity. Edited and translated by Ignacio Sánchez and James Montgomery. Oxford: Oxford University Press and the Institute of Ismaili Studies, 2014.

Ikhwān al-Ṣafāʾ. *On God and the World: An Arabic Critical Edition and English Translation of Epistles 49–51*. Epistles of the Brethren of Purity. Edited and translated by Wilferd Madelung, Cyril Uy II, Carmela Baffioni, and Nuha Alshaar. Oxford: Oxford University Press and the Institute of Ismaili Studies, 2019.

Ikhwān al-Ṣafāʾ. *On Life, Death, and Languages: An Arabic Critical Edition and English Translation of Epistles 29–31*. Epistles of the Brethren of Purity. Edited and translated by Eric Ormsby. Oxford: Oxford University Press and the Institute of Ismaili Studies, 2021.

Ikhwān al-Ṣafāʾ. *On Logic: An Arabic Critical Edition and English Translation of Epistles 10–14*. Epistles of the Brethren of Purity. Edited and translated by Carmela Baffioni. Oxford: Oxford University Press and the Institute of Ismaili Studies, 2010.

Ikhwān al-Ṣafāʾ. *On Magic: An Arabic Critical Edition and English Translation of Epistle 52, Part 1*. Epistles of the Brethren of Purity. Edited and translated by Godefroid de Callataÿ and Bruno Halflants. Oxford: Oxford University Press and the Institute of Ismaili Studies, 2011.

Ikhwān al-Ṣafāʾ. *On Music: An Arabic Critical Edition and English Translation of Epistle 5*. Epistles of the Brethren of Purity. Edited and translated by Owen Wright. Oxford: Oxford University Press and the Institute of Ismaili Studies, 2011.

Ikhwān al-Ṣafāʾ. *On the Natural Sciences: An Arabic Critical Edition and English Translation of Epistles 15–21*. Epistles of the Brethren of Purity. Edited and translated by Carmela Baffioni. Oxford: Oxford University Press and the Institute of Ismaili Studies, 2013.

Ikhwān al-Ṣafāʾ. *Rasāʾil Ikhwān al-Ṣafāʾ wa-Khullān al-Wafāʾ*. Edited by Buṭrus al-Bustānī. 4 vols. Beirut: Dār Ṣādir, 1957.

Ikhwān al-Ṣafāʾ. *Al-Risāla al-jāmiʿa: Tāj Rasāʾil Ikhwān al-Ṣafāʾ wa-Khullān al-Wafāʾ*. Edited by Jamīl Ṣalībā. 2 vols. Damascus: Maṭbaʿat al-taraqqī, 1949–1951.

Ikhwān al-Ṣafāʾ. *Al-Risāla al-jāmiʿa: Tāj Rasāʾil Ikhwān al-Ṣafāʾ wa-Khullān al-Wafāʾ*. Edited by Muṣṭafā Ghālib. Beirut: Dār al-Andalus, 1974.

Ikhwān al-Ṣafāʾ. *Risālat Jāmiʿat al-jāmiʿa li-Ikhwān al-Ṣafāʾ wa-Khullān al-Wafāʾ*. Edited by ʿĀrif Tāmir. Beirut: Dār Maktabat al-ḥayāt, 1970.

Ikhwān al-Ṣafāʾ. *Sciences of the Soul and Intellect, Part I: An Arabic Critical Edition and English Translation of Epistles 32–36*. Epistles of the Brethren of Purity. Edited and translated by Paul E. Walker, David Simonowitz, Ismail K. Poonawala, and Godefroid de Callataÿ. Oxford: Oxford University Press and the Institute of Ismaili Studies, 2016.

Ikhwān al-Ṣafāʾ. *Sciences of the Soul and Intellect, Part III: An Arabic Critical Edition and English Translation of Epistles 39–41*. Epistles of the Brethren of Purity. Edited and translated by Carmela Baffioni and Ismail K. Poonawala. Oxford: Oxford University Press and the Institute of Ismaili Studies, 2017.

Al-Kindī. "Risālat al-Kindī fī kammiyyat kutub Arisṭūṭālīs wa-mā yuḥtāj ilayhi fī taḥṣīl al-falsafa." In *Rasāʾil al-Kindī al-falsafiyya*, edited by Muḥammad Abū Rīda, 363–384. Cairo: Dār al-fikr al-ʿarabī, 1950.

Al-Kirmānī. *Rāḥat al-ʿaql*. Edited by Muṣṭafā Ghālib. Beirut: Dār al-Andalus, 1983.

Khosrow, Nāṣir-i. *Kitāb Jāmiʿ al-ḥikmatayn*. Edited by Henry Corbin and Muḥammad Muʿīn. Tehran and Paris: Shirkat golshan, 1953.

Pseudo-Majrīṭī. *"Picatrix:" Das Ziel des Weisen von Pseudo-Mağrīṭī*. Translated by Helmut Ritter and Martin Plessner. London: The Warburg Institute and University of London, 1962.

Pseudo-Majrīṭī. *Rutbat al-ḥakīm: The Book of the Rank of the Sage*. Edited by Wilferd Madelung. Zürich: Living Human Heritage Publications, 2016.

Pseudo-Majrīṭī. *Das Ziel des Weisen*. Studien der Bibliothek Warburg. Edited by Helmut Ritter. Leipzig and Berlin: B.G. Teubner, 1933.

Al-Ṣafadī. *Kitāb al-Wāfī bi-l-wafiyyāt*, vol. 15. 2nd ed. Edited by Bernd Radtke. Beirut: Institut der Deutschen Morgenländischen Gesellschaft, 1991.

Ṣāʿid al-Andalusī. *Kitāb Ṭabaqāt al-umam*. Edited by Louis Cheikho. Beirut: Al-Maṭbaʿa al-kathūlīkiyya li-l-ābāʾ al-yasūʿiyyīn, 1912.

Al-Sijistānī. *The Muntakhab Ṣiwān al-ḥikmah of Abū Sulaimān as-Sijistānī: Arabic Text, Introduction and Indices*. Edited by D.M. Dunlop. Hague, Paris, and New York: Mouton, 1979.

Al-Tawḥīdī. *Kitāb al-Imtāʿ wa-l-muʾānasa*. Edited by Aḥmad Amīn and Aḥmad al-Zayn. 3 vols. Beirut: Dār Maktabat al-ḥayāt, 1939–1944.

Secondary Literature

Adamson, Peter. "Abū Maʿšar, al-Kindī and the Philosophical Defense of Astrology." *Recherches de théologie et philosophie médievales* 69 (2002): 245–270.

Adamson, Peter. *Al-Kindī*. Oxford: Oxford University Press, 2007.

Adamson, Peter. "The Kindian Tradition: The Structure of Philosophy in Arabic Neoplatonism." In *The Libraries of the Neoplatonists*, edited by Cristina D'Ancona, 351–370. Leiden: Brill, 2007.

Adamson, Peter. "Philosophical Theology." In *The Oxford Handbook of Islamic Theology*, edited by Sabine Schmidtke, 297–312. Oxford: Oxford University Press, 2018.

Alí-de-Unzaga, Omar. "The Missing Link: MS 1040: An Important Copy of the *Rasāʾil Ikhwān al-Ṣafāʾ* in the Collection of the Institute of Ismaili Studies." In *Texts, Scribes and Transmission: Manuscript Cultures of the Ismaili Communities and Beyond*, edited by Wafi A. Momin, 81–136. London: I.B. Tauris, 2022.

Almutawa, Shatha. "Education through Narrative in *Rasāʾil Ikhwān al-Ṣafāʾ*." In *Knowledge and Education in Classical Islam: Religious Learning between Continuity and Change*, edited by Sebastian Günther, 389–403. Leiden and Boston: Brill, 2020.

Alshaar, Nuha A. *Ethics in Islam: Friendship in the Political Thought of al-Tawḥīdī and His Contemporaries*. New York: Routledge, 2015.

Awa, Adel. *L'esprit critique des "Frères de la Pureté": Encyclopédistes arabes du IVe/Xe siècle*. Beirut: Imprimerie catholique, 1948.

Baffioni, Carmela. "Citazioni di autori antichi nelle Rasail degli Ikhwan al-Safa: il caso di Nicomacho di Gerasa." In *The Ancient Tradition in Christian and Islamic Hellenism: Studies on the Transmission of Greek Philosophy and Sciences Dedicated to H.J. Drossaart Lulofs on His Ninetieth Birthday*, edited by Gerhard Endress and Remke Kruk, 3–27. Leiden: Research School CNWS, 1997.

Baffioni, Carmela. "Eléments platoniciens et néo-platoniciens de la représentation d'Adam dans la *Jāmiʿa al-Jāmiʿa*." In *Adorare caelestia, gubernare terrena: Atti del colloquio internazionale in onore di Paolo Lucentini (Napoli, 6–7 Novembre 2007)*, edited by Pasquale Arfé, Irene Caiazzo, and Antonella Sannino. Instrumenta patristica et mediaevalia, 99–112. Turnhout: Brepols, 2011.

Baffioni, Carmela. "Embryology in an Ismāʿīlī Messianic Context: The Manuscript Tradition of the Iḫwān al-Ṣafāʾ." *Studia graeco-arabica* 8 (2018): 167–180.

Baffioni, Carmela. "Epistle 48 of the Ikhwān al-Ṣafāʾ and Their Ismāʿīlī Commitment." In *Case Studies in Transmission*, edited by Jaakko Hämeen-Anttila, Ilkka Lindstedt, Raija Mattila, and Robert Rollinger, 11–21. Münster: Ugarit Verlag, 2015.

Baffioni, Carmela. "The Esoteric Meaning of the Sin of Adam in the Manuscript Tradition of the *Rasāʾil* Iḫwān al-Ṣafāʾ." *Studi Magrebini* 20, no. 1 (2022): 3–28.

Baffioni, Carmela. "Esoteric Shīʿism in the Additions to Ancient Manuscripts of the *Rasāʾil* Ikhwān al-Ṣafāʾ." In *L'ésotérisme shiʿite, ses racines et ses prolongements* (*Shiʿi Esotericism: Its Roots and Developments*), edited by Mohammad Ali Amir-Moezzi, Maria De Cillis, Daniel De Smet, and Orkhan Mir-Kasimov, 281–326. Turnhout and Bristol: Brepols, 2016.

Baffioni, Carmela. "An Essay on Terminological Research in Philosophy: The 'Friends of God' in the *Rasāʾil Iḫwān al-Ṣafāʾ*." In *Scritti in onore di Giovanni Oman*, edited by Carmela Baffioni, Giuseppina Igonetti, and Vincenzo Strika. Studi Magrebini, 23–43. Naples: Istituto universitario orientale, 2002.

Baffioni, Carmela. "La Figure d'adam dans la *Jāmiʿa al-Jāmiʿa*." In *Adam, le premier homme*, edited by Agostino Paravicini Bagliani, 115–134. Florence: SISMEL edizioni del Galluzzo, 2012.

Baffioni, Carmela. "The Figure of Iblīs in the Ikhwān al-Ṣafāʾ-Related Esoteric Literature." In *Raison et quête de la sagesse: Hommage à Christian Jambet*, edited by Mohammad Ali Amir-Moezzi, 113–141. Turnhout: Brepols, 2020.

Baffioni, Carmela. *Frammenti e testimonianze di autori antichi nelle Epistole degli Iḫwān aṣ-Ṣafāʾ*. Rome: Istituto Nazionale per la Storia Antica, 1994.

Baffioni, Carmela. “From Sense Perception to the Vision of God: A Path Toward Knowledge According to the Iḫwān al-Ṣafāʾ.” *Arabic Sciences and Philosophy* 8 (1998): 213–231.

Baffioni, Carmela. “Ibdāʿ, Divine Imperative and Prophecy in the *Rasāʾil Ikhwān al-Ṣafāʾ*.” In *Fortresses of the Intellect: Ismaili and Other Islamic Studies in Honour of Farhad Daftary*, edited by Omar Alí-de-Unzaga, 213–226. London and New York: I.B. Tauris and the Institute of Ismaili Studies, 2011.

Baffioni, Carmela. “Ikhwân al-Safâʾ.” In *The Stanford Encyclopedia of Philosophy*, edited by Edward N. Zalta, 2008.

Baffioni, Carmela. “L’influence des Ikhwān al-Ṣafāʾ sur la minéralogie de Ḥamīd al-Dīn al-Kirmānī.” In *Une lumière venue d’ailleurs: Héritages et ouvertures dans les encyclopédies d’Orient et d’Occident au Moyen Âge*, edited by Godefroid de Callataÿ and Baudouin Van den Abeele, 319–331. Turnhout: Brepols, 2008.

Baffioni, Carmela. “Ismaili Onto-Cosmological Doctrines in the Manuscript Tradition of the Ikhwān al-Ṣafāʾ.” *Shii Studies Review* 3 (2019): 33–58.

Baffioni, Carmela. “Al-Madinah al-Fāḍilah in al-Fārābī and in the Ikhwān al-Ṣafāʾ: A Comparison.” In *Studies in Arabic and Islam: Proceedings of the 19th Congress, Union Européenne des Arabisants et Islamisants, Halle 1998*, edited by S. Leder, H. Kilpatrick, B. Martel-Thoumian, and H. Schönig, 3–12. Leuven: Peeters, 2002.

Baffioni, Carmela. “Nāṣir-i Khosrow, Translator of the Ikhwān al-Ṣafāʾ?” In *Sources and Approaches across Disciplines in Near Eastern Studies: Proceedings of the 24th Congress, Union Européenne des Arabisants et Islamisants, Leipzig 2008*, edited by Verena Klemm and Nuha al-Shaʿar, 319–331. Leuven: Peeters, 2013.

Baffioni, Carmela. “Onto-Cosmology and Hierohistory in the Manuscript Tradition of the *Rasāʾil Ikhwān al-Ṣafāʾ*.” In *Intellectual Interactions in the Islamic World: The Ismaili Thread*, edited by Orkhan Mir-Kasimov, 193–218. London: I.B. Tauris and the Institute of Ismaili Studies, 2019.

Baffioni, Carmela. “The Religious Approach to Natural Sciences: The Case of Mineralogy in the Ikhwān al-Ṣafāʾ and in Ḥamīd al-Dīn al-Kirmānī.” In *Islamic Thought in the Middle Ages: Studies in Text, Transmission and Translation, in Honour of Hans Daiber*, edited by Anna Akasoy and Wim Raven, 181–194. Leiden and Boston: Brill, 2008.

Baffioni, Carmela. “Temporal and Religious Connotations of the ‘Regal Policy’ in the Ikhwān al-Ṣafāʾ.” *Mélanges de l’Université Saint-Joseph* 57 (2004): 337–366.

Baffioni, Carmela. “Uso e rielaborazione degli autori classici nella *Risāla al-Ǧāmiʿa*.” In *La diffusione dell’eredità classica nell’età tardoantica e medievale: Filologia, storia, dottrina: Atti del Seminario nazionale di studio (Napoli-Sorrento, 29–31 ottobre 1998)*, edited by Carmela Baffioni, 1–10. Alessandria: Edizioni dell’Orso, 2000.

Bausani, Alessandro. *L’enciclopedia dei Fratelli della Purità: Riassunto, con introduzione e breve commento dei 52 trattati o epistole degli Ikhwān aṣ-Ṣafāʾ*. Naples: Istituto Universitario Orientale, 1978.

Bergé, Marc. "Épitre sur les sciences d'Abū Ḥayyān al-Tawḥīdī: Introduction, traduction, glossaire technique, manuscrit et édition critique." *Bulletin d'études orientales* 18 (1964): 241–300.

Biesterfeldt, Hinrich. "Arabisch-islamische Enzyklopädien: Formen und Funktionen." In *Die Enzyklopädie im Wandel vom Hochmittelalter bis zur frühen Neuzeit: Akten des Kolloquiums des Projekts D im Sonderforschungsbereich 231 (29.11.–1.12.1996)*, edited by Christel Meier, 43–83. Munich: Wilhelm Fink, 2002.

Biesterfeldt, Hinrich. "Medieval Arabic Encyclopedias of Science and Philosophy." In *The Medieval Hebrew Encyclopedias of Science and Philosophy: Proceedings of the Bar-Ilan University Conference*, edited by Steven Harvey. Amsterdam Studies in Jewish Thought, 77–98. Dordrecht: Springer, 2000.

van Bladel, Kevin. *The Arabic Hermes: From Pagan Sage to Prophet of Science*. Oxford and New York: Oxford University Press, 2009.

Bosworth, C.E. "A Pioneer Arabic Encyclopedia of the Sciences: Al-Khwārizmī's *Keys of the Sciences*." *Isis* 54, no. 1 (1963): 97–111.

Burnett, Charles. "The Planets and the Development of the Embryo." In *The Human Embryo: Aristotle and the Arabic and European Traditions*, edited by G.R. Dunstan, 95–112. Exeter: University of Exeter Press, 1990.

de Callataÿ, Godefroid. "Astrology and Prophecy: The Ikhwān al-Ṣafā' and the Legend of the Seven Sleepers." In *Studies in the History of the Exact Sciences in Honour of David Pingree*, edited by Charles Burnett, Jan P. Hogendijk, Kim Plofker, and Michio Yano, 758–785. Leiden and Boston: Brill, 2004.

de Callataÿ, Godefroid. "Brethren of Purity." In *Encyclopaedia of Islam Three*, edited by Kate Fleet, Gudrun Krämer, Denis Matringe, John Abdallah Nawas, and D.J. Stewart. Leiden: Brill, 2013.

de Callataÿ, Godefroid. "The Classification of Knowledge in the *Rasā'il*." In *The Ikhwān al-Ṣafā' and Their Rasā'il: An Introduction*, edited by Nader El-Bizri. Epistles of the Brethren of Purity, 58–82. Oxford: Oxford University Press and the Institute of Ismaili Studies, 2008.

de Callataÿ, Godefroid. "From Ibn Masarra to Ibn 'Arabī: References, Shibboleths and Other Subtle Allusions to the *Rasā'il Ikhwān al-Ṣafā'* in the Literature of al-Andalus." In *Labor limae: Atti in onore di Carmela Baffioni*, edited by Antonella Straface, Carlo De Angelo, and Andrea Manzo, 217–268. Naples: Centro di studi maġrebini, 2014–2015.

de Callataÿ, Godefroid. *Ikhwan al-Safa': A Brotherhood of Idealists on the Fringe of Orthodox Islam*. Makers of the Muslim World. New York: Oneworld Publications, 2006.

de Callataÿ, Godefroid. "Magia en al-Andalus: *Rasā'il Ijwān al-Ṣafā'*, *Rutbat al-ḥakīm* y *Ġāyat al-ḥakīm* (*Picatrix*)." *Al-Qanṭara* 34, no. 2 (2013): 297–344.

de Callataÿ, Godefroid. "Philosophy and Bāṭinism in al-Andalus: Ibn Masarra's *Risālat al-Iʿtibār* and the *Rasāʾil Ikhwān al-Ṣafāʾ*." *Jerusalem Studies in Arabic and Islam* 41 (2014): 261–312.

de Callataÿ, Godefroid, and Sébastien Moureau. "Again on Maslama Ibn Qāsim al-Qurṭubī, the Ikhwān al-Ṣafāʾ and Ibn Khaldūn: New Evidence from Two Manuscripts of *Rutbat al-ḥakīm*." *Al-Qanṭara* 37, no. 2 (2016): 329–372.

de Callataÿ, Godefroid, and Sébastien Moureau. "A Milestone in the History of Andalusī Bāṭinism: Maslama b. Qāsim al-Qurṭubī's *Riḥla* in the East." *Intellectual History of the Islamicate World* 5 (2017): 86–117.

Carusi, Paola. "Alchimia islamica e felicità nella Risāla Ĝāmiʿa: Inalterabilità delle sostanze e pace dell'anima." In *Le felicità nel Medioevo: Atti del Convegno della Società italiana per lo studio del pensiero medievale (S.I.S.P.M.), Milano, 12–13 settembre 2003*, edited by Maria Bettetini and Francesco D. Paparella, 277–296. Louvain-la-Neuve: Fédération internationale des instituts d'études médiévales, 2005.

Carusi, Paola. "Alchimia islamica e religione: La legittimazione difficile di una scienza della natura." *Oriente Moderno* 19, no. 3 (2000): 461–502.

Casanova, Paul. "Une date astronomique dans les épîtres des Ikhwân aṣ-Ṣafâ." *Journal Asiatique* 5 (1915): 5–17.

Casanova, Paul. "Notice sur un manuscrit de la secte des assasins." *Journal Asiatique* 11 (1898): 151–159.

Corbin, Henry. *Histoire de la philosophie islamique, I: Des origines jusqu'à la mort d'Averroës (1198)*. Paris: Gallimard, 1964.

Daftary, Farhad. *The Ismāʿīlīs: Their History and Doctrines*. Cambridge: Cambridge University Press, 1992.

De Smet, Daniel. "L'auteur des *Rasāʾil Ikhwān al-Ṣafāʾ* selon les sources ismaéliennes ṭayyibites." *Shii Studies Review* 1, no. 1–2 (2017): 151–166.

De Smet, Daniel. "La doctrine avicennienne des deux faces de l'âme et ses racines ismaéliennes." *Studia Islamica* 93 (2001): 77–89.

De Smet, Daniel. "Esotericism and Exotericism." In *Encyclopaedia of Islam Three*, edited by Kate Fleet, Gudrun Krämer, and Denis Matringe. Leiden: Brill, 2015.

De Smet, Daniel. "Henry Corbin et les études ismaéliennes." In *Henry Corbin: Philosophies et sagesses des religions du livre: Actes du colloque "Henry Corbin", Sorbonne, les 6–8 novembre 2003*, edited by Mohammad Ali Amir-Moezzi, Christian Jambet, and Pierre Lory. Bibliothèque de l'École des Hautes Études, 105–118. Turnhout: Brepols, 2005.

De Smet, Daniel. "Ismāʿīlī Theology." In *The Oxford Handbook of Islamic Theology*, edited by Sabine Schmidtke, 313–324. Oxford: Oxford University Press, 2018.

De Smet, Daniel. *La philosophie ismaélienne: Un ésotérisme chiite entre néoplatonisme et gnose*. Paris: Les Éditions du Cerf, 2012.

De Smet, Daniel. *La quiétude de l'Intellect: Néoplatonisme et gnose ismaélienne dans l'œuvre de Ḥamîd al-Dîn al-Kirmânî*. Leuven: Peeters, 1995.

De Smet, Daniel. "La *Risāla al-Jāmiʿa* attribuée aux Ikhwān al-Ṣafāʾ: Un précurseur de l'ismaélisme ṭayyibite ?". In *Labor limae: Atti in onore di Carmela Baffioni*, edited by Antonella Straface, Carlo De Angelo, and Andrea Manzo, 269–298. Naples: Centro di studi maġrebini, 2014–2015.

De Smet, Daniel. "Yves Marquet, les Iḫwān al-Ṣafāʾ et le pythagorisme." *Journal Asiatique* 295 (2007): 491–500.

Dieterici, Friedrich. *Die Philosophie der Araber im IX. und X. Jahrhundert n. Chr. aus der Theologie des Aristoteles, den Abhandlungen Alfārābīs und den Schriften der Lautern Brüder: Erstes Buch: Einleitung und Mikrokosmos (die Welt)*. Leipzig: J.C. Hinrichs, 1876.

Diwald, Susanne. *Arabische Philosophie und Wissenschaft in der Enzyklopädie: Kitāb Iḫwān aṣ-Ṣafāʾ (III): Die Lehre von Seele und Intellekt.* Wiesbaden: Otto Harrassowitz, 1975.

Ebstein, Michael. *Mysticism and Philosophy in al-Andalus: Ibn Masarra, Ibn al-ʿArabī and the Ismāʿīlī Tradition*. Islamic History and Civilization, 103. Leiden and Boston: Brill, 2014.

El-Bizri, Nader (ed.). *The Ikhwān al-Ṣafāʾ and Their Rasāʾil: An Introduction.* Epistles of the Brethren of Purity. Oxford: Oxford University Press and the Institute of Ismaili Studies, 2008.

El-Bizri, Nader. "Prologue." In *The Ikhwān al-Ṣafāʾ and Their Rasāʾil: An Introduction*, edited by Nader El-Bizri. Epistles of the Brethren of Purity, 1–32. Oxford: Oxford University Press and the Institute of Ismaili Studies, 2008.

Endress, Gerhard. "The Cycle of Knowledge: Intellectual Traditions and Encyclopædias of the Rational Sciences in Arabic Islamic Hellenism." In *Organizing Knowledge: Encyclopædic Activities in the Pre-Eighteenth Century Islamic World*, edited by Gerhard Endress, 103–133. Leiden: Brill, 2006.

Endress, Gerhard. "Mathematics and Philosophy in Medieval Islam." In *The Enterprise of Science in Islam: New Perspectives*, edited by Jan P. Hogendijk and Abdelhamid I. Sabra, 121–176. Cambridge: MIT Press, 2003.

van Ess, Josef. "Encyclopædic Activities in the Islamic World: A Few Questions, and No Answers." In *Organizing Knowledge: Encyclopædic Activities in the Pre-Eighteenth Century Islamic World*, edited by Gerhard Endress, 3–22. Leiden: Brill, 2006.

Fierro, Maribel. "Bāṭinism in al-Andalus: Maslama b. Qāsim al-Qurṭubī (d. 353/964), Author of the *Rutbat al-Ḥakīm* and the *Ghāyat al-Ḥakīm* (*Picatrix*)." *Studia Islamica* 84, no. 2 (1996): 87–112.

Flügel, Gustav. "Ueber Inhalt und Verfasser der arabischen Encyclopädie رسائل اخوان الصفا وخلان الوفا d.i. die Abhandlungen der aufrichtigen Brüder und treuen Freunde: Nebst Andeutungen über die Einrichtungen des Bundes der verbrüderten." *Zeitschrift der Deutschen Morgenländischen Gesellschaft* 13, no. 1/2 (1859): 1–43.

Goldziher, Ignaz. "Über die Benennung der 'Ichwān al-Ṣafāʾ'." *Der Islam* 1 (1910): 22–26.

Gutas, Dimitri. *Avicenna and the Aristotelian Tradition: Introduction to Reading Avicenna's Philosophical Works: Second, Revised and Enlarged Edition, Including an Inventory of Avicenna's Authentic Works*. Leiden and Boston: Brill, 2014.

Gutas, Dimitri. "The Greek and Persian Background of Early Arabic Encyclopedism." In *Organizing Knowledge: Encyclopædic Activities in the Pre-Eighteenth Century Islamic World*, edited by Gerhard Endress, 91–101. Leiden and Boston: Brill, 2006.

Gutas, Dimitri. "Paul the Persian on the Classification of the Parts of Aristotle's Philosophy: A Milestone between Alexandria and Baġdâd." *Der Islam* 1 (1983): 22–26.

Hamdani, Abbas. "Abū Ḥayyān al-Tawḥīdī and the Brethren of Purity." *International Journal of Middle East Studies* 9, no. 3 (1978): 345–353.

Hamdani, Abbas. "The Arrangement of the *Rasāʾil Ikhwān al-Ṣafāʾ* and the Problem of Interpolations." In *The Ikhwān al-Ṣafāʾ and Their Rasāʾil: An Introduction*, edited by Nader El-Bizri. Epistles of the Brethren of Purity, 83–100. Oxford: Oxford University Press and the Institute of Ismaili Studies, 2008.

Hamdani, Abbas. "Brethren of Purity, a Secret Society for the Establishment of the Fatimid Caliphate: New Evidence for the Early Dating of Their Encyclopaedia." In *L'Egypte fatimide: Son art et son histoire*, edited by Marianne Barrucand, 73–82. Paris: Presses de l'Université Paris-Sorbonne, 1999.

Hamdani, Abbas. "An Early Fāṭimid Source on the Time and Authorship of the 'Rasāʾil Iḫwān al-Ṣafāʾ.'" *Arabica* 26, no. 1 (1979): 62–75.

Hamdani, Abbas. "The Ikhwān al-Ṣafāʾ: Between al-Kindī and al-Fārābī." In *Fortresses of the Intellect: Ismaili and Other Islamic Studies in Honour of Farhad Daftary*, edited by Omar Alí-de-Unzaga, 189–212. London and New York: I.B. Tauris and the Institute of Ismaili Studies, 2011.

Hamdani, Abbas. "Shades of Shiʿism in the Tracts of the Brethren of Purity." In *Traditions in Contact and Change: Selected Proceedings of the XIVth Congress of the International Association for the History of Religions*, edited by Peter Slater and Donald Wiebe, 447–460. Waterloo: Wilfrid Laurier University Press, 1983.

Hamdani, Husain F. "*Rasāʾil Ikhwān aṣ-Ṣafāʾ* in the Literature of the Ismāʿīlī Ṭayibī Daʿwat." *Zeitschrift für Geschichte und Kultur des Islamischen Orients* 20 (1932): 281–300.

Hein, Christel. *Definition und Einteilung der Philosophie: Von der spätantiken Einleitungsliteratur zur arabischen Enzyklopädie*. Frankfurt am Main: P. Lang, 1985.

Heinrichs, Wolfhart. "The Classification of the Sciences and the Consolidation of Philology in Classical Islam." In *Centres of Learning: Learning and Location in Pre-Modern Europe and the Near East*, edited by Jan Willem Drijvers and Alasdair A. MacDonald, 119–139. Leiden: Brill, 1995.

Irwin, Robert. "The Arabic Beast Fable." *Journal of the Warburg and Courtauld Institutes* 55 (1992): 36–50.

Ivanow, Wladimir A. *A Guide to Ismaili Literature*. London: Royal Asiatic Society, 1933.

Ivry, Alfred L. *Al-Kindi's Metaphysics: A Translation of Yaʿqūb Ibn Isḥāq al-Kindī's Treatise "On First Philosophy" (fī al-Falsafah al-Ūlā)*. Albany: State University of New York Press, 1974.

Kacimi, Mourad. "Edición crítica árabe y estudio de la *Risālat al-ŷāmiʿa ḏāt al-fawāʾid al-nāfiʿa, tāŷ Rasāʾil Ijwān al-Ṣafāʾ* (finales del s. IV H./X E.C., principios del s. V H./XI E.C.): Tomo I." Doctoral thesis, University of Alicante, 2015.

Kacimi, Mourad. "Reflexiones sobre la relación de la *Risālat ǧāmiʿat al-ǧāmiʿa* con los Iḫwān al-Ṣafāʾ." *eHumanista/IVITRA* 8 (2015): 328–338.

Kraemer, Joel L. *Humanism in the Renaissance of Islam: The Cultural Revival during the Buyid Age*. Leiden: Brill, 1986.

Kraemer, Joel L. *Philosophy in the Renaissance of Islam: Abū Sulaymān al-Sijistānī and His Circle*. Leiden: Brill, 1986.

Krinis, Ehud. "*Al-Risāla al-Jāmiʿa* and Its Judeo-Arabic Manuscript." In *Islam: Identité et altérité: Hommage à Guy Monnot, O.P.*, edited by Mohammad Ali Amir-Moezzi. Bibliothèque de l'École des Hautes Études: Sciences Religieuses, 311–329. Turnhout: Brepols, 2013.

Kukkonen, Taneli. "Al-Ghazālī on the Origins of Ethics." *Numen* 63 (2016): 271–298.

Madelung, Wilferd. "Ḳarmaṭī." In *Encyclopaedia of Islam, 2nd Edition*, 1071–1076. Leiden: Brill, 1998.

Madelung, Wilferd. "Maslama al-Qurṭubī's Contribution to the Shaping of the Encyclopedia of the Ikhwān al-Ṣafāʾ." In *Labor limae: Studi in onore di Carmela Baffioni*, edited by Antonella Straface, Carlo De Angelo, and Andrea Manzo, 403–417. Naples: Centro di studi maġrebini, 2014–2015.

Marquet, Yves. "910 en Ifrīqiyā: Une épître des Iḫwān al-Ṣafāʾ." *Bulletin d'études orientales* 30 (1978): 61–73.

Marquet, Yves. "La détermination astrale de l'évolution selon les Frères de la pureté." *Bulletin d'études orientales* 44 (1992): 127–146.

Marquet, Yves. "Les épîtres des Ikhwân as-Safâ', œuvre ismaïlienne." *Studia Islamica* 61 (1985): 57–79.

Marquet, Yves. *Les "Frères de la pureté", pythagoriciens de l'Islam: La marque du pythagorisme dans la rédaction des Épîtres des Iḫwān aṣ-Ṣafāʾ*. Paris: Textes et travaux de Chrysopaiea, 2006.

Marquet, Yves. "Iḫwān al-Ṣafāʾ, ismaïliens et qarmaṭes." *Arabica* 24 (1977): 233–257.

Marquet, Yves. "Ikhwān al-Ṣafāʾ." In *Encyclopaedia of Islam, 2nd Edition*, 1071–1076. Leiden: Brill, 1968.

Marquet, Yves. "Note annexe: À propos d'un poème ismaïlien dans les épîtres des Iḫwān aṣ-Ṣafāʾ." *Studia Islamica* 55 (1982): 137–142.

Marquet, Yves. *La philosophie des alchimistes et l'alchimie des philosophes: Jâbir Ibn Ḥayyân et les "Frères de la Pureté"*. Paris: Maisonneuve and Larose, 1988.

Marquet, Yves. *La Philosophie des Iḫwān al-Ṣafāʾ: Nouvelle édition augmentée*. Paris and Milan: S.É.H.A and Archè, 1999.

Mattila, Janne. "The Authorship of *al-Risāla l-Ǧāmiʿa* Re-Examined." *Arabica* 68, no. 5–6 (2021): 461–494.

Mattila, Janne. "*Fihrist* of the *Rasāʾil Ikhwān al-Ṣafāʾ*: Textual Variants and Their Relation to *al-Risāla al-Jāmiʿa*." *Micrologus* 33 (2025): 249–274.

Mattila, Janne. "The Ikhwān al-Ṣafāʾ on Religious Diversity." *Journal of Islamic Studies* 28, no. 2 (2017): 178–192.

Mattila, Janne. "The Philosophical Worship of the Ikhwān al-Ṣafāʾ." *Journal of Islamic Studies* 27, no. 1 (2016): 17–38.

Mattila, Janne. "Sabians, the School of al-Kindī, and the Brethren of Purity." In *Religious Identities in Antiquity and the Early Middle Ages*, edited by Ilkka Lindstedt, Nina Nikki, and Riikka Tuori. Studies on the Children of Abraham, 92–114. Leiden: Brill, 2021.

Mattila, Janne. "Utopia." In *The Routledge Companion of Islamic Ethics*, edited by Mohammed Ghaly. New York: Routledge, forthcoming.

Michot, Yahya Jean. "Misled and Misleading … Yet Central in Their Influence: Ibn Taymiyya's Views on the Ikhwān al-Ṣafāʾ." In *The Ikhwān al-Ṣafāʾ and Their Rasāʾil: An Introduction*, edited by Nader El-Bizri. Epistles of the Brethren of Purity, 139–179. Oxford: Oxford University Press and the Institute of Ismaili Studies, 2008.

Nasr, Seyyed Hossein. *An Introduction to Islamic Cosmological Doctrines: Conceptions of Nature and Methods Used for Its Study by the Ikhwān al-Ṣafāʾ, al-Bīrūnī, and Ibn Sīnā*. Cambridge: Harvard University Press, 1964.

Netton, Ian Richard. "Brotherhood Versus Imāmate: Ikhwān al-Ṣafāʾ and the Ismāʿīlīs." *Jerusalem Studies in Arabic and Islam* 2 (1980): 253–262.

Netton, Ian Richard. *Muslim Neoplatonists: An Introduction to the Thought of the Brethren of Purity (Ikhwān al-Ṣafāʾ)*. Islamic Surveys. 2nd ed. Edinburgh: Edinburgh University Press, 1991.

Nokso-Koivisto, Inka. "Microcosm-Macrocosm Analogy in *Rasāʾil Ikhwān aṣ-Ṣafāʾ* and Certain Related Texts." Doctoral thesis, University of Helsinki, 2014.

Organizing Knowledge: Encyclopædic Activities in the Pre-Eighteenth Century Islamic World. Edited by Gerhard Endress. Leiden: Brill, 2006.

Pines, Shlomo. "Shīʿite Terms and Conceptions in Judah Halevi's Kuzari." *Jerusalem Studies in Arabic and Islam* 2 (1980): 156–251.

Plessner, Martin. "Hermes Trismegistus and Arab Science." *Studia Islamica* 2 (1954): 45–59.

Poonawala, Ismail K. "Why We Need an Arabic Critical Edition with an Annotated English Translation of the *Rasāʾil Ikhwān al-Ṣafāʾ*." In *The Ikhwān al-Ṣafāʾ and Their Rasāʾil: An Introduction*, edited by Nader El-Bizri. Epistles of the Brethren of Purity, 33–57. Oxford: Oxford University Press and the Institute of Ismaili Studies, 2008.

Richter, Gustav. *Studien zur Geschichte der älteren arabischen Fürstenspiegel*. Leipzig: Heinrichs, 1932.

Rosenthal, Franz. *Aḥmad b. aṭ-Ṭayyib as-Saraḫsî.* New Haven: American Oriental Society, 1943.

Rudolph, Ulrich. *Islamische Philosophie: Von den Anfängen bis zur Gegenwart.* Munich: C.H. Beck, 2004.

Rudolph, Ulrich. "Reflections on al-Fārābī's *Mabādiʾ ārāʾ ahl al-madīna al-fāḍila.*" In *In the Age of al-Fārābī: Arabic Philosophy in the Fourth/Tenth Century*, edited by Peter Adamson, 1–14. London: The Warburg Institute, 2008.

Saif, Liana. "Ikhwān al-Ṣafāʾ's Religious Reform and Magic: Beyond the Ismaʿili Hypothesis." *Journal of Islamic Studies* 30, no. 1 (2019): 34–68.

Sorabji, Richard. *The Philosophy of the Commentators, 200–600 AD: A Sourcebook: Volume 1: Psychology (with Ethics and Religion).* Berkeley, Los Angeles, and London: University of California Press, 2004.

Stern, Samuel Miklos. "The Authorship of the Epistles of the Ikhwān aṣ-Ṣafāʾ." *Islamic Culture* 20 (1946): 367–372.

Stern, Samuel Miklos. "New Information About the Authors of the 'Epistles of the Sincere Brethren'." *Islamic Studies* 3 (1964): 405–428.

Stroumsa, Sarah, and Sara Sviri. "The Beginnings of Mystical Philosophy in al-Andalus: Ibn Masarra and His *Epistle on Contemplation.*" *Jerusalem Studies in Arabic and Islam* 36 (2001): 201–252.

Takim, Liyakat N. *The Heirs of the Prophet: Charisma and Religious Authority in Shiʿite Islam.* Albany: State University of New York Press, 2006.

Tibawi, Abdul Latif. "Ikhwān aṣ-Ṣafā and Their *Rasāʾil*: A Critical Review of a Century and a Half of Research." *Islamic Quarterly* 2 (1955): 28–46.

de Vaulx d'Arcy, Guillaume. "Aḥmad b. al-Ṭayyib al-Saraḫsī, réviseur de L'*introduction arithmétique* de Nicomaque de Gérase et rédacteur des *Rasāʾil Iḫwān al-Ṣafāʾ.*" *Arabic Sciences and Philosophy* 29, no. 2 (2019): 261–283.

de Vaulx d'Arcy, Guillaume. "The *Epistles of the Brethren of Purity* Edited by the Institute of Ismaili Studies: When the Re-Edition of a Text Can Be Its Destruction." *Mélanges de l'Institut dominicain d'études orientales* 34 (2019): 253–330.

de Vaulx d'Arcy, Guillaume. *Les Épîtres des Frères en Pureté (Rasāʾil Ikhwān al-Ṣafāʾ): Mathématique et philosophie: Présentation et traduction de six épîtres.* Paris: Les Belles lettres, 2019.

Wagner, Ewald. "Sprechende Tiere in der arabischen Prosa." *Asiatische Studien: Zeitschrift der Schweizerischen Asiengesellschaft* 48, no. 3 (1994): 937–957.

Walker, Paul. "Bāṭiniyya." In *Encyclopaedia of Islam Three*, edited by Kate Fleet, Gudrun Krämer, and Denis Matringe. Leiden: Brill, 2009.

Walker, Paul. *Early Philosophical Shiism: The Ismaili Neoplatonism of Abū Yaʿqūb al-Sijistānī.* Cambridge Studies in Islamic Civilization. Cambridge: Cambridge University Press, 2003.

Westerink, Leendert G. "The Alexandrian Commentators and the Introductions to

Their Commentaries." In *Aristotle Transformed: The Ancient Commentators and Their Influence*, edited by Richard Sorabji, 325–348. Ithaca: Cornell University Press, 1990.

Index of Epistles

Index of Ancient and Medieval Names

Index of Subjects

www.ingramcontent.com/pod-product-compliance
Lightning Source LLC
Chambersburg PA
CBHW071333230926
51849CB00002B/387
* 9 7 8 9 0 0 4 6 9 0 1 3 4 *